To

Yours

Andrew.

The Neutron and the Bomb

The Neutron and the Bomb

A biography of Sir James Chadwick

Andrew Brown

OXFORD • NEW YORK • TOKYO

OXFORD UNIVERSITY PRESS
1997

Oxford University Press, Great Clarendon Street, Oxford OX2 6DP

Oxford New York
Athens Auckland Bangkok Bogota Bombay Buenos Aires
Calcutta Cape Town Dar es Salaam Delhi Florence Hong Kong
Istanbul Karachi Kuala Lumpur Madras Madrid Melbourne
Mexico City Nairobi Paris Singapore Taipei Tokyo Toronto

and associated companies in
Berlin Ibadan

Oxford is a trade mark of Oxford University Press

Published in the United States
by Oxford University Press Inc., New York

A catalogue record for this book is available from the British Library

Library of Congress Cataloging in Publication Data
(Data available)

ISBN 0 19 853992 4

Typeset by OUP using Synthæsis
Printed in Great Britain by Biddles Ltd, Guildford and King's Lynn

For Jane, Jamie, Ben and Ollie

Preface

One spring afternoon in the early 1950s, an economist and a nuclear physicist were sauntering idly along the backs of the River Cam. Just as a couple of schoolboys on the same walk might have argued about the batsmanship of Denis Compton and Len Hutton, the two young dons passed the time making comparisons between academic figures. The economist asked the physicist for his opinion of two leading Cambridge physicists, Sir George Thomson and Sir James Chadwick. Both were Nobel Laureates and the Masters of Cambridge Colleges—Corpus Christi and Gonville and Caius, respectively. Sir George was of the Cambridge purple: the son of J.J. Thomson, himself a Nobel prizewinner and discoverer of the electron. In the physicist's opinion, George Thomson was a first class scientist, who thoroughly deserved the recognition he had received for his work on the diffraction of electrons passing through very thin films of metals. By contrast, Chadwick, the discoverer of the neutron, was quite simply one of the makers of modern physics.

By coincidence Chadwick's father shared J.J. Thomson's initials and even came from the same part of England, but worked in a cotton mill and died in his seventies, unmourned by his eldest son. James Chadwick was a very close man when it came to personal matters, and I have not been able to discover the reason for the estrangement from his family. He was a product of the late Victorian age, and the aspiration to better himself was a natural one, so that it cannot have been just the urge to shake off his working class origins that was responsible.

Chadwick's keen intelligence was nurtured by remarkably good state schools, and his scientific career blossomed under Rutherford's inspirational leadership, first in Manchester and then at the Cavendish Laboratory in Cambridge. A less fortunate legacy of being born an Englishman in the last decade of the nineteenth century was the duty to fight, and often die, for King and Country in the 1914–1918 war. As readers will see, Chadwick avoided this terrible opportunity by being interned in Germany, where he had been studying with Hans Geiger. Geiger along with many other German scientists went off to the trenches, as did their English counterparts, but the personal warmth and regard for former scientific colleagues never dimmed, as is made plain in some of Chadwick's and Rutherford's wartime correspondence. By the time of the second world war, knowledge in the field of nuclear physics had advanced so far, in large part due to the experimental work carried out and supervised by Rutherford and Chadwick at the Cavendish, that scientists were too valuable to risk in mere combat; by degrees, they came to realize that they held the potential to unleash a new weapon of unimaginable power on the world. Chadwick's life was emblematic of these changes: indeed he was unique in history as the only man to witness both the Kaiser announcing the outbreak of war to Berliners in 1914 and the explosion of the first atomic bomb in the New Mexico desert in 1945. Between these two cataclysmic events Chadwick discovered the subatomic particle that launched the atomic age; he was centrally involved in the bomb's development from the first months of the war.

There were no such possible ramifications in Chadwick's mind while he carried out his dogged search for the neutron. It is remarkable how the Cavendish Laboratory in the 1920s and 1930s managed to retain the carefree approach of happier times, with scarcely a thought given to the utility of any of its discoveries. In Chadwick's view they were competing with nature to unlock its secrets. Apart from some banter between him and Rutherford about alchemy, commercial considerations were entirely absent; nor were practical applications of major concern. As captain of the vessel, Rutherford set a bold course to explore the innermost depths of the nucleus, and Chadwick as his helmsman stuck to his directions tenaciously. The impenetrability of their chosen route led to criticism from more faint-hearted and worldly bodies, but they were not to be deflected. Their purity of purpose meant that the only ethical dilemma that confronted them concerned the means of their search—were they being scrupulously objective in their observations—and this was the subject of a spirited controversy with the school of physics in Vienna. There were no qualms about the outcome—knowledge could not be inherently dangerous.

Chadwick was abruptly thrown from this state of innocence when he realized that the energy stored in the uranium nucleus could be released in an explosion of devastating power. It seemed highly probable to him that his former colleagues in Germany would have reached the same conclusion, and so despite any personal misgivings, it was necessary to pursue this idea as vigorously as possible to the point of production of a weapon. His efforts towards this end both as a scientific leader and in the unaccustomed role of a trusted diplomat were immeasurable. When the first atomic bomb was successfully detonated at the Trinity test, Germany was no longer a combatant and the United States was clearly the only country in possession of the weapon, but Chadwick thought its use against Japan was justified. In the immediate aftermath, he held a press conference in Washington and warned an unheeding American public that the US monopoly in nuclear weapons would be short-lived.

Unlike Bohr, Oppenheimer, Teller or his own Liverpool research student Joseph Rotblat, Chadwick refrained from any other public statements on the politics of the bomb. In part this was because of innate shyness, but mainly because he regarded such policies as the prerogative of statesmen rather than of scientists. He once wrote that 'Rutherford's genius was in getting knowledge not in expounding it', and regarded his own talents in a similar, if lesser, light. On the question of controlling nuclear weapons, it was clear to Chadwick that the pitfalls were so cavernous and the stakes so high that there was never going to be an easy and guaranteed solution. He saw the problems of verification and international enforcement more clearly than many of his scientific contemporaries. There was a broad streak of realism, tainted with pessimism, running through his core. His lack of outward enthusiasm and unfailingly critical nature may have been the reason that General Groves seized on him as the most reliable scientist on the Manhattan Project. His unquestioned integrity enabled him to maintain the close relationship with Groves without losing the respect of his fellow scientists.

Even in the presence of such dominant personalities as Rutherford and Groves, Chadwick was always his own man. With Rutherford, any disagreements were short-lived, and caused Chadwick more anquish than they did his chief. Groves was a more arrogant and ruthless character than Rutherford, but Chadwick stood up to him quietly and implacably. Nowhere was this more evident than in their first meeting in 1943 in Washington, where Chadwick demolished the General's cherished compartmentalization approach to security. Chadwick's own attitude to security was complex and individual. He first confronted the question in Liverpool early in the war, where two of his key workers—Frisch and Rotblat—were classified as aliens. He thought the petty restrictions placed on their movements and activities were absurd and encouraged the two young men to flout them. He organized visits for Frisch to other universities so that he could share his thoughts on nuclear fission with other leading experts, and allowed Rotblat to include nuclear fission in lectures to undergraduates. Yet he stopped short of showing them the Maud Report he was compiling for the government on the prospects for an atomic weapon. Later in the war, he interviewed Frédéric Joliot-Curie after the liberation of Paris, and scrupulously observed the letter of the Quebec Agreement, which forbade any disclosure about the bomb to a non-Anglo-American.

Chadwick recognized that the fertilization of ideas depends on free communication, and he held to this philosophy even in the quasi-military setting of the Manhattan Project. After a while, General Groves seemed to appreciate his approach and asked to be supplied with some of Chadwick's correspondence, since he found it was the most efficient way of gleaning what important advances were being made. Chadwick's trust was betrayed by two members of the British contingent—the atomic spies Fuchs and Nunn May. The former seems to have fooled everyone, whereas from records I examined in the American National Archives, Groves suspected Nunn May, just from monitoring his movements between Montreal and the Argonne Laboratory. In her new biography of Lise Meitner, Ruth Sime refers to an evening in February 1946 which Meitner spent with Chadwick in Washington. Meitner was disappointed because he would not be drawn on questions about the bomb, and his sparse comments on the subject seemed to be intended to cut off any further questions. While this episode again reflected his adherence to the Quebec Agreement, his reticence was surely reinforced by the growing hysteria that week in DC, as the news of the Canadian spy ring became public; Chadwick knew that there would soon be more disclosures about Nunn May's role in particular. As was often the case during the war, Chadwick was preoccupied by sensitive information which he could not discuss with any of his scientific peers. Despite this, his natural predilection for open debate remained, and the following year he wrote a delicious letter of reprimand to the Cabinet Office in London for labelling as 'secret' a public lecture given by the head of the newly formed US Atomic Energy Commission.

Although I cannot pretend to retain an objective view of my subject, it would seem to me that he led an extraordinarily eventful life and that his stature as a physicist, not to mention his subsequent distinguished war service, easily merit historical study. Why is this the first biography of Chadwick to be written? He never sought the limelight and was always exceedingly modest about his own accomplishments. In the copious press stories last summer commemorating the fiftieth anniversary of the bomb, I noted Chadwick's name just once in the *New York Times*, and not at all in the British newspapers. As a man he was not easy to approach and the reputation he earned tended to be one of dourness and acerbity. This persona is dismantled by his own hand in correspondence, which apart from providing a concise and stylish record of contemporary events, also reveals a sharp wit and often surprising sensitivity about the feelings of others. The seventh Duke of Devonshire, founder of the Cavendish Laboratory, was described in an admittedly servile *Vanity Fair* portrait as: 'Shy and reserved, he has been unjustly accused of supercilious reserve and want of geniality, but in fact there is no man more ready than he to do a kind or a generous act'. I hope I will leave the reader with the impression that Chadwick was very much in this Cavendish mould.

The other problem with relating Chadwick's life is that he was a pioneer in nuclear physics, an incomprehensible subject to all but a few. I am not a physicist and have not presumed to teach physics during the course of the book to those who might understand even less than I. It is fortunate from my point of view that most of Chadwick's work was not mathematical in nature, and depended on Newtonian rather than quantum mechanics so that it can be described to a large extent in non-technical language. I have tried to present an accurate account of his scientific contributions, and if I have succeeded at all, it is because his original papers were so clearly written. The letter to *Nature* (see Appendix 1) announcing the discovery of the neutron illustrates this perfectly. I believe that even a non-scientist, with no knowledge of the fascinating story leading up to the discovery, could read this and recognize the incisiveness of the argument, constructed with all the drama of an end-game by a grand master in chess.

I have made free use of Chadwick's letters to fellow scientists, including Rutherford, Bohr, and Lawrence. Most of these have not appeared in print before, and they convey the excitement felt by some of the leading figures as the boundaries of nuclear physics were pushed back. Similarly his wartime letters to Oliphant, Peierls, Rotblat, Kinsey, Feather, and sundry civil servants remind us of the imminent dangers faced in wartime Britain and the obstacles (some remarkably petty) confronting all of them. The vital but often ramshackle British effort, which sparked American interest in the bomb, succeeded despite a shortage of scientists and skilled manpower with its participants doing without telephones, petrol, cash, uranium and sleep. Chadwick kept them productive, if not always happy, by adroit tact, sympathetic cajoling, occasional withering rebukes and his own ceaseless example.

His letters are scattered through various archives and piecing them together was a satisfying, but not always straightforward, process. I mentioned to Sir Rudolf Peierls that I had found some letters in which Chadwick suggested a search of German journals as a way of determining whether the German physicists were still publishing and lecturing at their usual universities, or whether there was an ominous silence that would suggest they were collaborating on a Nazi atomic bomb programme. Peierls replied that his recollection was that the examination of German periodicals was his idea. More thorough searching on my part revealed a letter that made it plain the same idea had occured to Chadwick and Peierls independently and at the same time. I was able to send this evidence to Sir Rudolf, who replied with great charm: 'It is of course of no consequence who first suggested the idea, only I found it disturbing that my memory should be so wrong, and I find it comforting that it was right after all. At the same time there are flaws in one's memory—I had completely forgotten that this search was done with the cooperation of Fuchs'. Sir Rudolf, who died last year, was just one of a distinguished cadre of senior academics who willingly provided me with invaluable remembrances of Chadwick, and my sincere thanks to each of them are recorded in the Acknowledgements.

While I have had access to all Chadwick's papers in the Churchill Archives Centre, his file at the Archives of the Royal Commission for the 1851 Exhibition at Imperial College, London, the Public Record Office in Kew, letters in the Niels Bohr Archives, Copenhagen, Cambridge University Library, and in the Bancroft Library at Berkeley, not to mention the National Archives near Washington, there are still documents and letters which remain classified. Some are retained by the UK Atomic Energy Authority at Harwell, and others at the Cabinet Office in London. Some of the wartime letters between Chadwick and Peierls that have never been released in England were available at the National Archives, but possibly as a result of the Gulf War, they were recently recensored by the US authorities.

Andover A.P.B
July 1996

Acknowledgements

Although I had this book in mind for some years, it was only after moving to the USA in 1990 that it became a realistic proposition. I started, and finished, my research with no qualifications as a physicist or as a historian. Almost everyone I approached for help, responded generously and with warm encouragement. My initial boost was from thc late Sir Nevill Mott, who was at the 1932 meeting of the Kapitza Club when Chadwick announced the discovery of the neutron. He arranged for an invitation to the Chadwick Centenary Meeting in Cambridge in 1991, through the good offices of Yao Liang and the Cavendish Laboratory, and Peter Gray, the Master of Gonville and Caius College. Peter put me in touch with Chadwick's daughters, Judy and Joanna; they have read each chapter as it appeared and helped with numerous details, as well as supplying photographs. The Caius old-boy network alerted Ernest Pollard, who worked as Chadwick's research student in the late 1920s, and he has fired off letters and phone calls to me to check on progress in the most positive way.

At the 1991 meeting, I was taken under the wing of Margaret Gowing, the historian, who introduced me to Sir Mark Oliphant and the late Sir Rudolf Peierls. These two remarkable men, apart from making wonderful after-dinner speeches, corresponded with me and read several chapters. Through Sir Rudolf, I came to know Lorna Arnold, the historian at the UK Atomic Energy Authority, who supplied me with invaluable pointers to material in the public domain, and lucidly conveyed her deep knowledge of wartime and postwar atomic history. A chance encounter in the queue for tea at the Cavendish led to a mostly one-sided exchange with Jeff Hughes, a historian of science who was finishing his Ph.D. thesis on the Laboratory. He saved me enormous amounts of time, and also kindly but firmly corrected some wayward tendencies in interpreting events. He directed me towards material that I might have otherwise overlooked—most importantly in the Royal Commission for the Exhibition of 1851 Archive, where Valerie Phillips could not have been more charming or helpful.

Charles Johnson, the current Lyon Jones Professor of Physics in Liverpool, put me in touch with David Edwards, who is a notable historian of science on Merseyside. He acted as my guide on a whirlwind tour of Liverpool, including a productive visit to the University Archive, and also introduced me to Dave King, who had just completed a thesis on the Liverpool synchrocyclotron. John Holt, an early student of Chadwick's in Liverpool and later professor there himself, gave me transcripts of various talks he had given about Chadwick and amplified other material I had gathered from the archive. Maurice Pryce and the late

Bernard Kinsey were two other members of the wartime department who wrote to me. Joseph Rotblat, who was so close to Chadwick in Liverpool and at Los Alamos, was eventually in London during one of my visits and gave me a three hour interview, which he supplemented by meticulous comments on material that I wrote.

Dorothy Dennis of the Bollington Public Library and George Longden unearthed local information that allowed me to flesh out the meagre facts at my disposal concerning Chadwick's childhood. The acquisition of other published information was accelerated by Bruce Kupelnick, whose intimate knowledge of the Widener Library at Harvard allowed the most obscure journals to be located in the stacks. It was Bruce, too, who alerted me to a lecture to be given at Harvard by Maurice Goldhaber, on his time at the Cavendish. Once again I had the thrill of presenting my account of events to a man who was there and recalled them in fascinating detail. My meeting with him led to further information from Hans Bethe.

Unexpected help came from two of my patients in New Hampshire. Ernst Gutbier translated Chadwick's 1914 paper on the continous β-spectrum for me. The late Heinz Nitke, who was just starting a physics Ph.D. in Berlin in 1932 when Chadwick discovered the neutron, eagerly straightened out some basic scientific misconceptions for me, and with equal tact improved my English on more than one occasion.

Lord Bauer is the economist mentioned in the introduction, and he patiently tried to explain the nuances of the Peasant's Revolt to me. Francis Crick wrote to say that he did not even know about the affair until it was over! David Shoenberg entertained me to lunch at Caius and provided insight into Chadwick's relationship with Kapitza, as well as recollections of his difficulties as Master. His fellow Caian, the medieval historian Christopher Brooke also talked to me about the mastership and helped me on diverse matters of college history. Sir Denys Wilkinson, wrote me a letter coruscating with anecdotes. Lord Sherfield spoke to me about life in the British Embassy in Washington during the 1940s. Aage Bohr recounted his wartime memories of Chadwick and the close friendship with his father.

Such oral and informal material was invaluable. On many occasions I could check it against historical documents and my admiration for the clarity and accuracy of my informants' memories for events of five and six decades ago is boundless. I apologize if I have forgotten any other contributors. The major collection of Chadwick's papers is held in the Churchill Archives Centre in Cambridge, where my research was constantly assisted by Alan Kucia and his staff. Anne Neary, the archivist at Gonville and Caius College, was always ready to help. Hilary Coote provided photographs from the Cavendish Laboratory. The Public Record Office in Kew were unfailingly efficient in producing wartime files, and I was made welcome by Jerry Lee at the Cabinet Office Historical and Records Section.

Anders Bárány persuaded the Nobel Committee for Physics to release material concerning Chadwick's nominations to me. Finn Aaserud, the Director of the Niels Bohr Archive in Copenhagen, has been both a stimulating critic and an invaluable expert; Felicity Pors has also helped on several occasions with material from the archive. The staff of the Niels Bohr Library at the American Institute of Physics, especially Joe Anderson, have trustingly sent me much valuable material. At the National Archives in Washington, I benefited from the encyclopaedic knowledge of John Taylor, like many before me, and he also put me in touch with Stanley Goldberg, who is writing a biography of General Groves. When the Manhattan Project files were moved to the Archives 2 in College Park, Maryland, Marjorie Chalante located some important documents for me. Peter Hanff and Bill Roberts of the Bancroft Library, University of California at Berkeley, provided copies of the cache of letters between Chadwick and Lawrence. Dale Mayer archivist at the Herbert Hoover Library, West Branch, Iowa, did some diligent searching for me. I would also like to record my gratitude to library staff at the Massachusetts Institute of Technology, Memorial Library Andover, and the University of Cambridge.

My good fortune extended to having Oxford University Press as my publishers. Donald Degenhardt, the commissioning editor, could not have shown more trust in a tiro author. He moved on to new pastures after handing over to Susan Harrison. I am indebted to her as well as Janet Walker and Keith Mansfield for their expertise and enthusiasm. I am grateful to Susan Marcus for preparing the index.

One of the enriching pleasures of writing a book, I have discovered, is that it brings one into contact with new people, like all those mentioned above. It also offers the opportunity to entice old friends to indulge in one's latest obsession. My father-in-law, John Ballantyne, and Michael St. Clair have been devoted readers of the unfolding manuscript. Penny Phipps freely applied her famous powers of persuasion. Broad support and valuable advice have come from Thelma Bates, Bleddyn Jones, Graham and Betty Hines, Tom and Nicki Shields, Charles and Annabel Merullo, Jane Maddox, Mary Catterall, Frank Bunn, Cathy Shaer and Ben U. My partners at Radiation Oncology Associates have feigned interest in radiological history, and permitted me to make sudden visits to England.

The drawbacks of authorship have been more apparent to my family. My wife has displayed a surprising degree of tolerance as papers spilled over from my desk onto hers. To our three sons, the answer to their most frequently asked question is 'twenty'. The book is dedicated to them with my love.

Copyright permissions

Copyright permissions

[illegible] We thank [illegible] and [illegible]

[illegible] the [illegible] Cambridge [illegible] UK.

[illegible] reproduced [illegible]

[illegible] the Physics Department [illegible]

[illegible] have been [illegible] of the [illegible] for permission to reproduce them in this volume.

Contents

The Plates are to be found between pp. 142 and 143

1 Obscure origins

Bollington lies in the lee of the Peak District, sheltered by hills from the north and to the east. The land on which it stands was carved from the Macclesfield forest, one of England's great medieval hunting reserves. A climb to the top of Kerridge hill to the east of the village is rewarded by a grand panoramic view. Looking to the south-west, you will see the old market town of Macclesfield about three miles distant, standing high on its own escarpment, and below you, Bollington cradled in a wide valley which broadens into the fertile Cheshire plain to the west. On this precipitous windswept ridge, the Gaskell family erected a beacon to commemorate the Battle of Waterloo in 1815. In fact it was more than a beacon—a white stone summerhouse built in a conical shape so that it resembled a giant beehive. The villagers christened it White Nancy and it became the symbol of Bollington.[1]

During the first hundred years of its existence, White Nancy witnessed radical changes in the village below as the Industrial Revolution transformed the appearance of the valley. By the beginning of the nineteenth century, all the necessary inventions had been made to mechanize the production of cotton, which until then had been a true cottage industry. Across the Pennines in Derbyshire, Arkwright had opened the first power driven mill in 1771.[2] His carding machine, which separated out the cotton fibres, and his water-frame, which spun the fibres into long thread, were also engines of social change and led to the spectacular expansion of the industry over the next seventy years. Lancashire became the cotton county, but Bollington had enough natural geographic advantages to become a prime site on its own.

The first requirement was water, both in a fast flowing river to provide power and a humid atmosphere to help the cotton fibres cling together by reducing the electrostatic charges created by friction. This permitted spinning to take place at higher speed without breaking the fibres—an important economic advantage. There had been a watermill in Bollington since the time of Edward III, and its annual rainfall made it as damp as anywhere in England. Coal was also plentiful with the Adlington and Poynton collieries a few miles to the north. During the first half of the nineteenth century, the steam engine replaced water as the primary source of power and Bagshaw's directory for 1850 tells us that the two together provided 'four hundred and fifty-eight horses power' to the Bollington mills. The early mills were constructed from the local Kerridge free stone, quarried from the adjoining sandstone hills. The stone was prized for its 'whiteness' by church stonemasons—it has a pale golden colour—and the huge cotton mills presented a noble appearance.

In 1831 the Bollington cotton business received another boost with the opening of the Macclesfield Canal. It provided a convenient artery for transporting coal to the mills; even though it was eclipsed in many ways by the coming of the railway in 1869, 95 per cent of the coal in 1880 still came by barge. The canal also gave the Bollington producers better access to Manchester, the major cotton marketplace 18 miles to the north-west. The industry prospered and further expansion took place in the 1850s with the building of the Adelphi and Clarence Mills by the Swindells family.[3] These were constructed on the banks of the canal and came to dominate the landscape. Massive buildings with flat roofs, they were buttressed by square towers (which contained much needed dust extractors) and enlivened by large, rectangular windows. Their tall tapering chimneys spewed black smoke that could shroud the valley in gloom. Within a few years, apart from the canal, they were also served by the Macclesfield, Bollington and Marple railway,[1] which ran past their front doors as it bisected the village in a south–north direction. The leading textile manufacturers had appointed themselves to the committee that determined the route the line should take through Bollington. Their influence ensured a direct and ready means of access to all their markets and indirectly led to the construction of a graceful stone-arched viaduct at Bollington which carried the track over the River Dean.

The success of the cotton industry meant that the population of Bollington trebled from fifteen hundred in 1811 to four and a half thousand in 1851.[3] Rows of stone cottages sprouted in the shadows of the mills to provide homes for the workers. The 1851 census showed that only one-quarter of the people living in the area closest to the mills were born outside the county of Cheshire. This undermines the local folklore that Bollington is 'a Derbyshire village, in Cheshire, peopled by Lancashire folk'. The increased numbers needed not only dwellings but places of worship and schools for their children which were provided, often on adjoining sites. In 1851, however, the legal minimum age for employment in a cotton mill was still only nine years.

With the exception of the cotton famine caused by the American Civil War (1861–1865), when half the families lost their entire income and two thousand depended on charity,[3] the second half of the nineteenth century was a time of gradually increasing prosperity and leisure for the Bollington mill workers. During this period cotton brought in about one third of Britain's overseas earnings, and the Lancashire mill owners boasted that they met the needs of the home market before breakfast and devoted the rest of the day to exports.[2] The Bollington mills had a reputation for supplying fine cotton to the lace makers of Nottingham, and Mr Oliver proudly told his workers that 'the goods they produced were worn by the very richest people, the most beautiful people, and he didn't think there was a crowned head in the civilized world who had not got in his household some of Oliver's cotton'.[3] From the earliest days the spinners and weavers had banded together in their own trade unions and now took leading roles in educational, religious and sporting societies. In July the mills shut for the Wakes week holiday and the town resounded to the noises of the fairground, steam organs and the Bollington Brass Band.

James Chadwick was born in this lively and industrious place on 20 October 1891. His parents, John Joseph and Anne Mary, were both in their late twenties. Anne, whose father was a gardener, worked as a domestic servant. James was her first child and was named after his paternal grandfather, who was a hand-loom, silk weaver. In later years, Chadwick liked to say that he was born in the country and that his parents were country people.[4] His birth certificate shows that he was born in Clarke Lane, on the southern outskirts of Bollington, where there were indeed several small farms. The family home was probably one of a row of rustic, stone cottages which were occupied by agricultural labourers, and quarry and mill workers. From there, his father, a cotton spinner, could walk to work at the Adelphi Mill along the canal towpath. Despite the general improvement in conditions, cotton spinning remained a deafening and dusty occupation. The workers were not particularly skilled and by the end of the century the British industry faced competition around the globe. In one cyclical downturn in 1895, so many Bollington cotton workers were out of work that it was suggested they might be usefully employed in road mending. It may well have been this slump that prompted John and his wife to leave Bollington and the mills to seek a better life in Manchester. James, as yet their only child, was left behind in the care of his grandparents.

As an old man, James Chadwick could remember, or was prepared to tell, little of his boyhood.[4] He absorbed from his grandfather a love of gardening, and remembered the enjoyment of climbing trees. His early schooling was at the Bollington Cross School,[1] which was built in the 1840s on land given by Samuel Greg Jnr., another cotton merchant. It was designed and used both as a school and a church. There was a distinctly ecclesiastical style to the building: the schoolroom, illuminated by stained-glass windows, contained an organ, choir stalls, lectern and pulpit. The headmaster was Herbert J. Sutton who doubled as the choirmaster and was also active in village life as secretary of the Horticultural Society. The headmaster's son, Herbert Jnr., was James' closest boyhood friend and shared the spoils of at least one early adventure. The two boys took a day off from school to go bird-nesting; seventy years later Chadwick retained a 'quite vivid memory of being caned very severely for it'.[4] The closing years of the century gave the villagers many opportunities for public celebrations, and James Chadwick would have surely witnessed or participated in some of them. In 1897 Queen Victoria's Diamond Jubilee was marked by huge tea parties for the schoolchildren, and bonfires were lit by White Nancy. The first reservists were given a joyful send off to the Boer War from the railway station in 1899, and the Relief of Mafeking in 1900 was the occasion for impromptu parties and processions.[1] This time a flag was placed on White Nancy which temporarily became the 'Spion Kop'* and the target of spirited attacks by the children.

* A mountain near Ladysmith which had been captured by the British and then lost to the Boers in a bloody counter-attack.

When in later years Chadwick occasionally spoke about his childhood in Bollington, it was usually to recall his grandmother with affection. Relations with his own parents, particularly his father, seem to have been less happy. He once remarked rather darkly: 'There are some things that I remember as a very small child but I don't tell them to anybody.'[4] At some time in the first years of the century, Chadwick left Bollington and joined his parents in Manchester. By then they had two other sons, Harry and Hubert, and had lost a baby daughter. It is easy to imagine why the transition was difficult for a sensitive ten year old boy to make: the simultaneous loss of his doting grandmother, his friend Sutton and the simple delights of the rolling, wooded fields of Cheshire. In return, a crowded struggling household in a grimy backstreet of a big city, where he as the eldest child was expected to take all sorts of responsibility for younger brothers whom he scarcely knew. In his teenage years James began to show an intellectual precocity that may have added further strain within the family and increased his sense of solitude. Looking back, he made the laconic admission 'I'm afraid I don't have a great family feeling.'[4] This lack of affection is borne out in the paucity of his remembrances.

John Chadwick had started a laundry in Manchester, but the venture did not prosper. James never seemed inclined to give his father any credit for initiative and, in later life, was scornful of this unsuccessful enterprise. He thought that John Chadwick must have borrowed the capital to start the business from relatives, 'probably of my mother's'. To him, with the benefit of hindsight, the seemingly inevitable failure was caused by his father's shortcomings. 'He didn't make a success of it. It wasn't the kind of thing he could do. He had to take a job on the railway as a storekeeper in a small way. We were very poor. Well, poor anyhow.'[4]

Although Chadwick said that he did not take school in Bollington Cross very seriously, he did well enough to be offered a scholarship to Manchester Grammar School. But he could not take it up because there were still small fees to be paid that were rather more than his parents thought they could afford. He went to the less prestigious, but new, Central Grammar School for Boys in Whitworth Street. Chadwick viewed this as only a slight misfortune and remembered the teaching as extremely good.[4] The curriculum, at least until the fifth or sixth form, was broad and his one regret was that there was no Greek, only Latin. He was clearly well taught in English to judge by his own concise literary style; he had an ingrained aversion to the grammatical slips of others. It was only in his senior school years that he displayed real ability in science—for mathematics and physics, not chemistry. His sixth form master, Mr Wolfenden, whose own strong interest in mathematics impressed Chadwick, helped this academic development. Chadwick remembered him as an encouraging teacher to whom he owed a very great deal.

By the age of sixteen, Chadwick seems to have exhausted the resources of the school. Although, on reflection, he thought his accomplishments in mathematics and physics at that stage were quite modest, they were clearly of an order to mark him out from his fellows. Wolfenden entered him for two university scholarship examinations, both of which he won. Mindful of his own immaturity and shyness,

Chadwick was reluctant to leave school, but Wolfenden persuaded him that there was nothing left for him to learn unless he went into a class of his own. Eventually convinced that such isolation would not be in his best interests, Chadwick, as he disarmingly put it, decided 'I had to go'.[4] He could keep only one of the two scholarships, a detail which mildly rankled him, and this was naturally the more valuable one awarded by Manchester Education Committee.

Poised on the verge of a university education, Chadwick had already come a long way from his humble beginnings. A generation or so earlier, such an ascent would have been much less likely. By Chadwick's day, the system had advanced far enough to ensure that a boy showing great promise stood a good chance of being selected for further education. The secondary schools in the North of England had a fine tradition of science teaching stemming from the dissenting academies of the eighteenth century, which took the place of universities for those who did not conform to the Church of England. Priestley and Dalton were two dissenters who had shown that there was more to education than classics.

Chadwick's poor background may have shaped his school career to a minor extent, but did not prevent him from receiving an excellent general education. Whatever their limitations, personal or financial, his parents allowed him to remain at school and did not insist that he leave at the earliest opportunity in order to supplement the household income. His childhood was dislocated by his father's ill-fortunes, and although he received practical parental support there does not seem to have been a close loving relationship. Chadwick never sought to make a virtue of his early hardships and tended to be unforthcoming about this period of his life, even with his own family. A writer[5] in the 1960s described the people of Bollington as being of 'independent nature . . . close-knit, dour, inbred, different.' Those acquainted with James Chadwick later in life might have recognized several of these characteristics in the man, even if, as was likely the case, they knew little of his background.

References

Charles Weiner's oral interview with Sir James Chadwick, conducted in Cambridge in April 1969, is a major source of information for this book. It is quoted with the permission of the Niels Bohr Library, Center for History of Physics, American Institute of Physics, College Park, Maryland.

1 Longden, G. and Spink, M. (1986). *Looking back at Bollington.* Willow Publishing, Timperley, Altrincham, Cheshire.

2 Aspin, C. (1986). *The cotton industry.* Shire Publications, Princes Risborough.

3 Wilmslow Historical Society (1976). *Cotton town: Bollington and the Swindells family in the 19th century.*

4 Weiner, C. (1969). Sir James Chadwick, oral history. American Institute of Physics, College Park, Maryland.

5 Ingham, M. (1964). *The happy valley.* Macclesfield Press.

2 Manchester—the nuclear nursery

Edwardian Manchester was a bustling, cosmopolitan city. As the mercantile centre of the cotton industry, its most characteristic scene was 'the great horse-drawn dray, slowly carrying bales of cotton to the warehouses'.[1] The merchants were independent minded in many respects, but shared a common regard for mammon. They wanted 'solid worth' so that 'the houses were well built, the women plump [and] the food substantial.'[1] The scale and success of their trading in cotton ensured a comfortable living for a supporting cast of lawyers, bankers and insurance brokers. The technical needs of the industry and the extensive transport networks that served it, brought to Manchester some of the finest engineers in the country. Many were local men who had acquired their skill and professional training at Owens College, which was founded by a bequest from a wealthy textile magnate in 1851 and was the forerunner of the Victoria University of Manchester. This was a remarkable period of symbiosis between commerce and higher education. Nineteenth-century science began with John Dalton, a Quaker schoolmaster, announcing his atomic theory of matter to the Manchester Literary and Philosophical Society: it closed with J.J. Thomson, a graduate of Owens College who had risen to become director of the Cavendish Laboratory in Cambridge, discovering the electron in 1897. In the years between, John Prescott Joule, the son of a Manchester brewer and a pupil of Dalton's, had been the first to establish the equivalence of mechanical, electrical, chemical and thermal energy.

The majority of its citizens had no idea that Manchester was the locus for such virtuous activities. They did know that they were toiling long hours for meagre reward and were living in cold, damp and cramped accommodation, where the death of a young child from infection was not uncommon. Nowhere in England was the inequality in income and material possessions more extreme, and the movement for social change gained important momentum from Manchester. This came to a head in the 1906 general election, which was precipitated by the resignation of the Conservative prime minister, A.J. Balfour. He had been Member of Parliament (MP) for Manchester East for over 20 years, and was justly regarded as the leading intellectual of his party. He had a genuine interest in science and this was, no doubt, reinforced by his sister's marriage to Lord Rayleigh, the first Englishman to win the Nobel Prize for Physics. Balfour resigned because his government had become hopelessly split over the issue of free trade. This was a subject which resonated powerfully in Manchester, where many businessmen were in favour of protectionism, but there was, equally, a strong *laissez-faire* tradition. A leading proponent of free trade, Winston Churchill, had deserted the Conservative Party over the issue in 1904 and was now standing as the Liberal candidate in Manchester North-West. His constituency contained some of the city's worst slums and while canvassing there, he was moved to remark to his private secretary: 'Fancy, living in one of these streets—never seeing anything beautiful—never eating anything savoury—*never saying anything clever!*'[2] The election was won in a landslide by the Liberals, and Balfour was its most notable casualty. The *Manchester Guardian* gave eloquent expression to the aspirations of the new voters and was easily the most influential newspaper of the brief age of the New Liberalism.

Life at the Victoria University mirrored the intellectual and political life in the city. A.J.P. Taylor wrote that 'Manchester is the only place in England which escapes our characteristic vice of snobbery . . . Many of the Burghers were German in origin and, having shaken off subservience to their own authorities, felt no awe of any other.'[1] Sir Arthur Schuster, Professor of Physics at the university and man of independent wealth, was descended from exactly this type of pedigree.[3] He had been born in Frankfurt into a prosperous Jewish family with a long established textile business. When Frankfurt was annexed by Prussia in 1870, his father moved the business and his family to Manchester. Arthur Schuster became the Langworthy Professor of Experimental Physics at Owens College in 1881 and transferred with this title to the new physics laboratories at Victoria University in 1900. By 1906 he found himself 'drifting away from the experimental side and [wanting] more leisure for working at a book'.[4]

Schuster decided that he would give up the Langworthy Chair providing that he could arrange for the greatest experimental physicist of the age, Ernest Rutherford, to succeed him. Rutherford was then Professor at McGill University in Canada, where he had made his name in research revealing the nature of radioactivity. His book, *Radioactivity*, first published in 1904, had quickly run to two editions—the second considerably longer than the first, reflecting the burgeoning nature of the subject.[5] Negotiations over the Langworthy Chair were conducted through a series of letters between the two men over a six-month period with almost no input from the university administration.[4] The correspondence involved honest disclosures, mixed with entrepreneurial tactics on Rutherford's side, and culminated in his arrival in Manchester in May 1907. Schuster handsomely sealed his selfless bargain by personally funding a new readership in mathematical physics, because he wished to encourage a closer relationship between experimental and theoretical research.

Research into radioactivity was the consuming interest of Rutherford's life, and he took care prior to accepting the appointment that nothing would interfere with this pursuit. The university promised him that it was 'strongly in favour of the development of research work'[4] and that he would not be expected to give more than five lectures a week. He inherited from Schuster a modern and well equipped department, a prince amongst laboratory stewards, William Kay, who 'always appeared to have an instinctive knowledge of the anatomy and pathology of any piece of apparatus',[5] but no radium. This deficiency was corrected after a generous gift of about 500 milligrams of radium bromide from an admiring colleague, Professor Stefan Meyer of the Radium Institute in Vienna.[4] So the Manchester school of radioactivity was launched and, under Rutherford's inspirational leadership, it would soon be publishing results that would revolutionize science.

James Chadwick, having secured his scholarship, came to the university for a preliminary interview prior to the autumn term in 1908. The interviews were held in a large assembly hall divided into sections by low benches. The physics and mathematics sections were close together and he sat on the wrong bench. He 'had no intention whatever of reading physics' but when he started answering questions on 'what I'd done and so forth, I found . . . that he was a lecturer from the physics department and I was entered for physics'.[6] Chadwick, too shy to admit to his mistake and touched by the genuine and sympathetic manner of his interlocutor, thus started his life as a physicist by default.

Like most students at Manchester, he continued to live at home. This was about four miles from the university and Chadwick walked there each day. The first year's physics class was large and unruly, comprising not only honours physics students but a lot of noisy engineers. Schuster had warned Rutherford that he at first had to struggle to maintain order in the elementary lectures, and Chadwick was struck by 'a large number not caring two hoots whether they heard what was happening or not, and deliberately trying to upset the lecturer'.[6] They need not have feared for the new professor, who could more than match any rumbustious student. He wrote to his friend Bertram Boltwood, Professor of Chemistry at Yale, soon after arriving in Manchester:

> I am now in full swing. Classes are all going. I lecture to one of 150. I have started a course on Radioactivity especially for the chemists & have a roll up of 70. This is a pretty active place and but for its climate has a number of advantages—a good set of colleagues, a hospitable & kindly people and no side any where . . . I find the students here regard a full professor as little short of Lord God Almighty. It is quite refreshing after the critical attitude of Canadian students. It is always a good thing to feel you are appreciated.[7]

Although unimpressed by the tenor of the first year course, Chadwick showed sufficient promise to be awarded a share of the Heginbottom Scholarship in Physics. In the second year the class was much smaller, and Chadwick had his appetite whetted by 'the first stimulating lectures I had ever had in physics'.[6] These were given on electricity and magnetism by Rutherford who, when he was forced to give an elementary talk, delighted in demonstrating an experiment to the audience with the able assistance of Mr Kay. For this course, Rutherford described one of his early experiments from New Zealand, in which he had magnetized thin iron wires by alternating currents acting at very high frequencies. He then dissolved the outer layer of the wire in hot nitric acid to see if the magnetization was more than skin deep. To Chadwick, Rutherford talked as though 'the whole thing was alive',[6] and for the first time Chadwick's imagination was seized by the physical process involved, not just the mathematics.

At the end of the second year, the bulk of undergraduates concentrated on the more practical aspects of chemistry or electrical engineering, and it was customary for the few honours physics students to embark on a research project of Rutherford's choosing; they received no further formal tuition. This policy was the embodiment of Rutherford's firmly stated belief[8] that:

> No university . . . is worthy of the name that does not do everything in its power to promote original research in its laboratories . . . The opportunity for investigation should not be confined to the teaching staff but should be extended whenever possible to the advanced student who shows promise of capacity as an original investigator. In this way a laboratory becomes a live institution instead of a dull, if somewhat superior, high school.

Rarely can such an imaginative and risky academic strategy have been pursued so happily as it was in Rutherford's Manchester department. There were drawbacks which its architect would have regarded as minor, if anyone had had the temerity to bring them to his notice. Chadwick[6] summed them up as follows:

> I had half an education in physics. There were whole aspects of physics that I knew little about. I read about them, of course, but I must confess I knew little about them, and certainly did not really understand them until I had to lecture in Liverpool.

His close friend and exact contemporary at Manchester, Harold Robinson,[5] felt the same sense of limitation:

> This scheme of undergraduate training . . . had the grave disadvantage of leaving wide gaps in our knowledge of some important sections of classical physics—and this is putting it euphemistically, for the word gap normally connotes at least something more or less solid in which a gap can exist.

The drawbacks were surely outweighed by the sense of privilege each young Manchester student felt from 'living very near the centre of the scientific universe'.[5] And as Sir Ernest Marsden,[9] who as 'a callow youth from Blackburn' began the undergraduate course in 1906, pointed out, it did not take so long to approach the frontier of new knowledge in those days.

Chadwick's third-year research project had illustrious origins. Just over a decade earlier, in 1898, the Curies through back-breaking effort had separated minute amounts of the highly radioactive element, radium, from 100 kg of pitchblende ore. At about the same time, Rutherford was studying the radiation from uranium by the typically simple device of wrapping the radioactive material in increasing layers of aluminium foil. This led him to the conclusion 'that the uranium radiation is complex, and that there are at present at least two distinct types of radiation—one that is very readily absorbed, which will be termed for convenience the α-radiation, and the other of a more penetrative character, which will be termed the β-radiation.'[10] He also examined the radiation from thorium and noted the very penetrating radiation that would soon be named γ-rays. In collaboration with the chemist Frederick Soddy at McGill University, again using thorium, Rutherford went on to show how the emission of alpha (α) or beta (β) particles accompanied the transmutation of one element (mother) into another (daughter).* The two men further demonstrated that such transmutation or radioactive decay follows an exponential time course. This introduced the concept of the 'half-life' of a radioactive element, that is the time it takes for the substance to lose half its activity. Each radioactive substance has a characteristic half-life (which can be from a few seconds to thousands of years), and if 100 units of activity are present initially, there will be 50 units after one half-life and 25 units after two half-lives, etc. Herein was the problem—there were no internationally agreed units of radioactivity (unlike, say, for length, weight and time) so the comparison of results obtained in different laboratories engaged on radioactivity research was uncertain, to say the least. The inherently unstable nature of radioactive substances precluded a simple solution, such as an agreed mass of a pure sample of a given element (because it would steadily decay into daughter and grand-daughter substances).

In September 1910, Rutherford was a British delegate to the International Congress of Radiology and Electricity in Brussels at which the question of a radium standard was at the top of the agenda. In the event, the Congress was a badly organized, multi-lingual fiasco culminating, according to Rutherford[11], in a 'regular bear garden' at the closing session, where the 'audience got out of hand and started whistling and boo-hooing'. The report of the International Radium Standards Committee (of which Rutherford was a member) was approved, nevertheless, and their suggestion that the standard should take the form of the amount of radioactive energy released by a gram of radium was adopted. The unit would be known as the *Curie*, and Madame Curie was charged with preparing pure source of about 20 milligrams which would be used for comparative measurements.

On his return to Manchester, Rutherford wanted to devise a method which could be used to make accurate intercomparisons of different radium sources. His idea was typically simple—to balance the ionization current due to the gamma (γ) rays from the radium source in one chamber against the known ionization current produced in a second chamber by the alpha and beta particles of uranium oxide. The physicist chosen to perform this internationally important work was James Chadwick, aged nearly 19 years, and on his own admission very shy and afraid of Rutherford. The tiro seems to have grasped his master's intentions immediately, and even identified a few impediments.

> I'd seen a perfectly obvious little snag in the arrangement that Rutherford had suggested as soon as I began to work, but I was so afraid of him that I daren't mention it. Of course, when I showed him some of the measurements he saw at once what the trouble was, and I think he was rather disappointed in me that I hadn't pointed it out. I had seen it, but I was so afraid of him that I thought he must have seen it himself and there must be some reason why he wanted it this way . . . In those early days in Manchester the only thing that

---

* For his work explaining the nature of radioactivity, Rutherford won the Nobel Prize for Chemistry in 1908.

mattered was getting on with the experiments that he was doing—the things he was particularly interested in himself . . . he was a pretty hard taskmaster.[6]

After this timorous start and having disposed of the early flaw in the method, Chadwick successfully completed the exercise, working without supervision except for occasional visits from Rutherford to 'see how things were getting on'. This period marked an exceptional level of activity in the Manchester department, with the ebullient professor leading the way. In October 1911, he wrote to Boltwood: 'I have been laying hard into experimental work and also into my book, and am making very good progress in both. The Laboratory is full, and work is in full swing. You will have seen that the last number of the *Phil. Mag.* was unusually radioactive.'[12] Rutherford was notoriously impatient for results and 'the young research student had often to exert some will-power to resist being unduly hurried'[9]. He liked to 'ginger up' the workers in his laboratory, and as there was an actual date by which Chadwick's technique had to be ready for international use, the ginger may have been quite hot. Except for rare, explosive outbursts, when Rutherford lost control of his temper, his impatience was a reflection of a burning intellectual curiosity and not unkindly meant. He always welcomed the eventual results with enthusiasm, and the anticipation of his praise made one young student 'understand exactly the feelings of a fox-terrier as, after killing a rat, he brings it into the house and lays it on the drawing-room carpet as an offering to his domestic gods'.[5]

Rutherford met Mme Curie again at the first Solvay Conference in Brussels in the autumn of 1911. The conference was the idea of Ernest Solvay, the Belgian industrial chemist and philanthropist, who invited the world's leading physicists to congregate at the Hotel Metropole with all expenses paid plus a generous one thousand francs per head. Mme Curie told Rutherford that she had prepared the radium standard, but wished to keep it in her laboratory in Paris for reasons which seemed to Rutherford partly sentimental and partly scientific. He pointed out to her that 'the Committee could not allow the international standard to be in the hands of a private person'.[13] Eventually a compromise was reached and Stefan Meyer in Vienna was commissioned to prepare a duplicate standard. The International Committee met again in Paris in March 1912 to compare the two standards, and although Rutherford was confident about the balance method that he and Chadwick had devised, he was still apprehensive about the outcome. He confided to Boltwood[14] on 18 March:

I have not much doubt but that the two standards will be found in very good agreement, but it will be a devil of a mess if they are not. That is one of the reasons why I must be there to act as arbitrator between the two parties. I have developed a balance method of a γ ray comparison, with which I think I can compare two nearly equal standards [to] an accuracy of 1 in 1000. I suppose, however, we shall not worry if the agreement of the two standards is within 1 in 300 or 400.

This account seems less than generous to Chadwick, who was not mentioned as his collaborator, whereas Robinson, Hevesy, Geiger, Russell, Marsden, Miss

Leslie, Florance, Bohr and Moseley all receive recognition in the same long letter for their work in the department. Chadwick's omission was almost certainly unintentional and of no consequence in a letter between friends, but the fact that he of all 'the boys' could have slipped Rutherford's mind reflects his quiet, self-effacing, nature.

The balance method was put to the test under the eyes of a distinguished panel of scientists who gathered in Paris to compare the two standard sources, and it was sensitive enough to demonstrate that they were equally radioactive—at least to within 1 part in 300. With relief all round, the Curie radium sample was then adopted by the International Bureau of Weights and Measures as the agreed world standard. The method and some of its applications formed the basis of Chadwick's[15] first published paper (jointly with Rutherford) in 1912.

Harold Robinson was a more self-confident young man than James Chadwick and made friends easily with other scientists in the department. Rutherford seemed to have a particular liking for him, if for no other reason than Robinson at six foot four inches was the only man in the laboratory to whom he had to look up. In 1942 Robinson[5] gave the first Rutherford Memorial Lecture and described those Manchester days thus:

> We saw him at his best and most inspiring at the physics colloquium, which met on Friday afternoons. The meetings, at which we were joined by large numbers of chemists and mathematicians who came over to hear Rutherford, were preceded by an enormous tea-party, generally presided over by Lady Rutherford. Rutherford always addressed the first meeting of the session, giving with obvious enjoyment a summary of the main researches carried out in the laboratory during the preceding year. I am fairly certain that this inaugural address was always called 'The progress of physics, 1907–1908', or whatever the session might be . . .

The preparation of these annual lectures cannot have been too taxing, since so many of the important advances in physics (or at least in radioactivity) took place in Rutherford's own department. The astounding discoveries of his Manchester years more than justify the famous retort to his friend and biographer, Eve[16], who once said to him: 'You are a lucky man, Rutherford, always on the crest of a wave!' The answer came, with a laugh: 'Well! I made the wave, didn't I?'.

Since his pioneering experiments in Canada, Rutherford had believed that α-particles were helium atoms carrying a positive double charge. He told the Royal Institution[17] in January 1908 that 'the determination of the true character of the alpha-particle is one of the most pressing unsolved problems in radioactivity, for a number of important consequences follow from its solution.' That month, the stock of radium arrived from Vienna and he made use of it straight away to attack this problem with the able young assistant, whom he had also gained from Schuster, Hans Geiger. Using a prototype of the electrical counter that would make his assistant a household name, they quickly made the first direct estimation of the number of α-particles emitted per second per gram of radium. They went on to show that the α-particle carried a double positive charge. The decisive proof that the α-particle was a charged helium atom came from an ingenious experiment by Rutherford and Royds[18], another young graduate scholar, in 1909. Their demonstration depended crucially on the exquisite skill of a local glass-blower, Mr Baumbach, who provided a tube with walls so thin (less than 1/100 mm) that α-particles could escape from a radium source within it into a surrounding low-pressure chamber. After allowing the α-particles to collect in the outer tube for a few days, an electric discharge was passed through the chamber which revealed the characteristic line spectrum of helium.

Rutherford, recognizing Geiger's outstanding dexterity and gluttonous appetite for work, shrewdly promoted him to Research Assistant and made him responsible for training research students in experimental radioactivity. Ernest Marsden was detailed to work with Geiger during his third year as an undergraduate in 1908–1909. One day during a discussion with Geiger, Rutherford turned to Marsden[19] and said, 'What about trying whether you can get α-particles reflected from a solid metal surface.' Within a week, Marsden was reporting to Rutherford that he had been able to observe just such an effect, and that about 1 in 8 000 α-particles incident on a platinum foil was reflected back, instead of passing through. Long afterwards, Rutherford was to say 'It was quite the most incredible event that has ever happened to me in my life. It was almost as incredible as if you had fired a 15 inch shell at a piece of tissue paper and it came back and hit you.'[20]

Geiger and Marsden[21] duly published their paper 'On a Diffuse Reflection of the α-particles', which was read to the Royal Society in June 1909. Rutherford, whose curiosity about the huge electrical or magnetic forces at play in the atom that could deflect the fast-moving α-particles led to the work in the first place, pondered the extraordinary findings, unable to explain them. It was not until Christmas 1910 that he entered Geiger's room, 'obviously in the best of spirits and told [Geiger] that he now knew what the atom looked like and what the large deflexions signified. On the very same day, [Geiger] began an experiment to test the relation expected by Rutherford between the number of particles and the angle of scattering'.[22] Geiger quickly confirmed the predicted relation, and Rutherford was ready to present his theory of atomic structure to the world.

His chosen forum was the Manchester Literary and Philosophical Society, known colloquially as the 'Lit & Phil'. He had already discussed his ideas with several members of the laboratory, but Chadwick as one of the young boys was not privy to them. Rutherford did, however, invite him to the March meeting of the Lit & Phil, mentioning that he had got some interesting things to say. Chadwick went along with about a dozen colleagues from the university physics department and found a place at the back of the room, containing 50 or 60 people. He never failed to be delighted by the memory of the evening.

> The first item contributed to the meeting was a statement by a fruit importer that he had found a rare snake in a consignment of bananas . . . the second item was the communication by Elliot Smith. He was a professor [of anatomy] in the University, also a great friend of Rutherford's . . . he produced a model of a skull.

> The cheek part went back to early man . . . [Robinson and I] decided that Elliot Smith's talk was extremely interesting, but in our opinion it was a mixture of Plaster of Paris and imagination. And then came Rutherford about the nuclear theory. In a sense, [it was] a most shattering performance to us . . . You couldn't escape from it. The experimental evidence which he only mentioned very briefly was so conclusive that there was no escape . . . It was to most people there, you see, something quite new, and of such consequence that one wanted time to think it over, while at the same time recognizing that a completely new idea had come in which was going to make things quite different.[6]

Rutherford[23], with his penetrating insight into physical processes, saw that the deflection of a heavy, fast-moving α-particle through a large angle, as observed by Geiger and Marsden, must be due to 'a single atomic encounter': it could not result from the summation of a large number of smaller deflections due to successive encounters with numerous atoms. From this physical intuition, 'a simple calculation shows that the atom must be a seat of an intense electric field'. Rutherford concluded that this electric charge had to be concentrated in a very small volume at the centre of the atom, and not dispersed in a large sphere as the theories of J.J. Thomson (1910) and Kelvin (1902) had proposed. Thus the enduring planetary or solar system model of the atom took shape, where there is a dense central (positive) charge in the nucleus around which (negatively) charged electrons orbit. When first presented, Rutherford's model was not completely worked out—for example, he could not be sure whether the intense field concentrated in the central nucleus was positive or negative. Despite the incontrovertible evidence supporting it, as Chadwick[6] pointed out 'some of the important consequences were not seen for quite a time. And you will find very few references indeed in the literature to the nuclear theory of the atom for about two years after the publication.'

In the summer of 1911, Chadwick graduated with first class honours. Surprisingly, it was not all plain sailing. He completed the written papers—'one or two perhaps reasonably well, others pretty badly'—then terror struck.

> I suddenly found out that we had to have a practical examination, and the external examiner was J.J. Thomson. I was terrified. It was not anything that he did. He came around to talk to me and I could hardly say a word. He was a kind of legend, you see . . . I was in a complete pickle. I just couldn't do anything.[6]

It defies belief that any undergraduate preparing for finals, and certainly not one of Chadwick's assiduous temperament, would really not have known in advance that he was going to have to take a practical exam. The attempt to rationalize his embarrassment, instead of being able to laugh at it, shows how deeply he took to heart any personal shortcoming. In this specific regard, it did make him sympathetic in future years, when he was examining a hapless student whose apparatus 'came to pieces in his hands'.[6]

Chadwick's real talents were obvious enough to Rutherford for him to be kept on as a graduate research student. By this time, the department was attracting researchers from other English universities and from overseas. The numbers had risen to 25 or 30 and space was at a premium, as was equipment. Geiger 'acted as a watch-dog over the research apparatus, and although he was a jealous guardian his popularity and prestige were high enough to enable him to do this without much friction.'[5] Andrade[24], who had previously studied in Heidelberg, was impressed by the Utopian spirit he found in Manchester, 'with the professor in closest touch with all his research men who, with little thought for their future living, were eagerly engaging themselves in obtaining results that seemed remote from any possible practical application.' Marsden, Robinson, Nuttall and Chadwick formed a quartet of home-grown demonstrators who shared 'a devotion to work and a high sense of humour'.[25] Their salary was £125 per annum, modest enough, but sufficient for Chadwick to start eating lunch again, after three years as an impoverished undergraduate when he would use the noontime break to read or to play chess. Some of the other young researchers came from grander lineages. Charles Darwin, son of Sir George Darwin, the Astronomer Royal, and grandson of the Charles Darwin became the second holder of the mathematical readership that Schuster had created for Rutherford's department. He was appointed in 1910 after his father had written to Schuster saying: 'I find that my son Charles would very much appreciate the studentship.'[26] At the same time H.G.J. 'Harry' Moseley was taken on by Rutherford as a lecturer and demonstrator, after graduating from Oxford. His father and both his grandfathers had been Fellows of the Royal Society, and he was the most brilliant and tragic English scientist of his generation.

Moseley[27], [28] will always be remembered for his X-ray spectrography work which led him to conclude that 'there is, in the atom, a fundamental quantity which increases by regular steps as we pass from one element to the next. This quantity can only be the charge on the central positive nucleus'. Thus, Moseley assumed that the fundamental quantity that he had discovered in X-ray spectra of atoms was the atomic number, but Chadwick would be the first to provide definite proof of the relationship a few years later, as we shall see. Moseley began his X-ray diffraction studies with Darwin in 1912; prior to this he had been engaged on radioactivity experiments, mainly on β-particles, that were more in the classical Rutherford tradition. During this period he worked in the room next door to Chadwick, but there was very little personal contact between the two men. For once, the reticence does not seem to have been Chadwick's alone, or at least there were others who approached Moseley with some trepidation. According to Russell:[25]

> Moseley from the first was recognised by his colleagues as the most original and gifted on the regular staff after the professor himself. He was a nice man but not hail-fellow-well-met. He was an Etonian not at all amiable to fools and time wasters. He rarely let any loose statement pass in conversation without either a total denial of its validity or a considerable and often pungent modification of the original formulation. He was one of the few who contradicted Rutherford.

An extract from a letter written to his sister in 1912 illustrates just how devastating Moseley[26] could be.

> I fear that I will not be able to put up with my excellent landlady much longer. Her temper is disturbing at times: this evening for example, when she took the opportunity of my being in my rooms to serve up my supper at 5.30, and I demanded

> it at its proper time two hours later. Landladies are all alike in one respect, that if you give in to them in trifles they think you weak kneed, and impose the more. However there is one prime virtue in the evil tempered. They generally preserve an icy silence, which is delightfully restful compared to the plump and garrulous.

Apart from the galaxy of resident physicists on the Manchester staff, there was a constant flow of foreign visitors. In March 1912 a post-doctoral fellow from Denmark, Niels Bohr, arrived and stayed for a few months. He was in his late twenties and had come to learn something of the techniques of radioactivity, but Chadwick[6] noticed that he did not talk much about his experiments. 'I don't think he was very interested in them, and . . . he gave them up fairly soon'. Bohr then spent most of his time away from the laboratory, working at home. He had noticed a false assumption in one of Darwin's papers, and this gave him the clue as to how quantum theory could be applied to Rutherford's model of the atom. His modification essentially was that the electrons surrounding the nucleus could occupy only certain discrete orbits. Each orbit corresponds to some definite energy level, and when an electron jumps from one orbit to another, a quantum or packet of energy is exchanged. Bohr's ideas were published in 1913 and to Chadwick[20] were 'one of the greatest triumphs of the human mind'. Despite a language barrier, Chadwick found Bohr approachable and they began an enduring friendship during Bohr's brief stay in Manchester.

Rutherford[29] once said he had never given a student a dud problem, and Chadwick's first assignment certainly supports that boast. He also had a favourite axiom that 'No man should ever have more than one problem given to him':[5] Chadwick, in turn, was equal to this stipulation. Having perfected the balance method for measuring gamma radiation, Chadwick[30] embarked on a wide-ranging investigation of the absorption of γ-rays in liquid air, carbon dioxide and hydrogen. As he stated in the introduction to the paper describing this work, until that time

> all information on the absorption of the γ-rays by gases is indirect, for the absorption coefficients are deduced either by assuming the density law or from the observed absorption coefficients of gases for β-rays. Values obtained in this way are, however, rather indefinite, and vary among themselves by about 50 per cent.

By simple but ingenious techniques, making measurements with the gases under pressures up to 90 atmospheres, Chadwick was able to make direct determinations of the various γ-ray absorption coefficients, with estimated experimental errors of a few per cent. As an aside, having established the absorption coefficient for air, he pointed out that the relatively high concentration of ions (charged atoms) found in the upper atmosphere could not be 'wholly due to the radiation from the radioactive material in the earth' and, by implication, must be the result of radiation from outer space. This work was published, with Chadwick as sole author, at the same time as his first joint paper with Rutherford in 1912, and bears the hallmarks of James Chadwick as a great experimenter. He had identified a difficult question of considerable practical importance and devised a direct attack on the problem, using various approaches. The precision of the experimental results and concision of the subsequent published account are perhaps the most characteristic Chadwick touches.

With these achievements, his stock in the laboratory rose steadily. Another member of the department, J.A. Gray, had just shown for the first time that when β-particles from radium E* impinge on 'external' materials, γ-rays are emitted, just as the impact of electrons on a metal target excites X-rays. Gray asked Chadwick to confirm his results, and this he did using aluminium, zinc, tin, lead and uranium as external absorbers of the β-particles from radium C. Chadwick found that the energy of the γ-rays produced increased with the atomic weight of the absorber. Looking back, he described it as 'a trivial thing',[6] but it was the work of a highly accomplished experimenter. He had purposely set himself a difficult task when choosing radium C. In Gray's original experiment with radium E, the amount of the excited γ-rays from the external target material was large compared with the amount of primary γ-rays emitted directly from the radioactive source. Radium C emits a much a higher ratio of γ-rays to β-particles, and Chadwick knew that the amount of gamma radiation excited by the β-particles would at most be only 1 or 2 per cent of the primary radiation. He undertook this exacting measurement by modifying the balance method that he had devised with Rutherford as a third year undergraduate. Chadwick[31] first introduced a lead screen which absorbed about 95 per cent of the primary γ-rays from the radium C source. The ionization produced by the remaining 5 per cent was then balanced against an equal and opposite ionization current due to a film of uranium oxide. The radium C source was positioned between shaped pole pieces of an electromagnet, and when the initial balance had been obtained, as described, the magnet field was activated. The shaped pole pieces of the magnet served to focus the β-particles on the target material, while having no effect on the uncharged γ-rays. Chadwick found that the γ-rays then excited by the β-particles caused about a 20 per cent increase in the ionization current that he had first observed from the attenuated primary γ-rays alone. This simple, direct manipulation of natural forces in order to measure a marginal effect (which had been looked for, but missed by other less talented experimenters) must have been admired by Rutherford as being recognizably in his own style.

Chadwick's steady progress continued and he received his M.Sc. in the summer of 1912. He was then appointed Beyer Fellow of the University and for his next work teamed up with A.S. Russell[25], 'an impressionable Scot from Glasgow'. He had trained as a radiochemist in his native city under Soddy, Rutherford's old collaborator from McGill, and Chadwick[6] remembered him at about this time proposing 'rules about the change in place in the Periodic Table of the radioactive substances with the emission of alpha and beta rays' (now known as the Law of Radioactive Displacement). The new collaborators[32] set out to examine the γ

* There is a radioactive series or family headed by uranium. Radium is a notable member of this series and decays with a half-life of 1620 years into the inert gas, radon. Radon in turn transforms into a succession of solid isotopes, originally labelled as Radium A, B, C etc. They are not, in fact, isotopes of radium but of various other elements such as polonium, bismuth and lead.

activity of α-emitters in a way that would be complementary to Chadwick's previous study of β/γ radiations. The most significant consequence of their partnership was that Chadwick[6] 'got a reputation with Rutherford as a chemist in some mysterious way'. Indeed he felt that the professor's opinion of him had risen considerably from the early days, 'when he was regarded as not showing very much promise'. Even Chadwick's sense of insecurity lifted when he was awarded an 1851 Exhibition Scholarship in 1913:[6]

> In those days it was a great distinction to get an 1851 and I never would have got it unless he [Rutherford] had recommended me quite strongly.

Chadwick was nominated by the Victoria University of Manchester for an 1851 Exhibition Science Research Scholarship in April 1913. Apart from his first class honours degree and M.Sc., his curriculum vitae also listed five research publications on various aspects of radioactivity—an impressive record for a 21 year old scientist. Indeed, the university[33] felt obliged to reassure the Commissioners of the 1851 Exhibition on the question of his maturity.

> Although somewhat younger than the candidates recommended in recent years we are of the opinion that he is sufficiently mature in his work to profit fully by the tenure of the scholarship. He is a young man of undoubted ability . . . His scientific ability is very highly spoken of by Professor Rutherford.

In fact, the only doubts about Chadwick being awarded the scholarship resulted from the powers at Manchester (Rutherford) seeking to keep him there, whereas the terms of the 1851 Exhibition stipulated that a scholar must work at a university other than the one nominating the student. The case was put that:

> Mr Chadwick himself would derive more advantage by continuing for part of his first year in the Physical Laboratories here under the supervision of Professor Rutherford. If, however, the Commissioners felt it necessary to insist on Mr Chadwick going elsewhere, arrangements could be made for him to study at Berlin or Vienna or at some other University on the Continent.[33]

The Commissioners[34] replied in uncompromising terms on 23 May 1913.

> Chadwick's nomination will be accepted only on condition that he spends his two years in another Institution to be stated by the applicant. If the authorities of the Victoria University think it undesirable that this course should be followed I would suggest that James Chadwick's name be withdrawn . . .

At this juncture, Rutherford caved in and the Registrar[35] of Manchester University replied to the Commissioners as follows:

> I am asked to say that Mr Chadwick is quite ready to fulfil the conditions and to study during his first year of tenure either in Berlin or in the University of Vienna—probably in the former.

No one could have foreseen the consequences of the Commissioners' adherence to protocol. James Chadwick was about to embark on one of his life's great adventures.

References

Letters and reports concerning Chadwick's 1851 Exhibition scholarship are quoted with the permission of the Archives of the Royal Commission for the Exhibition of 1851 (herein abbreviated to 1851 Archives) which are housed with the Archives of Imperial College, London.

Rutherford's letters are quoted with the permission of the Syndics of Cambridge University Library. Throughout the book, letters and documents are dated as follows: day/month/year.

1 Taylor, A.J.P. (1976). Manchester. In *Essays in English history*, pp. 307–25. Penguin Books, Harmondsworth.

2 Gilbert, M. (1991). *Churchill: a life*, p. 175. Henry Holt, New York.

3 Badash, L. (1979). British and American views of the German menace in World War I. *Notes and records of the Royal Society of London*, **34**, 91–122.

4 Wilson, D. (1983). Starting at Manchester. In *Rutherford: simple genius* pp. 216–37. Hodder and Stoughton, London.

5 Robinson, H.R. (1962). Rutherford: life and work to the year 1919. In *Rutherford at Manchester* (ed. J.B. Birks), pp. 53–86. Heywood & Co, London.

6 Weiner, C. (1969). Sir James Chadwick, oral history. American Institute of Physics, College Park, Maryland.

7 Rutherford, E. (20/10/07). Letter to Boltwood, B. In *Rutherford and Boltwood: letters on radioactivity* (ed. L. Badash, 1969), pp. 170–1. Yale University Press.

8 Wilson, D. (1983). The atom. In *Rutherford: simple genius*, pp. 268–307; Hodder and Stoughton, London.

9 Marsden, E. (1948). Rutherford Memorial Lecture. In *Rutherford by those who knew him*, reprinted by the Physical Society (1954), London.

10 Rutherford, E. (1899). Uranium radiation and the electrical conduction produced by it. *Philosophical Magazine*, **5**, 109–63.

11 Rutherford, E. (27/9/10). Letter to B. Boltwood. In *Rutherford and Boltwood: letters on radioactivity* (ed. L. Badash, 1969), p. 226. Yale University Press.

12 Rutherford, E. (21/10/11). Letter to B. Boltwood. In *Rutherford and Boltwood: letters on radioactivity* (ed. L. Badash, 1969), p. 255. Yale University Press.

13 Rutherford, E. (20/11/11). Letter to B. Boltwood. In *Rutherford and Boltwood: letters on radioactivity* (ed. L. Badash, 1969), p. 257. Yale University Press.

14 Rutherford, E. (18/3/12). Letter to B. Boltwood. In *Rutherford and Boltwood: letters on radioactivity* (ed. L. Badash, 1969), p. 254. Yale University Press.

15 Rutherford, E. and Chadwick, J.(1912). A balance method for comparison of quantities of radium and some of its applications. *Proceedings of the Physical Society*, **24**, 141–51.

16 Eve, A.S. (1939). *Rutherford*, p. 436. Macmillan, New York.

17 Rutherford, E. (1908). Recent advances in radioactivity. *Nature*, **77**, 422–6.

18 Rutherford, E. and Royds, T. (1909). The nature of the α particle from radioactive substances. *Philosophical Magazine*, **17**, 281–6.

19 Marsden, E. (1962). Rutherford at Manchester. In *Rutherford at Manchester* (ed. J.B. Birks), pp. 1–16. Heywood, London.

20 Chadwick, J. (1954). The Rutherford memorial lecture. *Proceedings of the Royal Society A*, **224**, 435.

21 Geiger, H. and Marsden, E. (1909). On a diffuse reflection of the α particles. *Proceedings of the Royal Society A*, **82**, 495.

22 Geiger, H. (1963). Some reminiscences of Rutherford during his time in Manchester. In *The collected papers of Lord Rutherford* (ed. J. Chadwick), pp. 295–8. George Allen and Unwin, London.

23 Rutherford, E. (1911). The scattering of α and β particles by matter and the structure of the atom. *Philosophical Magazine*, **21**, 669–88.

24 Andrade, E. (1962). Rutherford at Manchester, 1913–14. In *Rutherford at Manchester* (ed. J.B. Birks), pp. 27–42. Heywood, London.

25 Russell, A.S. (1962). Lord Rutherford: Manchester, 1907–19. In *Rutherford at Manchester*, (ed.J.B.Birks), pp. 87–101. Heywood, London.

26 Heilbron, J.L. (1974). *H.G.J. Moseley*. University of California, Berkeley.

27 Moseley, H.G.J.(1913). The high-frequency spectra of the elements. *Philosophical Magazine*, **26**, 1024–34.

28 Moseley, H.G.J. (1914). The high-frequency spectra of the elements. Part II. *Philosophical Magazine*, **27**, 703–13.

29 Blackett, P.M.S. (1962). Memories of Rutherford. In *Rutherford at Manchester* (ed.J.B.Birks), pp. 102–13. Heywood, London.

30 Chadwick, J. (1912). The absorption of γ-rays by gases and light substances. *Proceedings of the Physical Society*, **24**, 152–6.

31 Chadwick, J. (1912). The γ rays excited by the β rays of radium. *Philosophical Magazine*, **24**, 594–600.

32 Chadwick, J. and Russell, A.S. (1912). Excitation of gamma rays by alpha rays. *Nature*, **90**, 463.

33 Registrar of Manchester University (10/5/13). Letter to Commissioners of 1851 Exhibition. In Chadwick file, 1851 Archives.

34 Commissioners of the 1851 Exhibition, (23/5/13). Letter to University of Manchester. In Chadwick file, 1851 Archives.

35 Registrar of Manchester University, (29/5/13). Letter to Commissioners of 1851 Exhibition. In Chadwick file, 1851 Archives.

3 Germany and war

Although maturing rapidly as a scientist, Chadwick was still a shy and naive young man; he described himself as 'slow at growing up'.[1] His home circumstances did not afford him much chance to enjoy the cultural and social life of Manchester. He did not accompany others from the laboratory on their frequent forays to the music halls, he was not interested in sport, and seems to have had neither the opportunity nor the inclination to go courting. A tall, lean figure with smooth, dark hair, neatly parted to reveal a high forehead, he looked and was serious and single-minded.

As the Registrar of the University of Manchester had anticipated in his letter to the Commissioners of the 1851 Exhibition, it was simple to arrange for Chadwick to go to a continental university. He chose to spend the first year of his scholarship in Geiger's laboratory at the Physikalisch-Technische Reichsanstalt* in Charlottenburg, a suburb of Berlin. Geiger had returned to Germany in 1912, having played a major part in the experimental development of atomic theory at Manchester. While in Manchester, at Rutherford's insistence, Geiger had also taught a course in basic scientific German; this was the provenance of what little German Chadwick had at his disposal on arrival in Berlin in the autumn of 1913. He rapidly acquired a working knowledge of the language so that he could take part in the general discussions and meetings in the department. In January 1914 Chadwick[2] provided Rutherford with the following impression of his new linguistic encounters.

> First I would like to send my sincere congratulations on your knighthood. It was not a great surprise to me but Geiger was tremendously surprised. I have had to explain to him several times what it means. He has forgotten a fair amount of English and English customs, but he has recovered some since I came. From now on, no English is allowed except dictionary English. I get along reasonably well now but have trouble with the different accents. There are scarcely two people in the Reichsanstalt who have the same accent . . . I find his (Geiger's) German very difficult to understand, in my opinion he has the worst accent in the Reichsanstalt. By 'worst' I mean the farthest removed from the Berliner accent.

Geiger was solicitous towards the young Englishman and did his best to introduce him to other colleagues. He soon took Chadwick to the new Kaiser Wilhelm

---

* Imperial Institute of Physics and Engineering. This was the German equivalent of the National Physical Laboratory in Britain or the U.S, National Bureau of Standards.

Institute for chemistry to meet Otto Hahn and Lise Meitner.[1] Hahn was a radiochemist whose first important discovery, the active element radiothorium (^{228}Th), had been made in 1905 when it was dismissed by one disbeliever as 'a mixture of ThX (^{224}Ra) and stupidity'.[3] He would have been well disposed to any pupil of Rutherford's whom he revered after spending a year under his tutelage in Canada. Lise Meitner,[4] 'slight in figure and shy by nature', was born in Vienna in 1878, but was dubbed by Einstein 'the German Madame Curie'.

Chadwick got to know other leading figures by attending meetings of the Rubens' Colloquium, which met regularly in Berlin. These meetings were quite informal and would consist either of a visitor presenting results of his own work or a discussion of recently published scientific papers. Chadwick[2] also described these meetings to Rutherford.

> We go to Rubens' colloquium every Wednesday but I prefer ours. Here there is the advantage that every subject is discussed, but the abstracts are often carelessly done and the discussions are not at all stimulating. The Physikalische Gesellschaft is sometimes good, but is too long. Three hours on a hard wooden seat is too much for me, and if I really must keep awake I think I prefer to have my body comfortably provided for. Hahn I find extremely nice and he gives exceedingly clear papers.

There was another physicist from England in Berlin at that time who occasionally attended the Rubens' Colloquium, Frederick Lindemann (later Lord Cherwell). Chadwick[1] never spoke to him there, but remembered 'his making some very critical observations on one paper, perhaps on more than one occasion . . . [He] was very quick to seize on what he thought was any discrepancy, and he could calculate very quickly in his head, or appear to do so . . . He wasn't always right when you thought about it afterwards.'

Geiger's laboratory, the Magnetische Haus, was a small building in the garden of the Reichsanstalt that had originally been designed for magnetic measurements. Chadwick found the atmosphere friendly: he[2] told Rutherford 'working with Geiger is great fun and he helps me in every possible way'. Under Geiger's genial tuition, Chadwick learned a little glass-blowing and could make 'quite nice α-ray tubes'. He found a marked difference from the ever available Mr Kay in Manchester, when he did need to get apparatus made. At the Reichsanstalt there were centralized workshops where you had to put in a written order and wait your turn for two or three weeks. Chadwick[1] recalled a particular occasion when

> I wanted to make an ordinary simple electroscope—a tin box was all I wanted. I had to go to the tinsmith and give him written instructions. He was busy on making a seat for the lavatory of . . . a very important man (one of the directors of the institution). And so I couldn't get my electroscope until he'd made this lavatory seat through which warm water could be circulated in the winter. This story spread around the place and we were all interested what would happen when it was first used, and it was almost a disaster because it was too hot for Frau Herr Direktor. So I finally got my electroscope.

Chadwick[1] made friends and enjoyed getting to know 'people who were working on quite different things in the Reichsanstalt, some on the engineering side, some on the optical side . . . It was a very good atmosphere. They adopted me. I was much younger than most of them, of course.' The hours for the scientific staff were nine to three with no midday break. Geiger would bring in bars of chocolate to keep them going and

> then at three o'clock, down tools, and six, seven or eight of us would have lunch together. For most of them, that was the end of the day, but for some of us—certainly for Geiger and me, as a rule—it wasn't. We went back to the laboratory and went on with whatever we were doing. But sometimes if it were a particularly nice day and we felt like it, three or four of us would go for a walk in the country, and come back and perhaps have dinner together or not, depending upon the state of our pockets.

On the latter point, Chadwick's scholarship was worth 150 pounds a year, a little more than he had earned as a demonstrator in Manchester: he found he could manage quite well on this in Berlin, although 'no riotous living was possible'.

Chadwick's chosen area of work in the Reichsanstalt was β-radiation which would prove to be one of the most vexing and critical subjects for the development of atomic physics over a quarter of a century. The initial theories concerning the behaviour of β-particles were based by analogy on the more easily observed and simpler passage of α-particles through air. As we have seen, Rutherford proved that α-particles are helium nuclei—heavy particles which have a straight, finite course through air. As early as 1905, W.H.Bragg and his assistant Richard Kleeman concluded that 'Each α-particle possesses . . . .a definite range in a given medium the length of which depends on the initial velocity of the particle and the nature of the medium.'[3] They also showed that the range and initial velocity of any group of α-particles emitted by a particular radioactive element were constant and characteristic of that element—a sort of α-particle signature.

The analogous hypothesis was made for β-particles, that they were emitted with a single characteristic velocity. To verify this by direct observation was much more difficult because β-particles are electrons: very light charged particles that have no clear range through air because they are so readily scattered by the gas molecules. Hahn and Meitner began their scientific partnership in 1907 and their first aim was to demonstrate the characteristic β-particle signatures of various radioactive substances in the way Bragg and Kleeman had for α-particles. In their first joint paper a year later, things looked hopeful and they conjectured from their results that 'pure β-ray emitters only emit β-rays of one kind (i.e. one velocity), similar to what happens with α-rays.'[3] But exceptions to this tidy pattern began to appear and the theoretical basis of their method of investigation, the passage of electrons through matter, was called increasingly into question. The true complexity of β-particle absorption that made it an unsuitable way of studying the initial velocity of the particles was finally established in 1909 by William Wilson. Wilson, another of Rutherford's Manchester team, as part of this work also supplied a new technique for separating β-particles into a focused beam of homogeneous velocity by means of applying a magnetic field. This would become the standard technique for future experiments in β-ray spectrography.

Wilson's magnetic separation technique combined with a photographic method of β-particle detection was adopted by Hahn's laboratory and by Rutherford's. In May 1911, Rutherford[5] commented to Boltwood: 'You will have seen that Hahn is finding that the β ray problem is more complex than first appeared. Geiger and Kovarik have been investigating the matter in another way, and find that while there is some indication of order in the β rays, there are some very notable exceptions. The whole problem is very peculiar.' By this time all the interested parties, including Hahn and Meitner, had discarded the idea that the β-spectrum of a pure β-emitter is homogeneous or monochromic. Instead the magnetic separation–photographic detection paradigm had led them and the Manchester school to believe that a β-spectrum consisted not of one unique line or frequency, but a set of discrete lines.[3]

To James Chadwick these spectral lines suggested a means of exploring the poorly understood events of β-particle scattering. His way of doing this was to use Wilson's magnetic separation technique, but instead of using photographic plate as the detector, he employed a particle counter that Geiger had just developed. The first mention of this research was in his letter[2] to Rutherford, dated January 1914.

> I have not made much progress as regards definite results. We wanted to count the β-particles in the various spectrum lines of Ra (radium) B + C and then to do the scattering of the strongest swift group. I get photographs very quickly and easily, but with the counter I can't find even the ghost of a line. There is probably some silly mistake somewhere. I had hoped to settle it definitely this last weekend but the α-ray tube that we carefully filled with 40 mgms of emanation had a hole in it, and by the time I had carried it across to my room there wasn't one per cent left. And now I have to wait for a week or so till the emanation grows. In the meantime I have started a little experiment on the distribution of the secondary β-rays produced by γ-rays and it promises to go very well.

The emanation he refers to is radon gas which has a half life of about four days and decays by the emission of alpha particles into radium A, which quickly transmutes into the β-emitters radium B and C. He did not say whether the leaking α-ray tube was one of his own early efforts at glass blowing, but it seems likely. The last sentence gives an inkling of the fertility of Chadwick's mind: he was bursting with ideas for experiments, but rarely showed outward enthusiasm. Geiger, like Rutherford, was not misled by Chadwick's diffident manner and had no doubts about his extraordinary ability as an experimental scientist. In Chadwick's case, Geiger was always available for discussion, but knew that no direct assistance was required.

Chadwick repeated the experiment and convinced himself that there was indeed no silly mistake, and that β-radiation did produce a continuous spectrum, not a just a series of discrete lines. The paper was submitted for publication on 2 April 1914. Chadwick[6] was the sole author, but it seems likely that Geiger translated it into German. His original intention on embarking on β-particle work is clearly set out in the introduction to the paper.

> The purpose of the present investigation was to determine quantitively the intensity conditions by directly counting the β-rays in the individual groups [lines]. Thereafter it was intended to examine closely the laws determining the passage of β-rays through material by using the individual extremely homogeneous radiation groups. The counting of the the individual β-particles was carried out by the method proposed by Geiger.

Chadwick had intended to use the new counter to measure the lines in the β-spectrum more accurately than could be done with the usual photographic technique, and then to go on and use the 'strongest, swift group' in scattering experiments, in the hopes of unlocking more detail about the structure of the atom. The parallel with Rutherford's α-particle scattering experiments is obvious. What he found was that the lines in the spectrum were not as real as everyone had thought, but were rather artifacts of the photographic detection. He did find lines overlying the continuous spectrum but these were of very low intensity.

How did Chadwick know that his observations were correct and that the generally accepted picture painted by Hahn, Rutherford and others was wrong? He provided the answer after describing his own results.

> At first sight these results appear in part to contradict the photographic measurements. The difference is however explained by the fact that the photographic plate is extremely sensitive to small changes in the radiation intensity.

He went on to show that depending on the speed at which the photographic plate was developed, a line that was only 5% more intense than the continuous background could be made to look 'almost black on a clear background.'

Working virtually single-handed in a foreign department, Chadwick had made his first fundamentally important discovery, within about six months of arriving. Whatever his personal lack of confidence, he believed in his own scientific ability to the extent that he had no qualms about publishing the work. A man of lesser ability would never have made the observations in the first place, especially as they were not the primary objective of the experiment. A man of lesser integrity would have discarded them. The troubling implications of Chadwick's findings for those engaged in the thoeretical development of physics would not be resolved until Fermi's postulation of the neutrino, twenty years later. The first man to confront the paradox resulting from Chadwick's observation was Einstein, who visited the Reichsanstalt the spring of 1914. Chadwick[1], who had never met Einstein before, told him, in German, that he had just discovered 'the β-ray emission from the radioactive deposit had a continuous range of energy practically from zero up to a certain limit on which was superimposed these peaks'. Einstein's instant reply was: 'I can explain either of these things, but I can't explain them both at the same time.' Although Rutherford immediately accepted Chadwick's results, Lise Meitner for one, would be much slower to abandon her previous beliefs, as we shall see.

Despite Einstein's enigmatic response, the profound consequences of his findings do not seem to have been immediately obvious to Chadwick (nor perhaps to anyone other than Einstein). As he explained in a progress report to the 1851 Exhibition Commissioners, 'it was not practible to separate out these lines of homogeneous rays in order to use them to investigate the passage of particles through matter'.[7] Over half a century later, Chadwick would still refer to his discovery of the continuous β-spectrum in terms that suggested it was a nuisance rather than taking pride in being the first to identify the phenomenon.

As soon as the β-spectrum results were written up, he embarked on an experiment to study the scattering of β-particles by very thin gold foil. This work is very reminiscent of the Geiger and Marsden experiments in Manchester with α-particles, and was designed again to test Rutherford's single scattering theory. It was never completed and only presented later as a handwritten note to the Commissioners.[8] Even the title chosen by Chadwick for his report 'The scattering of β-particles through large angles' harks back to the famous 1913 paper of Geiger and Marsden 'The Laws of Deflexion of α Particles through Large Angles'. Chadwick's experiment gave a result for the nuclear charge of gold between $80e$ and $95e$, whereas the true value is $79e$. He was not entirely displeased and commented that 'considering the difficulties of the experiment, the results agreed very well with the theory'.

Another glimpse into his enquiring and imaginative mind in those days when he was occupied by β-particles and electrons comes from his memories of discussing new avenues of possible research with Geiger. Chadwick[1] suggested 'to him at that time that perhaps electrons might be scattered from a crystal surface in much the same way as X-rays. He [Geiger] said there was nothing in it, it was rather a silly kind of suggestion to make—as in fact of course it was. But I've often wondered what would have happened if we'd been able to do it and found a kind of wave nature of the electron long before our minds were ready for it.' (The same idea about the dual nature of electrons would occur to Duc Louis de Broglie[9] in 1923, 'after long reflection in solitude and meditation' and he would propose that electrons under certain circumstances would behave not as the familiar particles but as waves). In an eventful career, this would not be the only hunch that Chadwick failed to elaborate fully; it was obviously the one for which he quietly kicked himself the longest. Contrast the dignified mental process of the aristocratic Frenchman with Chadwick's[1] almost embarrassed dismissal of 'just one of those silly ideas that comes into one's mind'!

His brief, but notable, career in the Reichsanstalt came to an abrupt end at the beginning of August 1914. On the evening of Friday 31 July, Chadwick[1] had found himself caught up in an excitable crowd milling around in front of the palace, 'waiting for the Kaiser to come out and say something to them. So he did. He stepped out on to the balcony with what one would call military precision and delivered a few words most appropriate to the occasion, only quite few, and stepped back again'. Those few words were the declaration of *Kriegsgefahr* (Danger of War) and were the first public statement confirming the headlong rush to war. The following day Germany declared war on Russia and the Kaiser ordered a general mobilization of his forces. With this announcement, the public mood in Berlin became volatile and ugly; mobs roamed through the streets setting on 'suspected Russian spies, several of whom were pummelled or trampled to death in the course of the next few days'.[10]

Geiger was a reservist and had been called up sometime before the general mobilization. Before leaving, he had advised Chadwick[1] to go back to England and lent him 200 marks for the journey. Chadwick delayed until Saturday August 1st, when he went to Thomas Cook's, the travel agency, to see if he could get a ticket home via Holland. They told him this was impossible, but suggested a route via Switzerland. This would mean travelling through France which Chadwick thought too risky, especially as his friends told him that 'the worst thing that can happen here is that you'll have to stay for a month or two and then you'll be pushed out'. On the following Tuesday night, Chadwick dined with four or five friends in the city to 'find out what news there was'. The news was that the German Army had invaded Belgium; to the disbelief of Germany, Britain honoured the treaty they had both signed in 1839 recognizing Belgium as an independent and perpetually neutral state. The British declared war on Germany and instantly supplanted the Russians as the most hated foe in Berlin. Chadwick's small party witnessed a howling mob stoning the British Embassy; mounted police attempted to keep order but succeeded only in further inflaming the crowd who turned on them. Chadwick's friends, realizing the danger, took him to a safe house across the river. He would say later 'nobody could have been kinder than they were. They looked after me. It made no difference whatsoever after the Declaration of War. There were others in the laboratory who went out of their way to be kind to me whom I'd hardly ever seen before.'

The first unwelcome brush with the authorities came two weekends later.[1] He and a German friend were reading the latest news from a poster about the fall of Liège, the Belgian city, which had succumbed after a merciless bombardment lasting ten days. A woman reported to the police that one of the men had made the comment 'Die Saubande hat gesiegt'.* They were arrested, taken to a police station and told they were going to prison. To avoid the real danger of assault if they were taken there through the crowds handcuffed to a policeman, they were given the considerate option of paying for a carriage, which they did. 'We went in a kind of landau. We were on opposite sides, each with a policeman at his side gripping us, and the policeman on the opposite side holding his revolver.' Chadwick's accomplice was let out after a couple of hours, but he was detained for ten days on a diet of coffee and mouldy bread before intervention from the head of the Reichsanstalt secured his release.

For a short time afterwards, Chadwick returned to work in the Reichsanstalt; an understandable decision at the time when his colleagues there remained the only friendly faces in an otherwise hostile city, but one which later he would come to regard as quite wrong.[1] This equivocal interlude came to an end early on Friday 6 November with a knock at the door of his rooms. The police had been given instructions to round up all Englishmen and intern them as enemy aliens. Chadwick was

* A colloquial translation would be: 'The bastards have got away with it'.

marched with hundreds of other British civilians to a racecourse at Ruhleben,[11] about two miles west of Berlin, near the industrial suburb of Spandau. The racecourse was a bleak, windswept, patch of land reclaimed from the marshes of the River Spree. The new arrivals, cold, apprehensive and hungry, were marched off to one of the brick stables, each of which contained about thirty boxes. During the racing season, each box housed two trotting horses. A German sergeant slid back the door to one box and ordered Chadwick inside with five other men from the column of prisoners. The heavy iron door closed behind them, and they were left, standing in the darkness, holding their suitcases. The box was about ten feet square, with a concrete floor covered with straw; there was a pungent smell of horse dung and urine.

In the morning, a feeble light entered through a small window near the ceiling to give the new inhabitants their first look at each other. The guard issued them with a metal bowl, which was filled with cold coffee and breakfast was completed with a piece of black bread. The bowl would become the essential receptacle for coffee, soup, and gruel, as well as a washbowl. Cutlery was not provided, nor was soap, but could be purchased at the camp shop. Camp rations were meagre, and could only be collected after queuing at the cookhouse for two hours at each mealtime.[12] The prisoner then had to return to his box to sit on his bunk and eat what he could. After a few months, parcels began to arrive from home with food and clothing, which had an uplifting effect on morale; they also saved some lives from malnutrition and exposure to cold.

There were more than a dozen blocks of the racing stables and the camp soon held four thousand men, the largest contingent of whom were merchant seamen captured in Hamburg.[11] For Chadwick[1] it was the first opportunity to meet men from walks of life other than science and he got to know 'an Earl, various professional people of one kind and another, musicians, painters, a few racehorse trainers, a few jockeys', amongst others. The diversity amongst the internees was truly remarkable, and although about one-third were seafarers, there was a high proportion of businessmen who worked in Germany, as well as academics, and the artists and professional sportsmen, mentioned by Chadwick. The first winter on the northern German plain was physically hard, and brought the collective acceptance that internment was not going to be a temporary experience. With this realization, there came a tremendous unification of spirit amongst the men and a determination to make the best of their circumstances.[11]

Back in Manchester, following the outbreak of war, Anne Chadwick was naturally very concerned about her son's safety. She first wrote to Evelyn Shaw, secretary to the Royal Commissioners of the 1851 Exhibition, on 12 August 1914.[13]

> Dear Sir,
>
> James Chadwick is in Berlin as far as we know have not heard from him over a fortnight ago hoping for the best.
>
> Yours truly,
>
> A.Chadwick

Communications with Germany were now sparse, increasing the anxieties. Shaw wrote to Mrs Chadwick offering to make an advance on the scholarship stipend. Anne[14] replied on 23 September, thanking them very much for the kind offer. She had been told by the Foreign Office that it would be impossible to send money to any individual, so she did not feel justified in wasting Chadwick's meagre savings because he would need them when he got home. Of course, at that time no one was expecting a long war—it was meant to be over by Christmas. In October, Chadwick's old colleague from Manchester, Alex Russell,[15] also wrote to the Royal Commissioners to say that he had received a letter from James, 'telling me however nothing that I want to know about him, also giving no address.'

By December, the news of James' internment filtered through, and Mrs Chadwick's[16] tone had become querulous.

> 74 Glen Avenue,
> Blackey.
> Dec 13, 1914
> Dear Sir,
>
> I have been expecting to hear from you as Mr Russell informed me sometime ago. I am sorry to find you are not sending my son's scholarship as he wished me to send some money to a Mrs B de Hass in Holland he had to borrow before we could get any through and could not pay them back before he was interned as he had not enough. I am sorry Mr Russell has made a muddle of things, it is very hard when a young man works hard, denies himself all pleasures to gain a scholarship, is bound to go away to claim it, is detained through not fault of his there, his scholarship stops until he can take up his studies again. . . . Please excuse me I am a British mother and studies her son's welfare as I hoped British friends should feel for others in distress.

One wonders how she would have reacted if, instead of being interned, her eldest son had joined up and been sent to the trenches. Faced with this spirited but disjointed plea, Shaw[17] replied most diplomatically, again offering to send James' money to her. This seems to have mollified her, and she was bolstered by a postcard from James telling her how to send parcels to the camp. 'We are sending one tomorrow, he would have been pleased to join the Army had he been at home',[18] she added in reply to Shaw.

In March 1915, the Ruhleben prisoners were allowed to write their first letters home.[12] Chadwick's letter[19] is brimming with news about the extraordinary society that was evolving in Ruhleben. He[1] had met Carl Fuchs, a cellist from Manchester, who suffered the double misfortune of having his cottage in Lancashire attacked because he was German and being imprisoned in Ruhleben because he was English! Fuchs[11] played some unaccompanied cello suites at the first camp concert in December 1914, and with the encouragement of Baron von Taube, *Lageroffizier,* the concerts became weekly events. As Chadwick mentions in the letter below, there were many other internees from Manchester and its environs, and they banded together to form the Ruhleben Lancastrian Society—the first and most popular of the territorial associations. Another Mancunian founded the Literary and Debating Society which was enthusiastically supported by all sections of the camp. One snowy December night in 1914, nine newly captured fishermen were marched in, to find a hotly contested debate taking place on 'Should bachelors be taxed?'[11] As part of this social ferment, all university graduates were invited to a meeting in the grandstand to consider the educational needs of the camp; from this meeting sprang the Arts and Science Union, which would come to have a pervasive effect on the lives of many. Chadwick's letter also reflects the strain of life in the overcrowded stables, and the real privations. While the allocation to the boxes was virtually a random process, there grew up a strong loyalty to your stable or barrack. Some of the barracks were dominated by the merchant sailors, but Chadwick's stable housed a disproportionate number of businessmen and professionals. Barrack 10 was 'the glory of Ruhleben'.[11] It contained more dons, students and schoolboys than any other barrack and they were regarded as 'the public school crowd' by the rest of the camp. While Chadwick was not a prominent individual in this group, he did draw confidence from being amongst the acknowledged leaders and decided to employ his talents as fully as possible.

Absender: Engländerlager Ruhleben
J Chadwick
Baracke 10
Box 14: 23 March 1915
Write in pencil only! Nur mit Bleistift schreiben!

Dear Mother,

I have had your cards of the 2nd and 15th and Harry's of the 8th. So far I have had 60 Marks from you, two lots of 10 and one of 40. Let me know how much you sent. I have had 100 M. from the brother of Dr. Pogge so I am well provided now. I hope you have had my last letter. Don't forget to send parcels regularly and particularly the Hovis loaf. You might send a tin of mustard and two tubes of toothpaste. We have had very cold weather lately and some heavy falls of snow, but the weather improved yesterday and we seem in for a fair spell now. I hope so for I have had enough winter this year to last me for the next two or three. I think I wrote before that there are many Manchester people here. One in our loft comes from Heaton Park. I wish he was in our box. We have two old men in the box, one of them is nervous and very quiet and the other has all the ailments under the sun. They are generally in the way and expect us to be as quiet and dull as they are. Otherwise, we get along quite well. One of the prisoners is a conductor and he has got together quite a decent orchestra, which gives concerts nearly every Sunday evening. We formed an Arts and Science Union some time ago and have had several lectures. We also have a Dramatic Society. This is still young and has just finished the first play 'Androcles and the Lion'. A school is being started but this is not yet in working order. We are getting more space shortly, so we shall be able to play cricket and football.

Do send the spectacles as soon as you can. I want the same ones as before, that is, oval ones, but rims on the spectacles. I think the others are liable to break at the hole. My glasses have gone there. Anyhow, whatever the Optician says, get spectacles with rims. (Unless there is some other and very important reason).

I also want some strong boots. The only ones I have here are military boots which come nearly up to my knees and they will be far too warm for summer and also far too heavy to run in. If possible, get some English military boots. At any rate, get good leather. My size is large 7, I think, with square toes. Also send a pair of tennis shoes. That is all I can remember now. I think it will see me through the summer. I am alright without my overcoat. I have a sweater and with that and my old raincoat I feel quite warm enough.

Don't forget to write regularly and to send parcels regularly. Tell me when you send a parcel and the date on which you send it, so that I shall know if they all arrive. But don't write half a dozen times about the same parcel, particularly when I've had it a month before. I got tired of hearing about the last one. I am expecting a parcel every day from Prof. Scheel. Pack the glasses up very well so that there won't be the slightest possibility of their getting damaged and send a bit of chamois leather with them.

It is very difficult to fill the whole letter. I always forget what it was I wanted to write. Write to Aunt Pattie for me. It will be some time before I can write again. I hope she has recovered from her illness. I have nothing more to say except 'Don't forget the parcels'.

With best love,
James.
I suppose you can get my money.

James was not the only one to be frustrated by his mother. In March 1915, Shaw[20] wrote to the Secretary of Manchester University asking for his assistance in clarifying procedures. He suggested 'asking Mrs Chadwick to call upon you and to explain to what extent she has financed her son and how she proposes to finance him in the future.' Shaw closed by saying the 'letters which I have received from Mrs Chadwick are hardly intelligible'. The arrangements became a little clearer as a result of a letter from James' younger brother Harry. He[21] explained that James 'who is under the impression that the money is being paid to him monthly, has instructed us to supply Dr Pogge (a German on holiday in England) with money as he has received money from Professor Scheel, Dr Pogge's brother-in-law. As Dr Pogge has spent all the money he had by him and cannot obtain any from Germany, he wishes to borrow £10 from us. We, however, have none to pay to him so that if we do not receive my brother's money, Dr Pogge who is not interned must starve.'

Shaw did his best to establish the facts and took advice from the Secretary of the Prisoners of War Help Committee, who told him that the camp commandant would retain any large sum of money sent and pass on only five or ten shillings a week to a prisoner. He[22] therefore strongly recommended to Anne Chadwick that she should 'discontinue your practice of paying to Dr. Pogge a sum of money on the understanding that the equivalent is placed at your son's disposal in Germany, because there is no guarantee that this arrangement can be carried out and it is not possible that your son would require as pocket money what you are in the habit of paying for the maintenance of Dr Pogge'. The misapprehensions seem to have been settled, but Ann Chadwick remained loyal to Dr Pogge. She[23] wrote to the Commissioners in June to say that he had now been interned in England and 'it is quite right that all Germans should be interned Guilty ones or not Guilty but I still think Dr Pogge is quite honest'.

The spring and summer of 1915 saw a tremendous explosion in education within the Ruhleben camp,[11] and Chadwick began to pursue his scientific calling again, despite the many obstacles. Courses of lectures were given in the spring and summer, out of doors in the grandstand. Chadwick[12] lectured on Radioactivity twice a week on Fridays and Saturdays, and he also gave two talks a week on Electricity and Magnetism, which were popular. He wrote to Rutherford on 14 September describing his efforts.[24]

> Radioactivity is naturally not in a very flourishing condition here. A series of lectures was begun in spring by the Arts and Science Union and I gave some on Radioactivity. It did me no harm, and I hope it did the students none. I managed to show a few experiments with an electroscope and scintillations. I have made a neat little electroscope that I am very proud of and I hope to do some nice experiments with it. The chief difficulty here is that there is no room.
>
> We have formed a colloquium of a kind called the Science Circle, of which I am the humble and hardworked secretary. The meetings are held under such conditions that one can hardly hear one's self speak.

The Science Circle[11] was inaugurated in June 1915 and met at 6 pm on Wednesdays as a forum for biologists, chemists and physicists to read 'papers on subjects of mutual interest, to be followed by discussion'. It was just one of the specialist societies which appeared that summer in response to the prisoners' amazing thirst for intellectual activity. As a unifying and morale-boosting force, the range of subjects offered by the Arts and Science Union was second only to organized sport. One young prisoner who was attracted to science in Ruhleben, and heard Chadwick lecture on radioactivity was Charles Drummond Ellis.[25] He had entered the Royal Military Academy, Woolwich, as a cadet in 1913 after a glittering scholastic and athletic career at Harrow School. At the end of his first year at Woolwich, he had passed out first in order of merit and decided to spend the summer with four or five fellow cadets in Germany. They were all interned in Ruhleben, and it must have become increasingly apparent to them, with the passage of time, that they could never make up for their lack of active service in terms of a military career. Ellis, just twenty years old, had his interest kindled by the Science Circle and became Chadwick's star pupil and colleague. The two men would continue to work together in Cambridge after the war, and there Ellis established himself 'as a world authority on the nature and behaviour of β- and γ-rays'.

The members of the Science Circle approached the camp authorities in the autumn of 1915 for space to be used as a laboratory. The oldest and most squalid of the stables, barrack 6, was used to house orthodox Jewish prisoners. The large loft above the barrack had been condemned as unfit for habitation by the American Ambassador during a tour of inspection in March 1915.[11] The loft was cleared that summer and was now handed over to the Arts and Science Union for winter classes. The loft was partitioned, and the scientists gained a room that was 12 yards long, 3 yards wide and about 9 feet high in the centre, sloping down to about 4 feet high at the windows. Conditions were primitive with no water supply nor lighting. Even with some heat provided by fat lamps, the temperature of the room fell to – 10 °C in the winter; under the sloping roof, in the summer, it would reach a sweltering 35 °C.[8] Not surprisingly some chemicals were forbidden, chiefly poisons and phosphorus compounds, but through ingenuity, generosity and a little bribery, materials were soon built up. Chadwick was given some radioactive toothpaste, popular in Germany at the time, where it was advertised on posters by a young woman with a dazzling smile. Chadwick set to work in the New Year of 1916, using his handmade electroscope of wood lined with tinfoil; for laboratory beakers, he used beerglasses and teacups. He later wrote[26] to Rutherford that the toothpaste

> . . .resembles no known radioactive body. I don't feel very certain yet, for the material seems to consist of several bodies of such curious chemical properties that they are difficult to separate clearly. It seems like a mixture of rare earths. Everything points to a new family of radioactive bodies, but the complete solution will take a long time. I still feel very sceptical about it.*

In the summer of 1916 the interned scientists obtained permission to have an electric main put up, at their expense, and from this time had 2 amperes at 220 volts (alternating) for their use.[8] A young chemist in the camp told Chadwick about liquid crystals, something he had not heard of before, and he decided to study them in a magnetic field. Chadwick[1]

> . . .made a sketch of the magnet, the amount of iron and so forth, and gave it to a soldier and he arranged to supply me with it. With another man, I worked out how much copper wire I should want, what dimensions and so forth, something like 300 metres or so, and I got that through this other soldier. And I had to wind the coils by hand . . . it was a hot evening, and the other friends who were working in the same room had all gone to the theatre and left me alone. The temperature was about 36 °C . . . I got very hot and tired indeed doing all this by hand. But it was done.

A few months after making this effort, Chadwick[8] obtained some copies of the annual reports of the Chemical Society and discovered what he was thinking about had been quite well understood ten years earlier. 'So that was washed out, and I started on other things.'[1] Even in the conditions of Ruhleben, there was no intellectual satisfaction for Chadwick in repeating the work of another; nor was there any lack of confidence, on his part, in finding other fruitful subjects for research.

The cooperation of the camp authorities in these activities was remarkable and unparalleled. Chadwick[1] described the officers as extremely lenient. 'Indeed, we did a number of things that were quite contrary to all the rules, and the officers certainly knew that something was going on.' Inspectors from the War Office in Berlin turned a blind eye to the laboratory. 'Frl. Dr. Rolten of the Auskunft und Helfstelle [an organization similar to the Prisoner's Aid Society in London] made strenuous efforts to interest German manufacturers and Professors on our behalf,

---

* He later realized the toothpaste contained traces of thorium.

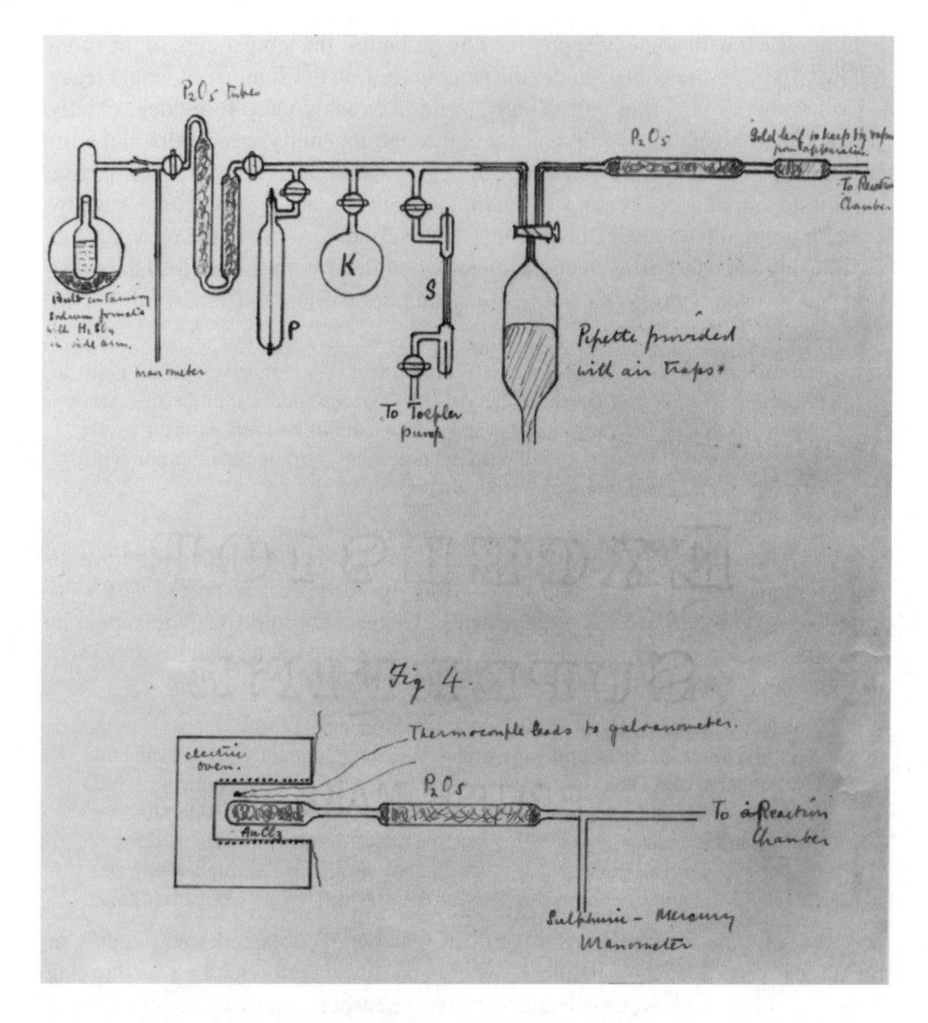

Fig. 3.1 Glass apparatus devised at Ruhleben by Chadwick for his experiments, taken from his unpublished 1918 report to the Royal Commissioners of the 1851 Exhibition.

and from various sources we received a great deal of apparatus.'[8] Chadwick and one of his friends, a chemist, were even allowed out of the camp on one occasion to visit two leading German scientists, Rubens and Nernst, who both offered practical help. One of the few unanswered requests Chadwick made was for an elementary book on inorganic chemistry. 'It was stopped, not by the Germans, but by our English censors, on the grounds that it might be giving information to the enemy. Anything more absurd one couldn't possibly think of.'[1]

Later, the camp scientists received a donation of about two hundred general science books from the publishers, Taubnitz. Journals were much harder to come by, but Chadwick did manage to obtain a copy of Einstein's 1915 paper on the general theory of relativity: 'I was probably one of the first English people to know about it, not that I could follow the mathematics. But there was a mathematician in the camp—I think he was a Rhodes scholar in Oxford—he was interested and he gave me a general description of the mathematics.'[1]

In the spring of 1917 the laboratory was transferred to a basement with more space and a water supply. The source of heat for glass blowing was an improvised gas lamp. Rancid butter was sprayed by a blast of air from a fine nozzle into an asbestos cone and ignited as it emerged as a jet from a small hole in the apex of the cone. Prior to the war, Chadwick had been used to a Bunsen burner where the airflow could be increased by foot-operated bellows. The butter burner had no such attachment and the bellows were Ellis' lungs. As he was kneeling on the floor one day providing the air supply, Chadwick's foot came down on his back, through force of habit, and he was exhorted to 'Blow, Charlie, blow!'[3] They worked on two main experiments together, both of which were chemical in nature and required the invention and construction of quite complicated, glass apparatus (shown in Fig. 3.1).

The first investigation carried out by Chadwick and Ellis was of the well-known phenomenon that 'the air in the neighbourhood of oxidising phosphorus conducts electricity'[8] (obviously the ban on working with phosphorus had been lifted). The second was to study the reaction of carbon monoxide and chlorine in light. Chadwick[27] wrote to Rutherford about the work on 24 May 1918.

> I have been wishing to write for some weeks, ever since I heard again from Geiger, but I put it off in order to be able to tell you of the experiments I am just beginning. Geiger sends you his kindest regards. At the beginning of March he was quite well and so far he has not been damaged at all. He carries a few books with him wherever he goes, but in spite of this he has forgotten even the radioactive constants. What really troubles him, however, is the scarcity of beer. What would he say if he was in the camp where the very name has lost its meaning and reminds one dimly of some past existence.
>
> During the winter I did a little work with Ellis on the ions produced during the oxidation of phosphorus. We had to break off the experiments owing to lack of apparatus but the few results we got were interesting. Apparently there are two distinct reactions; a slow one, accompanied by ions for which saturation was obtained as easily as for gaseous ions at the same pressure, and a second accompanied by very heavy ions. There was no evidence of the 'emanation' suggested by Schenck and Toreunig. I believe the mobile ions may be connected with the general process of oxidation, and we hope to return to this point later.
>
> We are now working, or rather about to work, on the formation of carbonyl chloride in light.
>
> More than a year ago we did some rough preliminary experiments which proved very suggestive. You see, we had no means of looking up the literature so we had first of all to find out the essential points of the reaction and to get a general idea of the magnitude of the various effects. Since then we have been making, buying and borrowing apparatus. Pure chlorine and carbon monoxide are troublesome but not very difficult to obtain. The greatest difficulty is to obtain a reasonable source

of light. We must use homogeneous light and measure directly or indirectly the amount of energy absorbed. We have hopes of borrowing a mercury vapour lamp. At present we use a small half watt lamp. This will enable us to do the kinetics of the reaction since the absorption of 1 $h\nu$ results in the reaction of between 10^2 and 10^6 molecules. All the photochemical reactions of chlorine are of this trigger type of reaction and are accompanied by the induction period ['Sauerstoff-Hemmung'*] etc., besides which chlorine sensibilises other photochemical reactions. The whole thing is really extraordinarily fascinating. Our idea was that the energy was always transferred by radiation but I am afraid we may have to give it up.

I have been working too much recently so I shall take a little exercise this summer. I shall probably play tennis. A game doesn't last very long and can be made fairly vigorous. Four long and dull years have given me a decided preference for the short and merry type of life. Within the last few months I have visited Rubens, Nernst and Warburg. They were extremely willing to help and offered to lend anything they could. In fact all kinds of people lend us apparatus.

I suppose you are not able to do much scientific work now. It is really very sad that at a time when even the papers write of the necessity of scientific work, science is almost forgotten.

There can be no doubt about Chadwick's need to do science nor his intellectual enthrallment with the results of his labours. In the setting of Ruhleben, the chance to be mentally stimulated and to have specific goals other than plain survival helped enormously. Conditions had been hard from the outset with limited rations, overcrowding, and, in the winters, numbing cold. Chadwick[1] recalled 'the agony when my feet began to thaw out about 11 o'clock in the morning'. From 1915 onwards the British Navy maintained a tight blockade on merchant shipping passing through the North Sea and by 1917 this was having a devastating effect on food supplies to Germany. Dairy products became unobtainable, and in 1916 the *Kriegswurst*, or war-sausage, was introduced. This was a revolting, soft mixture of bread soaked in blood and fat.[12] Although he received parcels from home, and, thanks to the Evelyn Shaw of the 1851 Exhibition, a little money that could be used to bribe the guards, Chadwick was severely undernourished and developed digestive problems, which would dog him for the rest of his life. While he put on a brave face in his letter to Rutherford quoted above, writing to his mother in August 1918, towards the end of the war, Chadwick[28] conveys some of the sense of weariness of a man who was down at heel, spiritually and literally, after four years of captivity.

I am afraid you are far too optimistic about the exchange. I think you will find that there are one or two difficult points to settle before the agreement is ratified and even after the ratification there are the transport arrangements. I don't think there is much chance of my coming home before Christmas. I don't like the idea of another winter here but I don't suppose it will be any worse than the previous winters. Can you send leather? I have only one pair of boots I can wear now and they need mending badly.

---

* Sauerstoff-Hemmung refers to the process by which oxygen slows down the photochemical reaction of chlorine. It is germane to the modern debate about the ozone layer.

Whatever the hardships he endured, Chadwick was spared the carnage of the First World War, when he might have become just another young man of genius to be cut down in his prime. This was precisely the fate of Harry Moseley whose cranium was shattered by a bullet fired by a Turkish soldier at Suvla Bay in the Dardenelles campaign of 1915. Chadwick[24] wrote to Rutherford at the time: 'I was very sorry indeed to see that Moseley has been killed. Still he has the satisfaction of knowing he has done his duty.'

The bathos might jar the modern reader, but James Chadwick was expressing his feelings with unaffected honesty. Just as with his boyhood, Chadwick rarely spoke about his wartime captivity in Ruhleben—to him it was a closed episode of his life. He was acutely aware in the post-war years that many men of his generation had suffered far more than he. There may have been a lingering, irrational, guilt that he had not served King and Country.

References

Chadwick's personal papers are held in the Churchill Archives Centre (CAC), Cambridge. Items from the collection are referred to as CHAD with the appropriate file number. In later chapters, I also refer to other collections at the CAC (Cockcroft, Feather, Meitner) and I am grateful to the Master and Fellows of Churchill College, Cambridge, for permission to use all these papers.

1 Weiner C. (1969). Sir James Chadwick, oral history. Niels Bohr Library, American Institute of Physics, College Park, Maryland.

2 Chadwick, J.(14/1/14). Letter to E. Rutherford. Rutherford papers, Cambridge University Library.

3 Pais, A. (1986) β-Spectra, 1907–1914. In *Inward bound*, pp. 142–162. Oxford University Press.

4 Frisch, O.R.(1970). Lise Meitner (1878–1968), *Biographical memoirs of fellows of the Royal Society*, **16**, 405–15.

5 Rutherford, E. (16/5/11). Letter to B. Boltwood. In *Rutherford and Boltwood* (ed. L. Badash, 1969), p. 249. Yale University Press.

6 Chadwick, J.(1914). Intensitätsverteilung im magnetischen Spektrum der β-Strahlen von Radium B + C. *Verh. d. Deutsch. Phys. Ges.*, **16**, 383–91.

7 Chadwick, J.(1914). Report to the Royal Commissioners, 1851 Exhibition. Chadwick's papers, 1851 Archives.

8 Chadwick, J.(1918). Report to the Royal Commissioners, 1851 Exhibition. Chadwick's papers, 1851 Archives.

9 de Broglie, L. (1924). Recherches sur la théorie des quanta. In *Inward bound* (see ref. 3), p. 252.

10 Tuchman, B.(1962). *August 1914*, p. 80. Papermac, London.

11 Ketchum, J. D.(1965). *Ruhleben: a prison camp society*. University of Toronto Press.

12 Sladen, D.(1917). *In Ruhleben*. Hurst & Blackett, London.

13 Chadwick, A. (12/8/14). Letter to E. Shaw. Chadwick's papers, 1851 Archives.

14 Chadwick, A. (23/9/14). Letter to E. Shaw. Chadwick's papers, 1851 Archives.
15 Russell, A.S. (5/10/14). Letter to E. Shaw. Chadwick's papers, 1851 Archives.
16 Chadwick, A. (13/12/14). Letter to E. Shaw. Chadwick's papers, 1851 Archives.
17 Shaw, E. (15/12/14). Letter to A. Chadwick. Chadwick's papers, 1851 Archives.
18 Chadwick, A. (18/12/14). Letter to E. Shaw. Chadwick's papers, 1851 Archives.
19 Chadwick, J. (23/3/15). Letter to A. Chadwick. CHAD II 3/1, CAC, Cambridge.
20 Shaw, E. (29/3/15). Letter to Secretary, Manchester University. Chadwick's papers, 1851 Archives.
21 Chadwick, H. (22/4/15). Letter to E. Shaw. Chadwick's papers, 1851 Archives.
22 Shaw, E. (22/4/15). Letter to A. Chadwick. Chadwick's papers, 1851 Archives.
23 Chadwick, A. (16/6/15). Letter to E. Shaw. Chadwick's papers, 1851 Archives.
24 Chadwick, J. (14/9/15). Letter to E. Rutherford. Rutherford papers, Cambridge University Library.
25 Hutchinson, W.K. Gray J.A. and Massey H.S.W.(1981). Charles Drummond Ellis (1895–1980). *Biographical memoirs of fellows of the Royal Society*, **27**, 199–233.
26 Chadwick, J. (31/3/17). Letter to E. Rutherford. Rutherford papers, Cambridge University Library.
27 Chadwick, J. (24/5/18). Letter to E. Rutherford. Rutherford papers, Cambridge University Library.
28 Chadwick, J. (31/8/18). Letter to A. Chadwick. CHAD II 3/1, CAC.

4 A new beginning

The internees were released from Ruhleben, without ceremony, following the armistice of November 1918. They had been spared the slaughter of the Great War in return for four dreary years of progressive hardships; no heroes' welcome awaited them and the prevailing mood was one of quiet relief rather than elation. Chadwick took advantage of the general disorganization to leave the camp on his own and made his way into Berlin. There he looked for pre-war friends and saw firsthand the social and economic damage that the long war had brought to the city. After a few days, he rejoined a batch of men from Ruhleben and began the long, slow journey home.

Their train meandered interminably through the Berlin suburbs and eventually headed north to the Baltic Sea. The group spent a few days on the island of Rugen before sailing to Copenhagen, where they changed ship and crossed the North Sea to Leith, the port of Edinburgh. Chadwick watched the ship dock, leaning over the taffrail at the stern of the boat. A man from the quayside shouted to him and asked for news of his son, who by chance had been a friend of Chadwick's in the camp. Chadwick later asked the man why he had picked him from the crowd hanging over the rail. The answer, which did not altogether please Chadwick,[1] was that he looked 'the university type'! James Chadwick was 27 years old, seriously malnourished and broke: he had eleven pounds and a few shillings to his name.

In a despondent frame of mind, Chadwick returned to his parents' house in Manchester. Although physically drained, he immediately set about picking up the pieces of his career. He wrote a long handwritten report to the Commissioners of the 1851 Exhibition on 17 December 1918, in which he gave a detailed account of all the experiments he had carried out in Ruhleben. One of his reasons for submitting this report so promptly was that he was anxious to impress the Commissioners that he had fulfilled the requirements of his scholarship despite the adverse circumstances; he hoped to persuade them to extend the period of their financial support. He did not, though, succumb to self-promotion and was careful to pay full tribute to the contribution of his junior assistant Charles Ellis. 'I may claim to have directed the work,' Chadwick[2] wrote, 'as Mr. Ellis had, previous to his internment, done no scientific work. Otherwise it is difficult to say what part of the work was done by Mr. Ellis and what by me.'

More uplifting than his return to the family home was Chadwick's first visit to the university at Christmas time. He was received by Sir Ernest Rutherford 'with even more than his usual kindness'[3] and took great heart from the warmth of the welcome. Rutherford had seen his brilliant Manchester school of physics dissolved by the corrosive effect of the Great War, and was trying to maintain a vestigial department, desperately short of staff. He immediately offered Chadwick a part-time job teaching in the laboratories, which was all that he felt able to manage given his physical debility. The offer was gratefully accepted, and they were soon discussing potential research projects.

One of the first things Chadwick saw on his return to the laboratory was Rutherford's work on artificial disintegration, which he was invited to witness. The origins of these atom splitting experiments (which represent Rutherford's third great contribution after the elucidation of radioactivity and the nuclear model of the atom) dated back to 1914, and a series of experiments carried out by Ernest Marsden, the talented young lecturer who had first brought Rutherford the astounding news that α-particles could be bounced off thin foil. In 1914, Marsden[4] set out to demonstrate recent theoretical predictions concerning the collision of α-particles with atomic nuclei. He chose to look at hydrogen because 'from Bohr's formula for the "velocity curve" of a charged particle, it can be deduced that in an end-on collision, the "H" particle will have about four times the range of the α particle producing it'. His apparatus consisted of a sealed tube about 1 m in length and 9 cm in diameter. At one end it was closed by a zinc sulphide scintillation screen, and mounted within the tube was a small α-ray tube which could be placed at various distances from the screen. When the tube was filled with hydrogen gas at atmospheric pressure, Marsden detected particles capable of producing scintillations (i.e. displaced H particles or protons) 'which can travel at least 3½ times as far as the α particles'.

Although Marsden had succeeded in making observations which were in line with theoretical expectations, he was dissatisfied with his experimental arrangement and thought it desirable to make more quantitative measurements. To do this, he redployed the apparatus that he and Geiger had originally used in their work on the scattering of α-particles through large angles. The effects he now observed[5] were 'several times larger than anticipated by the formula'. With air in his apparatus, the maximum range of the α-particles was 5.8 cm, but Marsden noticed scintillations 'similar in appearance to those produced by H particles' at a distance of 8 cm. By various further manipulations, it was shown that these long-range scintillations were indeed due to hydrogen particles and Marsden arrived at the following statement.

> These results would be explained if we were to assume that there is sufficient hydrogen either in the gas inside the α-ray tube, or in the material of the glass of the tube, or in any water-vapour or hydrogen occluded in the surface of the glass.

He took further steps that effectively excluded glass or water vapour as the source of the extra hydrogen, and finally concluded 'there seems a strong suspicion that H particles are emitted from the radioactive atoms themselves, though not with uniform velocity'.

The appointment to a chair in New Zealand (where he had no radium), quickly followed by the outbreak of war, prevented Marsden from completing his investigations; he wrote up his interesting results, jointly with Lantsberry his assistant, in a short paper submitted in May 1915. Rutherford, with his unfailing nose for a fundamental anomaly, wrote to Marsden,[6] who was by then serving in France, and asked if he 'minded' if he took over the work himself. Although busy with government work, in particular advising the Admiralty on the detection and location of submarines, Rutherford, assisted only by his faithful laboratory steward Mr Kay, spent long hours exploring this question. His technique was meticulous and in November 1917 he made the crucial observation that if he fired α-particles into nitrogen gas, hydrogen nuclei emerged at the far end. Rutherford knew intuitively that the effect was due to the bombarding α-particles splintering the nitrogen nuclei with the ejection of hydrogen nuclei. The extra H particles in Marsden's experiment had been produced in the same way from nitrogen in the air—a possibility so outlandish that he had overlooked it. When Chadwick returned to Manchester, Rutherford was putting the final touches to this long series of experiments, which by themselves would have guaranteed a place in the history of science for the man who brought them to fruition. In January 1919, Ernest Marsden spent a few weeks of precious leave back in the laboratory assisting Rutherford and was rewarded by an acknowledgement in one of the forthcoming papers (which seems to have been regarded as sufficient recognition by both parties). The work was published in a quartet of papers in the *Philosophical Magazine* of April 1919, and in a memorable summary to the last of the four, Rutherford[7] found it 'difficult to avoid the conclusion that the long-range atoms arising from collision of α-particles with nitrogen are not nitrogen atoms but probably atoms of hydrogen . . . If this be the case, we must conclude that the nitrogen atom is disintegrated under the intense forces developed in a close collision with a swift α-particle, and that the hydrogen atom which is liberated formed a constituent part of the nitrogen nucleus.'

Just as with the nuclear model of the atom, a surprising experimental effect accurately noted by Marsden had served as the vital clue which Rutherford worried and gnawed at until he produced a solution of universal importance. While Marsden was the first to split the atomic nucleus in his original experiment, he did not recognize what had happened, and the credit will always belong to Rutherford because his deep intuition enabled him to supply the crucial interpretation of the events. Chadwick's scientific fame would ultimately be established in a similar fashion.

One of the marks of Rutherford's greatness was the consistent way he strove for direct evidence concerning any problem: he would never allow himself nor his juniors to be content with indirect evidence, no matter how persuasive it seemed. To him, reliance on theoretical speculation without making the extra effort to obtain experimental verification was like 'drawing a blank cheque on eternity'.[8] In 1919 Rutherford was still bothered by the lack of definite proof for Moseley's conclusion (made on the basis of his work in 1913 on the X-ray spectra of the elements) that the charge on the nucleus is numerically equal to the atomic number of that element. He thought it was fairly certain that Moseley was correct, but it was such a fundamental relationship that Rutherford wanted direct proof. Now Chadwick was back, he asked him to think about the matter and to devise an accurate way of measuring the charge on the nucleus of different atoms. Chadwick's[9] thoughts are clearly set out in a handwritten paper to the Commissioners of the 1851 Exhibition, who were still supporting him.

> The physical and chemical properties of an element are determined by the charge on the nucleus, for this fixes the number and arrangement of the external electrons, on which these properties mainly depend. The mass of the nucleus influences the arrangement of the electrons only to a very small degree. The nuclear charge is thus the fundamental constant of the atom and the question of its actual magnitude of great importance for the development of atomic theory.

Chadwick next reviewed Rutherford's 1911 paper on the structure of the atom in which it was deduced that the charge on the nucleus of an atom is roughly ½ Ae, where A is the atomic weight and e the charge on an electron. Chadwick also considered the 1913 paper of Geiger and Marsden[10] which showed that this result was true 'to within about 20%'. He then came to the crux of the problem.

> Later, van den Brock [1913] suggested that the nuclear charge might be equal to the atomic number of the element, i.e. the number of the element when all the elements are arranged in order of increasing atomic weight. This proved to be in good agreement with the experiments on scattering [Rutherford's, and Geiger and Marsden's], but the importance of the atomic number was first shown by Moseley's work on the X-ray spectra of the elements. Moseley found that the X-ray spectra of the elements depended on the square of a number which increased by unity in passing from one element to the next of higher atomic weight. This number was not exactly equal to the atomic number, but was equal to $N - a$, where N is the atomic number and a a constant for the series . . . On the nuclear theory of atomic structure this characteristic number must be closely connected with the charge in the nucleus and Moseley concluded that the number gave in fundamental units the actual value of the charge . . . This is one of the most important generalisations in modern physics and gives a starting point for the development of the external structure of the atom. It is therefore necessary to be quite clear as to the nature of the evidence on which it rests. Moseley's discovery that the X-ray spectra of an atom are defined by a characteristic number is an empirical expression of his measurements, and is independent of any theories of atomic structure or origin of spectra. The identification of this number, however, is attained by an arbitrary choice of the constant a, and the statement that the atomic number gives the nuclear charge implies a theory of the origin of X-ray spectra. The fact that the characteristic number changes by unity from element to element is, of course, very suggestive on the nuclear theory, but there is no experimental proof of Moseley's conclusion.

Chadwick set out to establish the nuclear charge for different metals, using Rutherford's law of scattering for α-particles, as Geiger and Marsden had done. The Rutherford formula depends on knowing the number of α-particles in an incident pencil beam from a radioactive source and measuring the number of α-particles scattered through a certain angle after the beam passes through a target, typically a thin metal foil. The nuclear charge of the atoms in the foil can then be calculated and compared with the atomic number of that element. Chadwick pointed out that the scattered particles represent an extremely small fraction of those in the incident beam; Geiger and Marsden had used two different methods to measure the incident and scattered particles, which in Chadwick's view led to their large experimental error of 20%. Chadwick designed a unified measurement system where the scattering foils of platinum, silver and copper were in the form of an annular ring with the original beam passing through the hole in the middle. The direct beam was then reduced in intensity by a known factor so that the numbers of direct and scattered particles reaching the scintillation screen for counting were approximately equal. At Rutherford's suggestion, this was achieved by introducing a rotating glass wheel into the primary beam. The wheel had a narrow slit in it so that only a small, but known, fraction of the direct beam passed through to the screen behind. By altering the speed of revolution of the wheel, a manageable number of α-particles were permitted to hit the screen every minute so that they could be counted by eye. With this apparatus, Chadwick found values for the nuclear charges of platinum, silver and copper of 77.4, 46.3 and 29.3 compared with respective atomic numbers of 78, 47 and 29. He estimated that his probable experimental errors were in the range of 1–1.5 per cent. He[11] was able to conclude, therefore, that there can be 'little doubt that the nuclear charge does really increase by unity as we pass from one element to the next, and that its net value is given by the atomic number'.

By a further modification of his apparatus, he was able to show that the inverse square law of force holds to a high degree of accuracy between a heavy nucleus such as platinum and an α-particle when the two are within 10^{-11} cm* of each other. In the Rutherford-Bohr planetary model of the atom, the innermost orbit or shell of electrons around the nucleus had become known as the K ring and this was thought to be at about 10^{-10} cm from the nucleus. Chadwick had shown that the force between nuclei showed no discontinuity between 3×10^{-12} cm and 10^{-10} cm from the centre of the nucleus, and wrote in the final sentence of his paper: 'We may therefore conclude that no electrons are present in the region between the nucleus and the K ring.' Until now, this had been a generally but not universally held assumption, and again Chadwick was the first to furnish independent evidence that it was indeed the case. The intriguing question of the nature of the forces between nuclei at extremely close proximity would become the thrust of his next research.

The foregoing work on the charge on the atomic nucleus had its origin in Manchester, but Chadwick carried out the actual experiment at the Cavendish Laboratory in Cambridge. Rutherford had been appointed to the Cavendish Chair of Physics and Director of the Laboratory in April 1919, succeeding his teacher Sir

---

* 10^{-11} cm = 0.000 000 000 01 cm 10^{-10} cm = 0.000 000 0001 cm 10^{-6} cm = 0.000 001 cm.

J.J. Thomson. Thomson had been Cavendish Professor since 1884 and gave up the post only because he found he could not manage to combine it with his new duties as Master of Trinity College. The new Cavendish Professor asked two of his Manchester staff, Chadwick and William Kay, the laboratory steward, to join him in Cambridge. In Rutherford's[12] words, 'Kay first of all decided to come to me to Cambridge, but his wife was averse to leaving her friends so I advised him to stay where he was, as I did not want a disgruntled wife interfering with his work'. While the loss of his able and experienced lab steward was a disappointment to Rutherford, James Chadwick had no such encumbrance and readily agreed to follow his chief, eager to continue his research as well as to escape from the confinement of his parental home. He had still not recovered his health and remained financially vulnerable. To Rutherford, he had shown himself to be a sound and productive experimenter; perhaps he could now help to establish the new dynasty at the Cavendish by taking on the additional teaching and supervisory roles that Geiger had filled so successfully in Manchester before the war. Indeed Chadwick had been largely trained by Geiger and provided a happy link with the past.

Chadwick arrived in Cambridge in the late summer of 1919 and was smitten with the architectural beauty of the colleges and their grounds backing onto the River Cam; if Manchester University was Gothic by the yard, here was Gothic by the century. The Rutherfords were living in Newnham Cottage, which they leased from Gonville and Caius College (usually known as Caius, pronounced 'keys'); Sir Ernest persuaded Hugh Anderson, the Master of Caius,[13] and a great supporter of his, to nominate Chadwick for a Wollaston Studentship. This brought an income of 120 pounds per year plus rooms in the college.[1] Gonville and Caius College does not possess the external splendour of its grander neighbours, Trinity and King's; it occupies a tight, landlocked site and is likely to be ignored by a casual passer-by. If one resists the obvious attraction of emerging from the narrow shadows of Trinity Street into the glorious wide vista of King's Parade, and instead negotiates the unpromising passage into Caius, the quiet loveliness of Tree Court is an unexpected delight. Built in the style of chateaux of the Loire, the court contains avenues of mature trees and immaculate lawns, which according to College Rules may be stepped on only by fellows or their personal guests. Tree Court was designed by the noted Victorian architect, Waterhouse, in the late 1860s to provide much needed extra accommodation for the expanding number of students; it is connected to the medieval Caius Court by the Gate of Virtue, an imposing Renaissance arch. By long tradition, undergraduates leave Caius through the even more elaborate Gate of Honour, which dates from the late sixteenth century, to receive their degrees from the nearby Senate House.

The Cavendish Laboratory[14] was situated in Free School Lane about five minutes walk from Caius. The professorship and the laboratory itself were named after the family of the seventh Duke of Devonshire, who generously funded them both in the 1870s. A contemporary opinion[15] of the Duke held him to be 'a man of quite exceptional ability and acquirements, one of our few nobles who would have rendered the most plebian name distinguished and possibly illustrious. Had he not been a Duke he would have been a rare Professor of Mathematics; for when at Cambridge he was one of the best men of his time, and he left it a Master of Arts and Second Wrangler*.'

The new Cavendish Chair of Physics was held first by James Clerk Maxwell, then by Lord Rayleigh, who was succeeded in turn by J.J. Thomson. Maxwell, a modest man of wry humour who had already retired from academic life once at the age of 34 due to ill-health, was pressed by many Cambridge men to leave his Scottish estate in order to take up the new Chair of Experimental Physics. He made a number of fundamental contributions during his short life, none more important than his theory of electromagnetic fields published in 1864. 'This change in the concept of reality', wrote Einstein,[16] 'is the most profound and the most fruitful one that has come to physics since Newton'. The same expansive intellect that allowed him to propound a set of mathematical equations, synthesising the known properties of light and other forms of electromagnetic radiation into a coherent whole, made Maxwell a visionary founder of the Cavendish school.

In his inaugural lecture in October 1871, Maxwell[14] identified the need for both accuracy in experimental measurements as well as imaginative thinking, if the Cavendish were to flourish as a scientific institution.

> This characteristic of modern experiments—that they consist principally of measurements—is so prominent that the opinion seems to have got abroad that in a few years all the great physical constants will have been approximately estimated, and that the only occupation which will then be left to men of science will be to carry on these measurements to another place of decimals.
>
> If this is really the state of things to which we are approaching, our Laboratory may perhaps become celebrated as a place of conscientious labour and consummate skill, but it will be out of place in the University, and ought rather to be classed with the other great workshops of our country, where equal ability is directed to more useful ends.
>
> But we have no right to think thus of the unsearchable riches of creation, or of the untried fertility of those fresh minds into which these riches will continue to be poured. It may possibly be true that, in some of those fields of discovery which lie open to such rough observations as can be made without artificial methods, the great explorers of former times have appropriated most of what is valuable, and that the gleanings which remain are sought after, rather for their abstruseness, than for their intrinsic worth. But the history of science shews that even during that phase of her progress in which she devotes herself to improving the accuracy of the numerical measurements of quantities with which she has long been familiar, she is preparing the materials for the subjugation of new regions, which would have remained unknown if she had been contented with the rough methods of her early pioneers.

These prophetic words, together with mention of international cooperation and

---

* The Tripos examination in mathematics was an unimaginably arduous challenge for which Cambridge undergraduates were coached intensively. In the latter part of the nineteenth century, the Tripos reached its zenith and the top student, or Senior Wrangler, was almost a national celebrity. To be the Second Wrangler was to be the second placed student.

molecular structure, in many ways established the scope and philosophy that would make the Cavendish unique. The physical fabric of the laboratory took a few more years to arrive, and in October 1872 Maxwell[17] wrote: 'The Laboratory is rising, I hear, but I have no place to erect my chair, but move about like the cuckoo, depositing my notions in the Chemical Lecture Room in the first term, in the Botanical in Lent and in the Comparative Anatomy in Easter'. Maxwell's lectures were attended by only a handful of students, but by the end of the century demonstrations were attended by over two hundred; postgraduate researchers, like the young Ernest Rutherford, were attracted to the Cavendish from all over the world. J.J. Thomson was an enthusiastic proponent of the international flavour. 'The advantage gained by our students from this communion of men of widely differing training, points of view, and temperament can, I think, hardly be overestimated,' he[18] once wrote. 'In their discussions they become familiar with the points of view of many different schools of scientific thought, leading to a better, more intelligent, and more sympathetic appreciation of work done in other countries; they gained catholicity of view not merely on scientific but on political and social questions.'

In 1914 there were nearly 300 undergraduate students reading physics, but during the First World War the training of scientists virtually ceased. The Cavendish[17] became a billet for soldiers, and its workshops were given over to the manufacture of precision gauges, which were in great demand for the armaments industry. There was no concerted effort by the government to divert the activities of the laboratory to military ends. Indeed physics was not seen as having much relevance to warfare in 1914. Although some of the nation's leading figures such as Rutherford and J.J. Thomson were involved with the Admiralty's Board of Invention and Research from its inception in 1915, most young scientists from the Cavendish, like their contemporaries from Manchester and other universities, went off to fight. With demobilization there was an unprecedented demand for science training, and in 1920 the laboratory became seriously congested with twice as many students as ever before. When Chadwick[19] had written to Rutherford in 1918 that 'science is almost forgotten', he was preaching to the converted. Now Rutherford was intent not only on expanding the Cavendish facilities to meet the present teaching needs, but on guaranteeing the future supply of the brightest young research workers. In this he was helped by the fact that Cambridge University, after considerable internal resistance, finally offered the postgraduate Doctor of Philosophy degree in 1920. Rutherford[20] was also determined to establish beyond doubt the importance of well-trained scientists to the wealth and security of the nation. As he put it to the Vice-Chancellor, Sir Arthur Shipley, in a memorandum, 'The History and Needs of the Cavendish Laboratory—1919':

> The need of a laboratory specially devoted to training in research in Applied Physics is of pressing importance if we are to play our part in the researches required by the State, and in providing well-qualified research men for various branches of industry and for the scientific departments of the State.

The comprehensive memorandum concluded:

The most pressing and urgent needs for the department to make it thoroughly efficient are summarized below:

(1) Increased laboratory and lecture space for the teaching of Physics.

(2) Provision of new, well-equipped laboratories for Applied Physics, Optics and Properties of Matter.

(3) Provision of three additional lecturers of high standing, competent to direct advanced study in research in the new departments mentioned above.

(4) The endowment of another Chair of Physics in the University.

. . . It is estimated that the cost of the new buildings would not be more than £75,000 and that an additional sum of £125,000 will be necessary as an adequate endowment . . .

Sir Joseph Larmor, the Lucasian Professor of Mathematics at Cambridge, was an elector for the Cavendish Chair in 1919. He had first come to prominence in 1880 as Senior Wrangler in the Mathematical Tripos, when he beat J.J. Thomson into second place. Perhaps Rutherford's strongest supporter, Larmor had written to him at a time when he was experiencing mixed feelings about leaving Manchester:

The fame of the Cavendish ought to make it easy to beg for funds for an extension. Shipley [the Vice-Chancellor] and our other public men are past masters at importunate mendicancy, and would be available . . . I now see the possibilities of expansion that are practical if people here had a strong lead. They complain of J.J. that he not only did not give a lead but poured cold water over projects.

Rutherford certainly offered a strong lead, but there was no largesse forthcoming. His own attitude to money was ambivalent and he does not seem to have followed through after his initial demand—a lack of persistence which meant that he would never join the past masters of begging. Indeed, with one or two notable exceptions, during his tenure as director, any expansion of the Cavendish was piecemeal, and frugality was the watchword of the laboratory's economic management. In many ways, the consequences of this approach would be felt most acutely by Chadwick, whose responsiblity it gradually became to keep the Cavendish functioning from day to day and its research workers tolerably content.

Whatever his frailties as a department manager, Rutherford had none as a leader. He and Chadwick[1] agreed early on that the main thrust of research in the Cavendish should be aimed at understanding the structure of the atomic nucleus, and this policy endured for the period of Rutherford's directorship. Rutherford was mindful of the disruption that the war had brought to the international world of physics, and immediately set about restoring his close ties with former colleagues in Europe and further afield. He was a tireless correspondent and traveller, who did much to get cooperation and research moving again by his selfless advice and practical help. In part he was able to do this because Chadwick proved so effective as an organizer, releasing him from many mundane and time-consuming duties. With his reserved, unobtrusive personality, Chadwick[1] made a quiet start as Rutherford's research assistant. In his words,

Everything had been diverted to the war effort. There was in fact very little

> equipment . . . So, gradually, part of my duties in helping Rutherford—it was a matter between ourselves, it wasn't coupled to my studentship at all, there were no conditions of that kind attached to it, but it was understood between us that I was to do anything that I could to help him on the trivial matters. And I began to help in saying, well, we need this piece of apparatus, and ordering them.

Another reason for Chadwick to feel his way cautiously was that he was still relatively junior in scientific terms, and in 1919 the Cavendish contained some famous men other than its newly appointed director. Besides J.J. Thomson, who still came to the laboratory regularly with Rutherford's blessing, there was C.T.R. Wilson,[21] whose inspirational invention, the cloud chamber, had already proved to be one of the most useful pieces of apparatus ever devised in experimental physics. At the time of Chadwick's arrival, Wilson was 50 years old and would continue to work alone, 'a shy but enduring genius'[18]—he received the Nobel Prize for Physics in 1927. Edward Appleton[22] had started his research career at the Cavendish in 1914, but left to join the Army on the outbreak of war. He was recruited by the Army Signals Service and this shaped his future work on the thermionic valve and long-distance radio propagation. His subsequent discoveries represent a scientific hybrid, perpetuating both the Maxwell–J.J. Thomson tradition of electromagnetism and C.T.R. Wilson's interest in atmospheric research. Rutherford encouraged his investigations of radio transmission and appointed him an assistant demonstrator.

Francis William Aston,[23] the mechanically gifted son of a wealthy Birmingham metal merchant, was, like Wilson, a scientific loner. In 1919, he was busy perfecting his mass spectrograph with which he would soon show that neon, chlorine and other elements were mixtures of *isotopes.** Aston's work was quickly recognized with the Nobel Prize for Chemistry in 1922: Chadwick[24] later described him as one of the best experimenters he had known. Another great individualist appointed by J.J. was G.I. Taylor,[14] who became the country's leading expert on fluid mechanics. He had the misfortune to occupy the room next to Rutherford and Chadwick's laboratory, and all visitors to the director had to pass through Taylor's room. Rutherford unfairly compounded the inconvenience by treating Taylor as an unofficial guide, but he was of equable temperament and did not complain. Indeed there seems to have been no resentment towards the new professor and his assistant from any of these men; Aston and Taylor became two of Rutherford's regular golf partners in a Trinity College foursome.

Less well-known to the outside world and slightly more prickly was Dr. G.F.C. Searle.[14] He was a popular and revered figure in the Cavendish Laboratory, who had begun his fifty-five year career as an experimental demonstrator there before Chadwick was born. Searle excelled as a practical teacher and generations of undergraduates benefited from his wisdom and gentle humour. He openly opposed Rutherford's suggestion to the University that the intermediate degree of an M.Sc. should be introduced for those students who could not meet

---

* Isotopes are atoms of the same element, which differ in their atomic weight but are chemically identical.

the time or academic requirements of the three year Ph.D.; he[20] foresaw 'overcrowding of laboratories by students of little ability and resource' and feared that any further increase in numbers would lead to the 'discouragement of ordinary university teaching'.

While it is sometimes stated that Chadwick came to the Cavendish as Rutherford's research lieutenant, he was in reality the senior research student, given the extra duty of supervising the Radium Room. In Chadwick's[1] words, he had 'no authority or responsiblity except to Rutherford.' Rutherford, in the early days, kept a fatherly eye on him and was instrumental in strengthening his scientific credentials. In May 1920, Chadwick[25] wrote to Rutherford applying for the vacant Clerk Maxwell Studentship.[14] This was a University award, first made in 1890, after a bequest by Mrs Clerk Maxwell; previous holders had included C.T.R. Wilson and Aston. Mrs Maxwell's original intention had been to enable students not endowed with private means to continue in research. One of the conditions of the studentship was that it should not depend on the result of an examination, but on promise in research. In his application, Chadwick listed seven papers previously published in the scientific journals; in other times, this might have excluded him for being overqualified, but he was given the prestigious award. At about this time, Chadwick also registered for the new Ph.D. degree.

The first part of his thesis[26] was the important work already described, *The atomic nucleus and the law of force*. The second part explored more deeply the question of forces between nuclei at very close proximity, and the size and shape of the helium nucleus or α-particle. For this second experiment he was joined in the autumn of 1920 by Etienne Bieler,[27] an 1851 Exhibition Scholar visiting from McGill University, Montreal. Bieler* had no previous experience in radioactivity, but came to the Cavendish with a strong recommendation from A.S. Eve, Rutherford's friend and successor as Professor of Physics in Montreal. Under Chadwick's tutelage, Bieler learned the experimental techniques quickly, and Chadwick[28] found him 'a delightful companion both in and out of the laboratory'.

Part II of Chadwick's Ph.D. thesis began with the following words:

> When α-particles pass through hydrogen gas or a substance containing hydrogen, close collisions between an α-particle and a hydrogen nucleus occasionally take place. As a result of such a close collision, the hydrogen nucleus is set in swift motion and can be detected by the scintillation it produces on a zinc sulphide screen. Assuming that both the α-particle and the hydrogen nucleus can be regarded as points, and that the forces between them arise from their charges, C.G. Darwin [in 1914] calculated the number [of hydrogen nuclei] within any given angle to the path of the α-particle. Rutherford, [in 1919] however, found that the number and angular distribution of the projected H-particles did not agree with the simple theory, and he attributed the divergence to the complex structure of the α-particle.

---

* Etienne Samuel Bieler (1895–1929), born in Switzerland, educated in Paris and then Canada, after his father was appointed Professor of Theology at McGill University. After graduating in 1915, he enlisted with the Canadian Field Artillery and fought in Flanders, the Somme, Ypres and Vimy Ridge, where he was wounded. He came to Cambridge on Eve's recommendation and was awarded his Ph.D.in 1923. He died from pneumonia while leading a geophysical survey in Australia.

In making his original conclusion, Rutherford had based his argument on the then accepted theory that the α-particle, or helium nucleus, was composed of four hydrogen nuclei and two electrons. Chadwick pointed out that Rutherford's experiment was 'of a preliminary nature, and . . . not carried out to any high degree of accuracy. Recently, the optical condition of counting scintillations have been so greatly improved, that a more direct method of attacking the problem was possible.' What Chadwick had done was to study the catalogues of lens makers to find the microscope objective with the biggest aperture. He had selected 'a Watson holoscopic objective of 16 mm focal length and .45 numerical aperture combined with a low-power eyepiece. Compared with the old system, this increased greatly the brightness of the scintillations and gave at the same time a larger field of view, i.e. a larger number of particles, other conditions remaining constant.'[29] He also adapted the optically symmetrical apparatus that he had devised in the first experiment on the charge on the atomic nucleus. In order to employ this arrangement, the hydrogen target had to be held in the ring as the metal foils had been previously, and he decided to use paraffin wax, which is a combination of hydrogen and carbon atoms. Chadwick made the following, apparently sweeping, assumption:

> In these collisions it is immaterial whether the hydrogen is present in the form of hydrogen gas or in the combined state as paraffin wax.

Although he offered no evidence in support of this statement, Chadwick had begun to assist Rutherford in experiments to disintegrate light elements by α-particle bombardment, and they had found no evidence for the splintering of carbon nuclei. On this basis, the presence of carbon in the paraffin wax target should not have any influence on the observed results—although a few years later the behaviour of carbon would become the central issue in a celebrated scientific controversy. What Chadwick and Bieler found was that for α-particles of low velocity (which had insufficient energy to approach the hydrogen nucleus very closely) the inverse square law, as between two point charges, was obeyed. For the high-velocity α-particles the numbers of projected hydrogen nuclei were much greater, in some cases more than a hundred times as great, than would be expected according to the inverse square law. This was a startling finding and clearly perplexed Chadwick for sometime. He had a deadline to meet for the submission of his thesis, and the final Discussion section covered one side of paper! Its opening sentence was a masterpiece of scientific conservatism, and much more in keeping with his innate caution than the assumption, quoted earlier, that opened the Method section. The Discussion began:

> Until the fullest information which such experiments can yield has been obtained, it does not seem advisable to discuss in any detail the structure of the α-particle, or Helium nucleus, and the field of force around it.

The *Cambridge University Reporter* recorded that James Chadwick of Gonville and Caius College received his Ph.D. on 21 June 1921: he was one of the first to be conferred the degree. Five years later, Dirac[30] was awarded his Ph.D. for a thesis entitled '*Quantum Mechanics*', but few of the hundreds of dissertations submitted to Cambridge University down the years can have contained such profoundly important work as Chadwick's.

Chadwick and Bieler[29] soon arrived at more definite conclusions about their results and published them in a paper, *The Collisions of Particles with Hydrogen Nuclei.* This added some further results and a full Discussion, but was otherwise identical to the Ph.D. dissertation. The penultimate paragraph of the Discussion is reproduced below:

> As regards the structure of the α particle, it will be apparent at once that no system of four H nuclei and two electrons united by inverse square law forces could give a field of force of such intensity over so large an extent. We must conclude either that the α particle is not made up of four H nuclei and two electrons, or that the law of force is not the inverse square in the immediate neighbourhood of an electric charge. It is simpler to choose the latter alternative, particularly as other experimental, as well as theoretical, considerations point in this direction. The present experiments do not seem to throw any light on the nature or the law of variation of the forces at the seat of an electric charge, but merely show that the forces are of very great intensity.

It is now known that the α-particle does not consist of four protons and two electrons, and that the inverse square law does not account for the interactions within the atomic nucleus. Since the mid-1930s, physicists have believed the nuclear constituents are held together by the *strong interaction*, a novel kind of force different from gravity or electromagnetic attraction. The 1921 paper of Chadwick and Bieler has been cited 'as marking the birth of the strong interactions'.[31] When asked, years later, about his earliest notions of the strong forces, Chadwick's[1] answer was revealing:

> That was clear about 1921 or '22. But we had no explanation of it. I put it this way: Any idea one might have about the structure of the nucleus—particles had to be held together somewhere. So that in addition to the repulsive force between the positively charged particles, there had to be an attractive force somewhere. And I played around with various forms of the force with an attraction varying as the inverse fourth power of the distance. And I remember I had to brush up my mathematics a bit in order to solve some of the equations. And it was also clear, I think in Bieler's mind, too. I remember our making lantern slides showing the potential well and the peak and so forth, and they were used in lectures.

During the first two decades of the century, experimental physicists like Chadwick and Rutherford were attempting to unlock the innermost secrets of the atom by observing phenomena that they contrived to create in a controlled way. Their observations were subjective to a degree: the information did not present itself in a digital or mathematical form, it depended on the vagaries of human perception. As Hughes[32] has recently pointed out, disagreements between physicists during the 1920s often 'were not over matters of interpretation of facts; they were disputes about what the facts *were*.' The lack of any comprehensive theory of atomic structure beyond the Rutherford–Bohr model hampered the elucidation of experimental results; as we have already seen in Chadwick's work, it was frequently necessary to fly in the face of accepted wisdom in order to present new findings. Atomic physics, or *radioactivity* as its Cambridge practitioners would have termed it, was still an empirical subject, where experimental evidence was gathered and used to shape theoretical developments; in a later phase, the new quantum mechanics became such a powerful tool that specific predictions could be made and then tested experimentally.

In June 1920 Sir Ernest Rutherford[33] was accorded the unusual honour by the Royal Society of being invited to deliver his second Bakerian Lecture. Entitled 'Nuclear constitution of atoms', it encapsulated much of the state of knowledge on the subject, including a brief review of work completed during the previous decade and important speculations about the future. Early in his talk Rutherford made the following observation: 'The question whether the atomic number of an element is the actual measure of its nuclear charge is a matter of such fundamental importance that all methods of attack should be followed up'. He then stated that 'the results so far obtained by Mr Chadwick strongly support the identity of the atomic number with the nuclear charge . . .' He also referred to his own work on α-particle scattering by hydrogen atoms in a section headed 'Dimensions of Nuclei', but did not mention Chadwick and Bieler in this context, suggesting that their experiment had not yet begun. Rutherford deduced that 'the law of inverse squares no longer holds when the nuclei approach to within a distance of 3×10^{-13} cm of each other . . . in such close encounters there were enormous forces between the nuclei, and probably the structure of the nuclei was much deformed during the collision'. This suggests that Rutherford was already postulating a new force field in the immediate vicinity of the nucleus.

In contrast to the above passage, which can be construed as anticipating scientific discoveries of the 1930s, as Rutherford got to the heart of his lecture he stated his thoughts on the constitution of the nucleus in the following words:

> From a study of radioactivity we know that the nuclei of the radio-active elements consist in part of helium nuclei of charge $2e$. We also have strong reason for believing that the nuclei of atoms contain electrons as well as positively charged bodies, and that the positive charge on the nucleus represents the excess positive charge.

These views were very much of their time and had held sway in Rutherford's mind for some years. He had been the first to recognize the emission of α- and β-particles from radioactive substances, and later correctly interpreted their production as nuclear events. Taking the argument a simple but misleading step further, he[34] had written in a 1914 paper: 'The helium nucleus is a very stable configuration which survives the intense disturbances resulting in its expulsion with high velocity from the radioactive atom, and is one of the units of which possibly the great majority of the atoms are composed'. The second part of the hypothesis, as presented by Rutherford[35] in a discussion at the Royal Society in March 1914, seems utterly convincing on its face: 'The general evidence indicates that the primary β-particles arise from a disturbance of the nucleus. The latter must consequently be considered as a very complex structure consisting of positive particles and electrons . . .' But, for once, Rutherford's celebrated powers of physical intuition, which allowed him to visualize events happening at a subatomic level, were leading him astray. Although he was correct that radioactivity is a nuclear phenomenon, it does not follow that α- and β-particles are stable constituents of the nucleus. During the 1920s, even he would find that the paradoxes of physics were no longer susceptible to intuition alone, and mathematical abstraction rather than descriptions based on mechanistic images would become the subject's leading language. The exploration of the atom could not be accomplished by referring to mental charts that had been drawn up from the study of more commonplace physical surroundings and events.

This is not to say that his days as an innovative thinker were over, and the audience at his Bakerian Lecture were introduced to another revolutionary idea in the following passage. The word *proton* had not yet been coined by Rutherford to denote the basic positively charged particle in the nucleus and he still referred to hydrogen nuclei or charged hydrogen atoms instead.

> Under some conditions, however, it may be possible for an electron to combine much more closely with the H nucleus, forming a kind of neutral doublet. Such an atom would have very novel properties. Its external field would be practically zero, except very close to the nucleus, and in consequence it should be able to move freely through matter. Its presence would probably be difficult to detect by the spectroscope, and it may be impossible to contain it in a sealed vessel.

This was the first inkling, at least in public, of the *neutron*, but according to Chadwick[1]

> . . .it had been in his (Rutherford's) mind for some considerable time. He had asked himself, and kept on asking himself, how the atoms were built up . . . the general idea being at that time that protons and electrons were the constituents of an atomic nucleus. One could imagine easily enough how a proton and an electron might under certain circumstances get close together, although the presence, the existence of hydrogen as a stable element made it a bit difficult. But how on earth were you going to build up a big nucleus with a large positive charge. And the answer was a neutral particle.

In his thoughts about the 'neutral doublet' quoted above, Rutherford had already recognized that its lack of charge would make it difficult to detect. He did not use the word 'neutron' in the lecture, and instead persisted in a lax terminology of calling it an atom. His ideas about proving its existence were rudimentary and he confined himself to the following remarks:

> . . .it is to be expected that they may be produced, but only in very small numbers, in the electric discharge through hydrogen, where both the electrons and H nuclei are present in considerable numbers. It is the intention of the writer to make experiments to test whether any indication of the production of such atoms can be obtained under these conditions.

According to Chadwick,[1] Rutherford asked J.L. Glasson, a research student, to undertake such an experiment 'immediately after the Bakerian Lecture'. Glasson's work was sent for publication by Rutherford about one year later under the title,

Attempts to Detect the Presence of Neutrons in a Discharge Tube. The following extract from Glasson's[36] introductory section gives a clear idea of the thinking in the Cavendish at the time.

> In the ordinary atom of hydrogen we have a single electron separated from the nucleus by a distance of the order of 10^{-8} cm. It is here contemplated that a more intimate union of the two is possible, such as would be obtained if the electron fell into the nucleus, so that the separation became of nuclear instead of atomic dimensions. Such a particle, to which the name *neutron* has been given by Prof. Rutherford, would have novel properties. It would, for instance, greatly simplify our ideas as to how the nuclei of the heavy elements are built up. This building-up process is apparently at work in the evolution of stellar systems from the nebular state.

Glasson applied a range of voltages from 2000 to 50 000 V across his discharge tube containing hydrogen without conjuring up any trace of the neutral particle he was hoping to find. Nor was a similar experiment by another Cavendish researcher, J.K. Roberts, in 1922 any more successful, and he later described his attempt as 'a crazy idea of the Prof. or Chadwick on which I wasted a lot of time'.[24]

At the end of his Bakerian lecture, Rutherford had sounded a cautionary note. Talking about 'the strain involved in counting scintillations under difficult conditions,' he warned that 'further progress is not likely to be rapid'. The scintillation method of counting α-particles and protons was a technique indispensible to much of Rutherford's research in radioactivity. Small wonder that he[37] thought of it as a method of 'very great delicacy and power'; its immediacy and directness appealed to him, and fired his imagination.

> In a dark room the surface of the screen is seen as a dark background dotted with brilliant points of light which come and go with great rapidity. This beautiful experiment brings vividly before the observer the idea that the radium is shooting out a stream of projectiles each of which causes a flash of light on striking the screen.

Scintillation depends on the presence of a trace of copper or other metal impurity in small zinc sulphide crystals and arises from the transformation of the energy of the incident α-particle into light energy. Each scintillation lasts about one tenth of a millisecond and its detection by the human eye is a demanding test. The observer has to prepare for the task by sitting in a darkened room for twenty minutes or so to dark adapt the eye. Looking intently at the screen, waiting to count multiple instantaneous flashes is a little like peering into the summer night's sky hoping for a shooting star. Flashes in quick succesion are easy to miss; in quiet patches, non-existent flashes are easy to imagine. Despite his graphic description of the process, Rutherford often left the tedious counting to other, younger eyes. As early as 1908, he[38] made the following admission in a letter: 'Geiger is a demon at the work and could count at intervals for a whole night without disturbing his equanimity. I damned vigorously after two minutes and retired from the conflict.' In his 1919 paper on the disintegration of the nitrogen atom, Rutherford expressed his 'thanks to Mr. William Kay for his invaluable assistance in counting scintillations'. At the end of his Bakerian Lecture, there was an acknowledgement to Mr.

J. Chadwick for the same duty 'in some of the later experiments'.

In fact Chadwick[1] was more than another pair of eyes—he brought some discipline to the process.

> I also reduced the counting of scintillations to a drill which was very necessary, that is (and Rutherford agreed), a count for one minute at a time only, and have a little rest, count again, and so on. And not do more than a certain amount in a day because one's eyes got tired. Also, a much more important thing really, although the drill helped, was to see only the scintillations that were there and not to imagine scintillations which is very easy to do.

Chadwick also took upon himself to inculcate the research students with this routine so that they would count scintillations with consistency. Their training took place in the 'Nursery', a poorly ventilated attic room in the Cavendish, where some of the most talented young physicists of their generation were introduced to equipment and laboratory techniques during the summer vacation before they embarked on postgraduate work. Even the glow of nostalgia did not imbue the experience with joy in most students' memories, and the recollections of de Bruyne[39] are typical.

> Most of my contemporaries had spent the long vacation in 'The Nursery' which was the attic, learning the arts and crafts of radioactivity under the mournful eye of James Chadwick. They were assigned a research task and more or less left to get on with it.

Professor David Shoenberg,[40] a later research student, wrote in similar vein:

> I had first to undergo the traditional introduction to research in the Cavendish, which was a few weeks of practical work in the summer on vacuum techniques under Chadwick's supervision. The equipment provided was rather primitive—only a hand-operated Fleuss pump and a home-made diffusion pump—and the work consisted mainly of chasing leaks in poorly constructed glassware. It was a very hot summer and it was easy to get discouraged, but Chadwick would come by occasionally and encourage us with some gloomily sympathetic remark. He had made glass joints with a candle flame when he was interned in Ruhleben during World War I and felt we were lucky to have gas.

All Rutherford's great discoveries were made with benchtop instruments, and Chadwick was merely enforcing the tradition, going back to Maxwell, that students should use home-made apparatus. There was an advantage in experimental physics to the researcher having an intimate feel for the equipment which he would use to confront the mysteries of nature. Unlike Rutherford, who would immediately let young researchers know whether he was pleased or disappointed by their efforts, Chadwick was undemonstrative and the lack of feedback unnerved even the brightest, until they came to know him better (which not all did). Allibone,[41] who was to follow in Chadwick's footsteps as a Wollaston Scholar at Gonville and Caius, described how 'Dr. Chadwick would listen to your account, say nothing and then leave you so that you had no idea whether what you had reported pleased him or not'.

Rutherford was an intensely sociable man and Chadwick[42] soon realized that apart from any technical help he was expected to lend at the workbench, he also provided companionship. Rutherford 'wanted company to support the tedium of counting in the dark—and to lend an ear to his robust rendering of "Onward, Christian Soldiers".' In exchange for this dubious musical pleasure, Chadwick[1] was treated to the great man's thoughts on whatever subject took his fancy.

> Before the experiments, before we began to observe in these experiments, we had to accustom ourselves to the dark, to get our eyes adjusted, and we had a big box in the room in which we took refuge while Crowe, Rutherford's personal assistant and technician, prepared the apparatus. That is to say, he brought the radioactive source down from the radium room, put it in the apparatus, evacuated it, or filled it with whatever, put the various sources in and made the arrangements that we'd agreed upon. And we sat in this dark room, dark box, for perhaps half an hour or so, and naturally talked . . . it was these conversations that convinced me that the neutron must exist.

George Crowe[14] was the estimable laboratory steward Rutherford had the great good fortune to inherit when he came to the Cavendish, and he proved to be the perfect successor to Mr Kay. He had been hired as a teenager in 1907 and gave stalwart service to the laboratory until forced to retire in 1959. 'Cloud' Wilson had taught him the art of glass-blowing and he was a skilful carpenter, as befits the son of a boat-builder. One of Crowe's first tasks for the new professor in 1919 was to help put the precious supply of radium brought from Manchester into solution again so that radioactive sources could be prepared as required. He would regularly pump off the radon gas from the radium solution; as a precaution the Radium Room was situated right at the top of the building so that any stray radon gas, or emanation, would escape to the atmosphere and not permeate the laboratory. Precautions taken for personal protection were rudimentary and Crowe often ignored even these, prefering to work with bare hands rather than cumbersome gloves. After a few years, his hands started to show changes of radiation dermatitis, with necrosis of the fingers.

Apart from the talented men appointed by J.J., the early post-war influx brought other young students to the Cavendish who would become the foremost physicists of the next generation. One of these was already well-known to Chadwick: his fellow internee and protégé Charles Ellis[43] entered Trinity College as an undergraduate in 1919. The rules regarding the Tripos examination were relaxed during this period in recognition of the time students had already given to the war, and Ellis was able to take Part I of the examination in two terms with a First Class in mathematics. He completed Part II in 1920 with a First in physics, and immediately entered the Cavendish as a research student to resume the study of β-rays.

In January 1919 the Admiralty sent a contingent of four hundred young naval officers to Cambridge to spend six months broadening their education. Among them was a tall, handsome, wavy-haired Lieutenant who had fought at the Battle of Jutland in 1916 as an 18-year-old midshipman. His name was Patrick Blackett[44] and after three weeks at Cambridge he resigned his commission to become an undergraduate. After obtaining a First in physics in Part II of the Tripos,

Blackett took his place as one of about 30 research students in 1921. His patrician manner combined with obvious brilliance seems to have brought him to Rutherford's attention straightaway. Blackett[45] gave the following succinct account of his first project, which would set him on the road to a Nobel Prize.

> In 1921 Rutherford had set me the problem of attempting to use C.T.R. Wilson's beautiful cloud chamber method to photograph the disintegration of nitrogen nuclei, which he had observed by means of scintillations. This meant making an automatic cloud chamber which would take a very large number of photographs. After three years' preliminary work mainly concerned with the development of an automatic cloud chamber, I succeeded in 1924 in taking within a few months some 25 thousand photographs showing the tracks of 400 thousand α-particles, and amongst these tracks discovered six which clearly represented the process of atomic disintegration discovered previously by Rutherford. The novel result deduced from these photographs was that the α-particle was itself captured by the nitrogen nucleus with the ejection of a hydrogen atom, so producing a new and then unknown isotope of oxygen, ^{17}O.

The new constellation was forming rapidly and in the high summer of 1921 its most mercurial member, Peter Leonidovich Kapitza,[46] was first seen. The circumstances of his arrival in Cambridge were born of personal tragedy: a year earlier, he had lost his two-year-old son to scarlet fever, and then, within one month, his wife, infant daughter and father to the Spanish influenza epidemic that swept Europe. In an attempt to assuage his grief, A.F. Joffé, the founder and head of the Petrograd Physico-technical Institute appointed the young man to a new commission of the Russian Academy of Sciences which had been set up with the express purpose of renewing scientific relations with other countries. Joffé set out on an international tour in February 1921 to buy foreign equipment and to establish contact with other physicists across Europe. Eventually he managed to obtain permission from the Soviets for Kapitza to join him, only to find that Germany, France and Holland would not admit him, fearing that he might prove to be a young communist agitator. England was a little more self-confident, and in May granted Kapitza an immigration visa. Soon after arriving in London, he and Joffé were taken up by the Fabians and invited to dinner by H.G. Wells, as well as meeting George Bernard Shaw and Lord Haldane. The two men started their tour of British scientific departments in June, and it culminated in a visit to Cambridge on 12 July to see Rutherford.

There is an apochryphal story concerning Kapitza's acceptance into the Cavendish. It runs as follows: Kapitza asked Rutherford whether he might work in the Cavendish as a research student for some months. Rutherford demurred, pointing out the difficulty of accommodating an extra worker in an already overcrowded laboratory. Kapitza, unabashed, countered by asking what accuracy Rutherford aimed for in his experiments. The puzzled reply was '2 or 3%', to which Kapitza cheekily pointed out that as he would be one among 30 researchers, he would come within the experimental error and therefore not be noticed.

Such an exchange would be typical of Kapitza, but according to Chadwick[47], while the story was 'most amusing . . . it is quite wrong'. If Kapitza had spoken to him so impudently, 'Rutherford would never have admitted him'. In Chadwick's more prosaic version, he and Rutherford showed the two visitors round the Cavendish; it was natural that the professor should address himself to Joffé, while it fell to Chadwick to talk to Kapitza. Soon afterwards Kapitza made his request to work in the laboratory. 'Rutherford was not very happy about it and he consulted me, partly because he himself had had little talk with P.K. I could assure him that I thought P.K. would be an acquisition.'[3] But Rutherford, like the governments of Western Europe, was worried that he might be accepting a communist troublemaker into his territory. Chadwick suggested taking Kapitza on the proviso that he never talked politics. Rutherford took his assistant's advice, admitted Kapitza to the Cavendish and, on the day he arrived, declared that no communist propaganda would be tolerated in the laboratory. After his initial misgivings, Rutherford was captivated by Kapitza's zest for life and deeply impressed by his ability in the laboratory. Like many young scientists before him, Kapitza worshipped Rutherford, and soon nicknamed him the Crocodile,[46] for:

> In Russia the crocodile is the symbol for the father of the family and is also regarded with awe and admiration because it has a stiff neck and cannot turn back. It just goes straight forward with gaping jaws—like science, like Rutherford.

Chadwick's research achievements were further recognized in November 1921, when he was made a Fellow of Gonville and Caius College. This made little difference to his academic activities, apart from requiring him to carry out some teaching in college in the evenings. It did, at the age of 30 years, give him some sense of security, and an identifiable position in an institution that he came to love. He could now walk on the grass of Tree Court and take his place at High Table for dinner with the other senior college figures. There, seated at the long table set across the end of the vaulted dining hall and elevated symbolically a few inches above the other tables, he was initiated into the cryptic exchanges of wit and opinion, gradually forming friendships and alliances. The Master, Hugh Anderson,[13] had been a prominent neurophysiologist and was now an adroit university administrator. He was a man of quiet charm, who had a great knack for getting to know and encouraging younger Caians; he naturally took a close interest in the new Fellow in Physics, who worked with his friend Rutherford. Chadwick[1] was drawn to Anderson, and regarded him as 'the best Master Gonville and Caius has ever had since John Caius'. The personal debt he felt to Anderson, and his loyalty towards the College would never be extinguished.

References

1 Weiner, C.(1969). Sir James Chadwick, oral history. American Institute of Physics, College Park, Maryland.

2 Chadwick, J. (1918). Report to the Royal Commissioners, 1851 Exhibition. Chadwick's papers, 1851 Archives.

3 Chadwick, J (?1971). Undated notes re Oliphant's biography of Rutherford. Chad II 2/1, CAC.

4 Marsden, E. (1914). The passage of alpha particles through hydrogen. Philosophical Magazine, 27, 824–30.

5 Marsden, E. and Lantsberry, W. C. (1915). The passage of alpha particles through hydrogen. Philosophical Magazine, 30, 824–30.

6 Marsden, E. (1962). Rutherford at Manchester. In *Rutherford at Manchester* (ed. J. Birks), pp. 1–16, Heywood & Co., London.

7 Rutherford, E. (1919). Collision of alpha particles with light atoms. IV. An anomalous effect in nitrogen. *Philosophical Magazine*, **37,** 581–7.

8 Chadwick, J. (22/10/59). Letter to N. Feather. Feather papers, CAC.

9 Chadwick, J.(?1920). Undated notes to Royal Commissioners, 1851 Exhibition. Chadwick's papers, 1851 Archives.

10 Geiger, E. and Marsden, E. (1913). The laws of deflexion of alpha particles through large angles. *Philosophical Magazine*, **25,** 604–23.

11 Chadwick, J. (1920). The charge on the atomic nucleus and the law of force. *Philosophical Magazine*, **40,** 734–46 .

12 Rutherford, E. (2/11/20). Letter to B. Boltwood. In *Rutherford and Boltwood* (ed. L. Badash, 1969), p. 335. Yale University Press.

13 Brooke, C. (1985). *A history of Gonville and Caius College*. Boydell Press, Woodbridge.

14 Crowther, J.G. (1974). *The Cavendish Laboratory, 1874–1974*. Science History Publications, New York.

15 Junior, J. (1874). The Duke of Devonshire. *Vanity Fair*, 301.

16 Einstein, A. (1934). Clerk Maxwell's influence on the evolution of the idea of physical reality. In *Essays in science*, pp. 40–5. Philosophical Library, New York.

17 Wood, A. (1946). *The Cavendish Laboratory*. Cambridge University Press.

18 Larson, E. (1962). *The Cavendish Laboratory: nursery of genius*, p. 38. Edmund Ward, London.

19 Chadwick, J. (24/5/18). Letter to E. Rutherford. Rutherford papers, Cambridge University Library.

20 Wilson, D. (1983). Cambridge and the Cavendish. In *Rutherford: simple genius*, pp. 406–52. Hodder and Stoughton, London.

21 Blackett, P.M.S. (1960). Charles Thomson Rees Wilson, 1869–1959. *Biographical memoirs of fellows of the Royal Society*, **6**, 269–95.

22 Ratcliffe, J.A. (1966). Edward Victor Appleton, 1892–1965. *Biographical memoirs of fellows of the Royal Society*, **12**, 1–17.

23 Hevesy, G. (1947). Francis William Aston, 1877–1945. *Biographical memoirs of fellows of the Royal Society*, **5**, 635–50.

24 Oliphant, M. (1972). *Rutherford—recollections of the Cambridge days*. Elsevier, Amsterdam.

25 Chadwick, J. (30/5/20). Letter to E. Rutherford. Rutherford papers, Cambridge University Library.

26 Chadwick, J. (1921). Unpublished Ph.D. thesis, University of Cambridge. CHAD II, 3/2, CAC.

27 Bieler, E. (1895–1929). Unpublished papers, 1851 Archives.

28 Chadwick, J. (6/9/29). Letter to A.S. Eve. Bieler's unpublished papers, McGill University Archives, Montreal.

29 Chadwick, J. and Bieler, E.S. (1921). The collisions of alpha particles with hydrogen nuclei. *Philosophical Magazine*, **42**, 923–40.

30 Matthews, P.T. (1987). Dirac and the foundation of quantum mechanics. In *Reminiscences about a great physicist: Paul Adrien Maurice Dirac* (ed. B. Kursunoglu and E.G. Wigner), pp. 199–227. Cambridge University Press.

31 Pais, A. (1986). *Inward bound*, p. 240. Oxford University Press.

32 Hughes, J. (1993). Unpublished Ph.D. thesis, University of Cambridge.

33 Rutherford, E. (1920). Nuclear constitution of atoms (Bakerian Lecture), *Proceedings of the Royal Society A*, **97**, 374–400.

34 Rutherford, E. (1914). The structure of the atom. *Philosophical Magazine*, **27**, 488–98.

35 Rutherford, E. (1914). Discussion at the Royal Society, 19/3/14, *Proceedings of the Royal Society A*, **90**, 462.

36 Glasson, J. (1921). Attempts to detect the presence of neutrons in a discharge tube. *Philosophical Magazine*, **42**, 596–600.

37 Rutherford, E., Chadwick, J. and Ellis C.D. (1930). *Radiations from radioactive substances*, pp. 54–5. Cambridge University Press.

38 Rutherford, E. (1908). Letter to H.A. Bumstead. In *Rutherford: simple genius*, (Wilson, D., 1983), p. 287, Hodder and Stoughton, London.

39 de Bruyne, N. (1984). A personal view of the Cavendish 1923–30. In *Cambridge physics in the thirties*, (ed. J. Hendry), pp. 81–9. Adam Hilger, Bristol.

40 Shoenberg, D. (1987). Teaching and research in the Cavendish:1929–35. In *The making of physicists* (ed. R. Williamson), p. 101. Adam Hilger, Bristol.

41 Allibone, T.E. (1987). Reminiscences of Sheffield and Cambridge. In *The making of physicists* (ed. R. Williamson), pp. 21–31. Adam Hilger, Bristol.

42 Chadwick, J. (1954). The Rutherford Memorial Lecture, 1953. *Proceedings of the Royal Society A*, **224**, 435–47.

43 Hutchinson, W.K., Gray, J.A. and Massey, H.S.W. (1981). Charles Drummond Ellis (1895–1980), *Biographical memoirs of fellows of the Royal Society*, **27**, 199–233.

44 Lovell, B. (1975). Patrick Maynard Stuart Blackett, Baron Blackett, of Chelsea, 1897–1974. *Biographical memoirs of fellows of the Royal Society*, **21**, 1–115.

45 Blackett, P.M.S. (1962). Memories of Rutherford. In *Rutherford at Manchester* (ed. J. Birks), pp. 102–13. Heywood & Co., London.

46 Shoenberg, D. (1985). Piotr Leonidovich Kapitza, 1894–1984. *Biographical memoirs of fellows of the Royal Society*, **31**, 327.

47 Chadwick, J. (?1970). Undated draft of a letter to D. Shoenberg. CHAD III 4/4 CAC.

5 Sparks don't fly

By the early 1920s, the post-war bulge of ex-servicemen had passed through Cambridge, and the university slipped back effortlessly into its former ways as a citadel for the privileged. Many of the undergraduates had no academic pretensions, preferring to channel their talent into sport and other social activities, and spent a blissful three years before emerging with an Ordinary BA degree. They would still conform to the formalities of dress, wearing black undergraduate gowns for what few lectures they did attend, as well as to college dinner in the evenings. Their laughter and vitality suffused a college like Gonville and Caius, and no opprobrium attached to their lack of intellectual ambition nor to their behaviour, unless certain clear rules were breached.

Chadwick envied the young students their self-assurance and even perhaps their frivolity. For the first time in life, Chadwick found that he too could indulge in some pleasures. For the most part, these were simple things like visiting Trinity in the evenings to have dinner with Rutherford or Ellis. He took up fly-fishing and spent several holidays with the Rutherfords. He also struck up a friendship with Peter Kapitza. After the warning from Rutherford about communist propaganda, many members of the laboratory 'shunned this young man. They were afraid of compromising themselves if they befriended him.'[1] This was not true of Chadwick, who remembered how he had felt on first arriving in Berlin, not speaking much of the language, and the kindness he had received from his German friends. Kapitza[2] wrote to his mother on 5 December 1921 about a period production of 'The Beggar's Opera':

> I went with Dr Chadwick, Rutherford's assistant, and this was my first visit to the theatre in Cambridge. The theme of the opera was very silly . . . I have never seen applause so readily given by an audience as here in England—on every appropriate and inappropriate occasion.

Kapitza[3] was soon introducing Chadwick to more adventurous pursuits.

> One Sunday I went for a spin on my motorbike, taking Chadwick, one of the young Cavendish scientists, with me. I was stupid enough to let him drive, and he managed to upset the machine while going at a good speed and sent us both flying. I landed awkwardly, directly on my chin and split it open. Chadwick landed on his side and bruised it badly.

Kapitza also sent his mother a vivid impression of a gathering at which Chadwick may well have presided. The annual dinner of the Cavendish Physical Society was, by modern standards, a gargantuan feast with soup, a fish course, game, roast meat, pudding and a savoury—each course accompanied by a different wine. These Christmas feasts had been held since the turn of the century, and J.J. Thomson[4] revelled in them, especially the postprandial songs 'composed for the occasion by members of the Laboratory [dealing] with topical matters, personal or scientific . . . set to some well-known air'. To Kapitza[5] the experience must have seemed another example of the emotional immaturity of the English male. He captured the unruly spirit of the occasion, but, as in many of his letters, remained sufficiently detached from events to impart a sardonic edge to their description:

> Cambridge, 16 December 1921
>
> . . . The holidays will soon be here and the lab closes for two weeks. I asked the Crocodile to allow me to work but he said he wanted me to have a rest, because everyone has to have a rest. There has been a striking improvement in his attitude towards me. I now work in a separate room, which is a great honour here. I have obtained some results but they are not yet final; however, I now have a good chance of success.
>
> I must tell you about the dinner of the Cavendish Physical Society, which was very amusing. The members of this society are all the research workers in the laboratory. The dinner is an annual event and Professors Thomson and Rutherford are invited as guests, with a few professors from other universities—this year Professors Barkla and Richardson. Some 30 to 35 men were there, seated around a U-shaped table with one of the young physicists presiding and the guests on either side of him. First of all we ate and drank, and although the drinking was only moderate, the English get drunk very quickly. You can soon see this from their faces which lose their stiffness and become lively and animated.
>
> After coffee the port was circulated and toasts were proposed, first of all to the King and then to the Cavendish Laboratory, proposed by a young scientist and replied to by one of the professors. The toasts were largely in humorous vein for the English are very fond of jokes and witticisms. The third toast was to 'Old Students' and the fourth to 'The Guests'. Between the toasts songs from a collection written by the physicists were sung, in which the lab, physics and the professors were all serenaded in comical terms. Everyone joined in singing to tunes borrowed from operettas. These customs date from Clerk Maxwell's time.
>
> You could do anything you liked at the table—squeal, yell, and so on—so the general picture was rather wild and quite unique. After the toasts everyone stood on their chairs, crossed arms and sang a song recalling old friends and so on [Auld Lang Syne]. It was very funny to see such world famous luminaries as J.J. Thomson and Rutherford standing on the chairs and singing at the tops of their voices. Finally we sang 'God save the King' and at midnight we broke up, but I didn't get home until three in the morning. This was because some of the diners had to be taken home, but I can assure you that I was among the takers and not the taken. The latter were doubtless the happier but my Russian belly seems better adapted to alcohol than the English one. There were no ladies at the dinner.

One wonders how the supposedly fragile Chadwick constitution contended with such excesses. He still found the ordinary meetings of the Cavendish Physical Society an ordeal if he had to present a paper. These meetings were held fortnightly and were similar to the departmental meetings that used to occur in Manchester, with Lady Rutherford pouring the tea and her husband giving an annual 'State of the Laboratory' review in October. Chadwick's[6] first presentation of his own work was marred by nerves:

> I didn't do it well. In fact, I did it rather badly. And Rutherford made no bones whatever about his opinion on it afterwards . . . In effect he said 'That won't do. I will find an occasion for you to give another address.' . . . That was to give me another chance to rehabilitate myself, give me some confidence . . . I did do it quite well, and he was quite satisfied. But many people would not have put themselves out in that way . . . He was kind, and he was considerate, but in a somewhat peculiar way, because in his general manner he was rather hearty and robust, and one didn't see at first what was really behind it all.

Rutherford had long recognized Chadwick's consummate skill as an experimenter and saw through his diffident manner to the brilliance that lay beneath. As we have seen, Chadwick had already helped with some of the work reported by Rutherford in his 1920 Bakerian Lecture, and after the lecture was asked to join with him on a systematic study of the disintegration of the lighter elements by α-particles. For the first year Chadwick was assisting Rutherford, he was also carrying out the research for the second part of his Ph.D. thesis. Charles Ellis was called on to count scintillations from time to time, and recalled the Rutherford–Chadwick experiments thus:

> A radioactive source provided the alpha particles, which fell on a foil of the material to be disintegrated, and the resulting protons, hitting a zinc sulphide screen, were detected by the scintillation method. The whole apparatus was contained in a small brass box and the scintillations were viewed with a microscope. I can well remember being surprised, in fact mildly shocked, that the apparatus was not more impressive, yet these experiments, so simple on the surface, required the highest experimental skill to make them yield dependable results . . . The experiments started about four in the afternoon and we went into his laboratory to spend a preliminary half an hour in the dark to get our eyes into the sensitive state necessary for counting. Sitting there drinking tea, in the dim light of a minute gas jet at the further end of the laboratory, we listened to Rutherford talking of all things under the sun. It was curiously intimate but yet impersonal and all of it coloured by that characteristic of his of considering statements independently of the person who put them forward.[7]

Rutherford's 1919 publication describing the splitting off of hydrogen particles from nitrogen nuclei had been widely reported in the popular press, and he himself regarded it as a revolutionary event. His famous apology to the international anti-submarine warfare committee for missing a meeting—'If, as I reason to believe, I have disintegrated the nucleus of the atom, this is of greater significance than the war'[8]—might seem breathtaking in its arrogance, but leaves one in no doubt as to the writer's opinion of its importance. Although his experiments were described in great detail in the 1919 papers and no one had seriously questioned his conclusions, Rutherford was still troubled by the possibility raised by Marsden[9] in 1915 'that the H particles are emitted from the radioactive atoms themselves'.

This nagging concern about possible alternative sources of hydrogen surfaces in a letter to his friend Bertram Boltwood at Yale University on 19 August 1920, when Rutherford[10] described the extension of his investigations to other light elements in his pursuit of the structure of the nucleus.

> Recently I believe I have found that carbon and possibly fluorine break up in the same way as oxygen, but it will take a great deal of experiment to prove this definitely. Apparently neither sulphur nor aluminium are disturbed by the shocks. I wish I had a live chemist tied up to this work who could guarantee on his life that the substances were free from hydrogen. With this little detail set on one side, I believe I could prove very quickly which of the lighter elements give out hydrogen, but it is very difficult to do so without the chemical certainty as the effect is so very small . . . You can see that things are moving pretty rapidly on questions of atomic structure. My own work is rather difficult and takes a good deal of time, but I have now got a couple of good counters on whom I can rely. I hope in the next few years to form some idea of the constitution of the nucleus of the lighter atoms, but on account of the impossibility of getting the α particles close to the nucleus of the heavy atoms, I do not think it likely that we shall get much beyond neon.

Rutherford still clung to the notion he had formed in Manchester regarding Chadwick's talents as a chemist, and this was another reason for their partnership: 'he [Rutherford] expected difficulties in proving that any effects which might be found were not due to the presence of hydrogen, and he thought I was enough of a chemist for the purpose'.[6] The two men made steady advances and six months later, Rutherford[11] wrote to Boltwood again:

> You will be interested to hear that I have made some progress on the general question of the disintegration of elements. Chadwick and I have been hard at work and we find that a number of elements like nitrogen, boron, aluminium, fluorine and phosphorus emit long range particles under an alpha ray bombardment. These particles go much farther than ordinary H. atoms under the same conditions, showing the effect cannot possibly be due to the presence of any hydrogen or hydrogen compound in material under examination. The range of these particles is very great; in the case of aluminium it is at least three times that of the normal H. atom. If these, as they probably are, prove to be hydrogen nuclei their energy of motion is even greater than that of the incident alpha particle. I shall probably publish a brief account of this in *Nature* shortly . . .

A few days later, Rutherford and Chadwick[12] did just that and their letter, *The Disintegration of Elements by α-Particles*, appeared in *Nature* on 10 March 1921. Using α-particles from radium C, they showed that the maximum range of any protons (hydrogen nuclei) liberated from hydrogen gas or a hydrogen compound was 29 cm. To exclude all such protons from their observations, they asked their assistant George Crowe to prepare very thin strips of mica to insert in front of the scintillation screen. Crowe's dexterity was such that he could cut a piece of mica that would precisely match the absorption of a given thickness of air. Thus for these experiments they made use of mica that was exactly equivalent to 32 cm of air, so that any protons arising from 'natural' hydrogen would not pass through to be counted on the screen. In this way, they 'obtained definite evidence that long-range particles are liberated from boron, fluorine, sodium, aluminium, and phosphorus, in addition to nitrogen'. All these elements gave rise to protons with ranges of over 40 cm in air and in the case of aluminium the range was 'certainly not less than 80 cm'. They also reported a list of elements where no such disintegration was observed, viz. 'lithium, beryllium, carbon, oxygen, magnesium, silicon, sulphur, chlorine, potassium, calcium, titanium, manganese, iron, copper, tin, and gold'.

Crowe[13] had his own memories of these exhaustive and difficult experiments carried out in the cluttered twilight of Rutherford's laboratory. The Professor's impatience, born of uncontainable excitement at the prospect of an imminent discovery, is palpable. Crowe suggests that Rutherford was expecting a certain effect, but was unsure of its magnitude.

> He was shooting protons out of light atoms by means of alpha particles; carbon would not play. The next day was the turn of aluminium, and on some theoretical grounds Rutherford thought that the protons would have a high velocity and a long range. 'Now, Crowe, have some mica absorbers ready tomorrow with stopping power equivalent to 50 cm of air.' 'Yessir.'
>
> On the next day: 'Now, Crowe, put in a 50 cm screen.' 'Yessir.' 'Why don't you do what I tell you—put in a 50 cm. screen.' 'I have, sir.' 'Put in 20 more.' 'Yessir.' 'Why the devil don't you put in what I tell you, I said 20 more.' 'I did, sir.' 'There's some damned contamination. Put in *two* 50s.' 'Yessir.' 'Ah, it's all right, that's stopped 'em! Crowe, my boy, you're always wrong until I've proved you right! Now we'll find their exact range!'

Chadwick's sense of excitement was no less intense but not openly displayed; his quiet unflappable manner was the perfect foil to Rutherford's rumbustious approach. According to Oliphant,[14] Chadwick's 'clear-headedness prevented him from being carried away by evanescent enthusiasms'. A pattern became apparent in the experimental results, namely that 'no effect is observed in "pure" elements the atomic mass of which is given by $4n$, where n is a whole number. The effect is, however, marked in many of the elements the mass of which is given by $4n + 2$ or $4n + 3$. Such a result is to be anticipated if atoms of the $4n$ type are built up of stable helium nuclei and those of $4n + a$ type of helium and hydrogen nuclei'. Rutherford's conviction that the α-particle was a stable component of atomic nuclei was growing. The results were presented in more detail in a long paper, *The Artificial Disintegration of Light Elements*,[15] in November 1921.

Two additional observations were added to those reported in the earlier letter to *Nature*. Firstly, the number of protons liberated increased rapidly with the velocity of the bombarding α-particles. Secondly, in the case of aluminium, protons were ejected in all directions, largely independent of the direction of the impinging α-particles: nearly as many were expelled backwards as forwards. The latter finding 'suggests at once the possibility of an atomic explosion in which the energy of the α particle plays the part of a detonator, and the escaping H particle gains most of its energy from the nucleus'. This possibility was rejected, a few lines later, by the authors because of the first observation that the 'energy of escape appears to be nearly proportional to the energy of the incident α particle', and the force of an explosion does not depend on how fast the detonator is applied. The suggested solution, which was almost certainly largely due to Rutherford, was that for those elements whose atomic mass is given by $4n + 2$ or $4n + 3$, there was a central nucleus built up of helium nuclei around which the two or three additional protons orbited as satellites. By applying the laws of conservation of energy and momentum to such a model of the aluminium nucleus, they found that their results were consistent with theory. They were left with the problem of what would hold the positively charged protons in tight orbit around the massive central nucleus, also positively charged. This was the same question of the nature of attractive forces within the nucleus which confronted Chadwick in the paper he was writing at the time with Bieler; he had now arrived at the same impasse by two different routes.

Etienne Bieler, the 1851 scholar visiting from Canada, helped Rutherford and Chadwick to count scintillations in the disintegration experiments just described, and was therefore privy to the wide ranging discussions that took place in the darkened laboratory. Under Chadwick's supervision, he continued to explore the apparent breakdown of the inverse square law in the immediate vicinity of the nucleus. Using Chadwick's annular ring technique, Bieler studied the scattering of α-particles by aluminium and magnesium. He found that at large angles of scattering, there was a consistent departure from the inverse square law of force in close collisions, with less scattering observed than expected. In order to explain his findings, Bieler postulated an attractive force that would hold sway over the electrostatic repulsive force within the nucleus. His Ph.D. thesis[16] contained theoretical curves showing the attractive force varying as the inverse fourth power of the distance, and Rutherford and Chadwick would find these ideas useful in interpreting future experiments of their own. Through their scattering experiments, the Cavendish group explored the strength and the range of the nuclear potential with unmatched determination.

Aside from serving as an indentured scintillation counter in these efforts, Charles Ellis, Chadwick's protégé from the Ruhleben camp, was becoming involved in a lively disagreement in the scientific journals with Lise Meitner, the 'German Madame Curie'. When Ellis began his own research career in 1920, Rutherford had suggested to him that he should start by re-examining the spectrum of β-rays (electrons) arising from certain heavy metals, when they were irradiated by γ-rays from radium B and C. While in Manchester, Rutherford and two co-workers[17] had published preliminary findings on this subject in 1914, but the war had interrupted their line of enquiry. In 1921, Ellis[18] produced his own first paper which confirmed the validity of Einstein's photoelectric equation at much higher energies than previously tested, and also introduced a completely new physical process which became known as *internal conversion*. This is an inner photoelectric effect whereby a radioactive nucleus emits γ-rays which then 'hit' some of the surrounding electrons on their way out of the atom. A number of these electrons will be ejected from the atom at a characteristic velocity. Ellis argued that the lines superimposed on the continuous β-ray spectrum from radium B were due to this internal conversion.

Meitner[19] responded in 1922 with her first paper since the war, in which she seemed to ignore Chadwick's 1914 paper on the continuous spectrum, and instead suggested that all β-rays start out from the nucleus with a unique energy which may be partly converted into γ-rays of characteristic frequency. These γ-rays might subsequently eject surrounding electrons so that the entire β-ray emission from the radioactive atom consists of a series of groups of electrons with definite and characteristic energies, i.e. a line spectrum. Although she made no mention of Chadwick's 1914 paper in her publication, she was known to doubt his results and said that his 'method of counting electrons deflected by fixed angle in a variable magnetic field had not enough resolution to distinguish a continuous spectrum from a number of separate lines . . . But the fundamental reason for her scepticism was her conviction that the primary electrons, just like α-particles, must form a group of well-defined energy, in the spirit of the quantum theory'.[20] Ellis met her head on in both the English and German literature,[21] stating plainly that 'the simple analogy between α- and β-decay cannot be maintained'. He disputed her measurements and was concerned that her 'theory gives no possibility of explaining the general [i.e. the continuous] spectrum of β-rays, in fact it appears to deny its existence'.

Although Ellis was fast becoming the acknowledged β- and γ-ray expert at the Cavendish, it was Chadwick's original discovery that was being challenged, however indirectly, by Meitner, and it was natural that the two men should reunite in an effort to settle the matter. Their response was ready by September 1922 and quickly appeared in the *Proceedings of the Cambridge Philosophical Society*. After briefly summarizing the opposing Meitner and Cavendish positions, Chadwick and Ellis[22] forcefully presented the outcome of their investigation.

> These two theories are fundamentally distinct and involve entirely different interpretations of the continuous β-ray spectrum. On Meitner's view it is presumably held to be a fortuitous occurence due perhaps to scattering of the homogeneous groups, whereas on our view the continuous spectrum consists of the actual disintegration electrons. The test of the independent existence of the continuous spectrum thus gives a ready means of deciding between the two theories.
>
> In this paper we shall describe briefly some measurements on the intensity distribution in the β-ray emission of radium B and C by an ionisation method, and it will be seen that our results provide strong evidence in support of the following statements: Firstly, that the continuous spectrum has a real existence which is not dependent on the experimental arrangement and that any explanation of it as due to secondary causes is untenable; and secondly, that our view of the origin of the continuous spectrum is consistent with the magnitude of the observed effects.

Frl. Meitner was still not convinced, and it took another five years of hard work of exceptional technical difficulty by Ellis and his collaborators before she was. In 1927 Ellis and Wooster[23] demonstrated with a micro-calorimeter that the average energy released per individual β-decay was equal to the mean energy of the continuous spectrum and that secondary processes, as called for by Meitner, did not exist. In their words, 'the long controversy about the origin of the continuous spectrum of β-rays appears to be settled.'

The originator of the long controversy, James Chadwick, took no further active part in the dispute after 1922, for the simple reason that he was convinced by his own findings and Ellis'. He did, however, have one more contribution to make to the study of β-particle physics, which was published in 1925. The genesis of this work clearly dated from his time in the Reichsanstalt, when it had been his original intention to study the single scattering of β-particles—he had been working on this very experiment when war broke out and presented his preliminary results to the 1851 Commissioners in 1918. Now he offered the problem to a visiting research student from Switzerland, P.H. Mercier, as a project. The nature of the investigation is set out with typical clarity in the introduction to their paper.[24]

> In single scattering the deflexion of a charged particle through a certain angle is due to an encounter with a single atom, the effect of the other atoms in the scattering material being very small compared with this single large deflexion. In compound scattering, on the other hand, the final deflexion of the particle is the resultant of a very large number of small deflexions of the same order of magnitude. Both these types of scattering can be easily realized in the case of α-particles . . . While the experiments on the scattering of α-particles are of a simple and straightforward character, the corresponding experiments with β-particles are attended by numerous difficulties, some of which arise from the nature of the available sources of β-particles, while others are due to the small energy of the particles.

They went on to list 'the numerous difficulties': the fact that β-radiation is not homogeneous, the presence of strong γ-radiation from many β-sources and the absence of a simple, reliable method of measurement. With considerable technical ingenuity, these obstacles were overcome to the extent that 'the conditions for single scattering were approximately fulfilled'—an achievement of no lasting scientific importance, but one never previously realized.

Although the Cavendish was without question the most prestigious physics laboratory in the world during the 1920s, its administration and funding were distinctly amateurish by modern standards. According to Chadwick,[6] the only report that the Cavendish had to furnish for the University was a set of annual accounts. This was 'a drudgery which Rutherford shrank from', and was undertaken voluntarily by one of the lecturers, Henry Thirkill. At least there was more order than when J.J. Thomson held the laboratory cheque book himself, and would pull it out of his pocket to pay a member of staff he met on King's Parade or use it as a scientific notebook. It seems that after submitting his original memorandum to the university authorities in 1919, in which he had mentioned capital sums amounting to two hundred thousand pounds, Rutherford was content to let the annual budget run at a few thousand pounds. For example, in 1921, the total university expenditure on experimental physics, including salaries and wages, was £8764, of which £7710 was recovered in fees from students.[14] The cost of materials and apparatus for the year was £2548; this did not include half a gram of radium that Rutherford managed to extract from the Medical Research Council after persistent lobbying. A modest extension to the laboratory was built in the early 1920s and some space gained from the Engineering Department.

Research students were attracted to Rutherford like moths to a flame, and he was always tempted to take them, regardless of practical considerations. This led to considerable difficulties for Chadwick, who had to provide space, apparatus and guidance for them. He became concerned about the 'very serious' shortage of equipment and lack of money to spend. By the summer of 1922, Kapitza was producing transient magnetic fields of very high intensity, and his efforts and personality had completely won over Rutherford. Kapitza wrote to his mother on 17 August:[25]

> Preliminary tests of my new experimental set-up have been completely successful. I hear that the Crocodile can talk of nothing else. I've been given a large amount of space in addition to the room in which I work and for the full scale experiment I have permission to spend a fairly large sum of money [£150].

To Chadwick, the Russian must have started to resemble a cuckoo in the nest, and it seems likely that he remonstrated with Rutherford. In October 1922, Kapitza[26] gave the following account.

> Rutherford is really exceptionally good to me. On one occasion he was not in a good mood and he told me I must economize. I showed him that I was doing everything very cheaply and in the end he couldn't deny it and said: 'Yes, yes, that's all true, but it is part of my duty to talk to you like this. Bear in mind that I am spending more on your experiments than on all the other work of the laboratory put together'. And you know , that is true, for our experiment fairly runs off with his pennies.

Rutherford's half-hearted effort at cost containment had no effect, and only six months later Kapitza[27] proudly confided in his mother that he had been 'allocated £1200 for my experiments (all this must remain strictly within the family)'!

One of the most common requirements in the experimental studies in radioactivity undertaken at the Cavendish was to be able to evacuate the air from apparatus as completely as possible. The laboratory possessed a few old Gaede mercury pumps and force pumps, which were, in Chadwick's words,[6] 'small, turned by hand and very time-wasting'. There was a much better motor-driven pump, the Hyvac, on the market, but funds were so limited that he could afford to buy only a few each year. Picture Chadwick's chagrin, when alone one evening in the Combination Room of Caius College with the Master, Sir Hugh Anderson, he was suddenly asked: 'Why wouldn't Rutherford accept the grant I had arranged for him?'[14]

Chadwick replied that he knew nothing whatever about it, and the Master then told him the full story. Anderson was a wealthy man whose family were connected with the Orient Steam Navigation Company;

> . . .he thought the grant made by the University to the Cavendish for research expenses was not sufficient to enable Rutherford to pursue his work properly . . . Accordingly, he approached a number of his wealthy friends and persuaded them

> to offer Rutherford a private grant of £2000 a year for some years to use at his discretion. He told Rutherford what he had arranged. The position was that Rutherford had only to say that he needed this money for researches in the Cavendish, and it would be forthcoming. But Rutherford did nothing, to Anderson's astonishment and indeed exasperation.

At the time, Chadwick[14] 'could offer no explanation for Rutherford's refusal to say that he needed more money for research. I knew so well how hampered and restricted Rutherford was for lack of equipment and technical assistance, and his attitude was quite incomprehensible to me'. There had been a disagreement between the two men over the need to acquire the capability to produce high voltages, which Chadwick had pressed for.

> [Rutherford] showed no interest, I got no response at all. He may have had good reasons for not undertaking such work, but I could not understand his lack of interest. His reasons were probably 1/ lack of money, 2/ lack of space and perhaps 3/ lack of confidence in my ability to supervise such work.[28]

There was also a broader 'difference of opinion about how the laboratory should be run'.[6] Chadwick thought that they were accepting too many research students which was causing overcrowding and, more importantly, deflecting the laboratory from what he saw as its prime function to stay in the van of nuclear research. There is no question that Rutherford shared the latter intention, but he thought it could still be done while teaching graduates from Britain and the Empire. Chadwick's suggested solution[6] was:

> to have a few rooms set aside in which various pieces of equipment could be ready for any examination we wanted to make in a hurry, such as a cloud chamber, counting apparatus of different kinds, which only meant a few rooms; but it meant shutting off these rooms from the occupation of these students. It meant letting in fewer research students. Also, there was some difficulty in keeping the apparatus in order because we had very little technical assistance. And I suppose Rutherford was quite right in saying we shouldn't do that. Well, anyhow we didn't do it.

Rutherford's commitments outside the Cavendish were ever increasing. In 1923 he served as the President of the British Association, and his year in office culminated in a meeting in Toronto to which he tacked on a vacation in the Rockies and a lecture tour in the US, so that he was away from Cambridge from July to October 1924. He was elected President of the Royal Society in 1925, a position he would hold until 1930. It was perhaps inevitable that such absences should cause some resentment, and for a time, there was a distinct coolness between the director and his first assistant. In Chadwick's mind, Rutherford was experiencing a period of poor health which exacerbated things. 'I wasn't the only one affected by it. For a time Charles Ellis and I and one or two others did not attend the meetings of the Cavendish Physical Society.'[6]

Chadwick remembered these years as a time of frustrations. In 1923, Arthur Compton, working in Chicago, had described what soon became known as the Compton effect—an interaction of X-rays with matter that depended on the X-ray being thought of as a particle with quantum energy and momentum, rather than as a waveform. At 'about that time or immediately following', Chadwick[6] 'began to think about the collisions of quanta with electrons'.

> And I wanted to observe these collisions in an expansion [cloud] chamber. None was available. I spoke to Rutherford. He was completely uncooperative. It was difficult, but I realized he was really ill. I did build up an expansion chamber, for which I paid myself. I didn't have much money at that time. But I paid for it, bought my own lenses. There were some things in the laboratory I could use, but I couldn't do it properly. And just as I was about to start—I'd enlisted the help of a young student, K. Emeléus, . . . and told him what I was going to do—when one day C.T.R. Wilson came around. I told him what we were about to do. 'Ah, that's just what I'm doing'. Well, it was and it wasn't, because he was doing it with X-rays at quite low levels . . . I was interested in the higher energy. And I knew that he could take far better photographs than I could. In any case, he was doing it. So we dropped it, and we then did some very trivial experiments on the delta rays, as they were called, produced by alpha particles in their passage through gases—that is, the collisions of alpha particles with electrons. But I wasn't interested. It was perfectly obvious what was going to happen. There was no difficulty about it. But I had to give Emeléus something to do.

Any estrangement between Chadwick and Rutherford was short-lived and had no lasting consequences. At meetings in London, Rutherford would often find himself under attack from other members of the scientific establishment, who thought the concentration on nuclear research in his department was too narrow and likely to be unproductive. He could always look to Chadwick for reassurance on his return.

> He used to come back to Cambridge between half past five and six, and usually come back to the Cavendish on his way home to talk about what had been happening during the day. And if he didn't come in, then I generally went to his house. which was only a few minutes away, to tell him how the experiments had gone. But then, you see, these criticisms were fresh in his mind and perhaps weighing more heavily upon him then than they would have the next day. But he certainly was a little perturbed at times.[6]

For all his fame and achievements, Rutherford was not immune to the encouraging effect of a pat on the back, and Chadwick[6] came to realize this.

> I think he was always surprised at what he had done. He gloried in it. Oh, he used to enjoy anything written in his praise. He or I would come across something, say, in German publications he didn't read. He didn't know German at all well . . . so he would ask me to translate them for him. And he enjoyed it. But the one word which he enjoyed particularly was 'bahnbrechend' (pioneering). When that came he would repeat it and glory in it . . . I knew him fairly well, you see, because it wasn't only in the laboratory that I knew him. I knew him at home. And on at least three occasions I had spent holidays with him and with Mrs. Rutherford . . . So I felt that I knew him in off-guarded moments when he was not being the great physicist.

An example of Chadwick's close relationship with Rutherford comes from one short holiday in the spring of 1924, whenLady Rutherford had flu and was not fit for the rigours of driving to Somerset in an unheated, blowy motor car.

> I don't think that we were away more than a week, if that. Rutherford and I had long talks in the evenings after dinner. During one of these talks, Rutherford suddenly said—I say suddenly because it had little connection with our talk on the writing of papers for publication—'You know, Chadwick, I sometimes look back on what I have done and I wonder how on earth I managed to do it all.' . . . His words and manner showed no boasting, some satisfaction at most; what impressed me so much was his modesty, even a degree of humility in looking back on his achievements.[14]

Chadwick's impression of Rutherford's personality crystallized back at the Cavendish. It had been some time since Rutherford had inspected the radium room at the top of the laboratory, and in particular Chadwick wanted him to see the material that had been loaned by the Medical Research Council. After some persuasion, Rutherford made the climb, and Chadwick[6] used the visit to try to impress on him how they were still short of material for the numerous research projects that were going on.

> We could do with a great deal more than we had. I said, 'What a pity it is that somebody didn't give you a gram of radium as the women of America had given to Madame Curie'. We were then just coming down the staircase, and he stopped, patted me on the shoulder and he said, 'My boy, I am damned glad that nobody ever did. Just you think, every year I should have to think of the import of what I had done with a whole gram of radium, and I should find it impossible to justify it.'

This conversation had a lasting effect on Chadwick and for him encapsulated the essence of Rutherford's character, which could never be appreciated by those who only saw the bluff, larger than life figure from a distance.

> I had, of course, long realized that Rutherford was modest about his achievements, notwithstanding his eager enjoyment of his reputation, his almost boyish delight in any laudatory references to his work, his susceptibility to flattery . . . But I had not realized how deeply ingrained his modesty was until I pondered over this remark. And it threw some light on some arguments concerning the spending of money on research, especially his own work. It was then that I had an explanation for his refusal to ask for the research subsidy which Sir Hugh Anderson had arranged for him; he did not feel that he could justify spending so much money—for it was a great deal in those days—on himself and his research students. And this in spite of his unique position in the scientific world, his extraordinary achievements in the past, and the urgent need to press on with the establishment of nuclear physics as a new branch of enquiry.[14]

Although Sir Ernest was often unavoidably occupied by outside activities, and Chadwick became increasingly responsible for administration, the director's irregular tours of the laboratory, pipe in hand, continue to galvanize the research efforts. Much of Rutherford's zeal, and his favourite German word, are preserved in a letter he wrote to Boltwood,[29] regretfully declining an invitation to give a series of lectures in the USA.

> It is a pity but I cannot go. I am bound to the wheel into my Laboratory, Royal Institution Lectures & prospective President of the British Association at Liverpool in August 1923. I should very much like to be free to spend a few weeks in Yale again & regain some of the enthusiasm of my lost youth, but life for me is very busy in these days & I have to drive the 'boys' along. I am, however, fitter than I have been for years & bear up nobly under the strain. Cambridge suits me well alround & even the presence of the Master of Trinity [J.J. Thomson] in the Lab adds a fillip to life. He is as keen as a youngster still & is still searching for a 'bahn-brechende' scientific event as in the old days. I have got a very good group of researchers together—but too many of them—& it is a business to get sufficient apparatus & ideas to keep them going strong. The whole family including Fowler [his son-in-law] go off for 3 weeks in the Dolomites on June 20 & we hope to have a pleasant holiday. The young people are very keen to go while I am of course quite blasé, but will bring up the tail of the procession with my wife at the head.

Chadwick's tenure of the Clerk Maxwell studentship ended in the summer of 1923 (he was succeeded by Kapitza). At about this time, Sir William McCormick, the Chairman of the Advisory Council to the Department of Scientific and Industrial Research (DSIR) made one of his regular visits to the Cavendish Laboratory. The DSIR had been set up in the Great War to encourage the application of scientific research to British industry, and J.J. Thomson and Rutherford both had contact with it from its inception. McCormick would come to the Cavendish once or twice a year and Rutherford would take him on a tour of the laboratory, generally with Chadwick in tow.[30] On this occasion, McCormick sought Chadwick out and told him that he thought Rutherford had too much to do. He was concerned about the day-to-day supervision of laboratory research and wondered whether Chadwick would be interested in an expansion of his own role. Chadwick was 'naturally delighted' and within a short time was offered the post of Assistant Director of Research—the first in Cambridge. He now had a formal position in the hierachy of the university and some say in the admission of research students, as well as a secure salary. His stipend of £416 13s 4d was 'without cost to the university', and helped the laboratory finances so that the net deficit in 1924 was only £58 on a total expenditure of £9257. Chadwick's DSIR appointment marked 'the beginning of a great expansion' in his duties, and he would oversee an increase in the numbers of researchers from 20 to 40 or 50 which, in his view, were 'more than we could really look after'.[6] The Ph.D. course was for three years, but Cambridge would make an allowance of one year for research previously undertaken at another university. This became a commonly applied device for helping with the overcrowding.

> And it was almost a condition of admission that a man from outside should have done at least one year of research at his own university before we admitted him. We didn't want him for three years, but we didn't want to have to start from the beginning. We had to do that for our own men. There might be six or so of our own Cavendish people from the degree course staying on for research.[6]

Although they might have arrived with a little experience, very few students had a particular project in mind. Therefore, Chadwick and Rutherford

> . . . had to think of problems which would occupy that time and which were big enough for them to get their teeth into and keep them busy and give some kind of positive and reasonable results . . . Our procedure was to jot down during the course of the year various problems that we thought must be looked into. We did that independently. Then in the summer we knew what research students we were going to have, although one or two might turn up later . . . we sat down together and exchanged our views and decided what could be given, what would provide suitable work for a research student reading for the Ph.D. degree. Well, that meant, you see, that quite a number of interesting things couldn't be done. We didn't have the time or equipment.[6]

Each intake of new students seemed to contain one or two destined for scientific fame, and Chadwick would often be the first to recognize their outstanding ability, while they were completing his 'nursery' course. In 1924, John Cockcroft,[31] who had started out as an electrical engineer and then obtained a mathematics scholarship to Cambridge, began his physics career under Chadwick's watchful eye. Two years later, it was the turn of another Yorkshireman, the dapper Norman Feather,[32] whose Ph.D. 'involved the construction of a cloud chamber to study the relation between photographic density and ionization density in α-particle tracks'.

There was an unwritten admissions policy that a certain number of research places would be reserved for students from the Dominions. Chadwick[6] was an enthusiastic supporter of this idea, not least because Rutherford was 'owed' to New Zealand. Many came on overseas scholarships from the Commissioners of the 1851 Exhibition; most notable amongst these was a cherubic Australian named Mark Oliphant, who arrived in 1927. With his natural, friendly manner, he became a special favourite of Rutherford's and also a life-long friend of James Chadwick's.[33] He recalled their first meeting:

> I found Chadwick and introduced myself. The lean man in a neat, dark suit looked at me over the top of his glasses, with an intense stare which I came to know well over the next 47 years. He was silent for an embarrassing few moments; then his face broke into a rare smile as he said: 'Oh yes, I've been expecting you. I'll show you where you are to work.' He led me to a large room near Rutherford's office, explaining that one quarter of its area would be mine. With a limp handshake, characteristic of Rutherford and Chadwick, he left me. Such was my first meeting with a man who was to become a close friend, but who then scared me stiff.

Oliphant also offers the following exchange which occured during his early days in the laboratory, while Chadwick was making his daily round inspecting progress. Apart from providing a glimpse of the kindliness which was often completely obscured by an austere manner, the incident also shows that the laboratory was still woefully short of useful equipment.

> Chadwick: 'Is there anything you need?'
>
> Oliphant, who was vigorously turning the handle of a force pump: 'Yes, a Hyvac pump would help enormously.'
>
> Chadwick: 'Why?'

> Oliphant: 'Because I'm fed up with turning the handle on this one.'
> Chadwick, fixing him with a steely glance: 'Well, you just can't have one.'

Oliphant, who was close to tears, left for a long lunch. When he returned he found a Hyvac pump, which Chadwick had purloined from the director's own room, sitting on his bench. From that day on, Oliphant realised 'Chadwick's heart belied that stare'.[33] On occasion, the exuberance of the research students would defeat even Chadwick's severe countenance, as for example, when Edward Bullard (later a leading geophysicist), was using Harrie Massey (later secretary of the Royal Society) as a human counterweight on a weighing scale he had rigged up. He expressed the weight of a solenoid of wire in units of *Masseys*. 'Chadwick's grin of appreciation of Bullard's ingenuity, reminiscent of the techniques used by him and Ellis in the internment camp, burst into a chuckle—an almost unheard of event'.[33]

Another responsibility that Chadwick took upon himself was to scrutinize all papers published from the Cavendish.

> He was a severe critic of English usage, and seemed to carry in his head the whole of Roget's *Thesaurus of English Words and Phrases.* He was equally censorious of poor scientific argument, or of sloppy conclusions from inadequate experimental results, not even sparing Rutherford's own papers. Despite his inspiration and continuous help, Chadwick did not claim any credit for his part in much of the most important work done in the Laboratory. Under his guidance many young men achieved reputations not wholly deserved, as was shown when they moved away from his tutelage.[33]

Although there were occasional women students at the Cavendish, Caius College was an exclusively male preserve and married fellows were not allowed to live in residence. This lack of female company must have come to seem the norm to Chadwick, for the last time he had encountered members of the opposite sex of his own generation on a daily basis was at Bollington Cross School. Life suddenly changed one night, when he was introduced to a vivaceous, self-assured young lady who was staying with friends of his in Cambridge. Her name was Aileen Stewart-Brown and, by whatever law of the attraction of opposites, she had a devasting effect on the dry, reserved, Dr Chadwick. Kapitza,[34] writing to his mother in June 1925, remarked 'Chadwick is up to his ears in love and the Crocodile growls that he is not working enough.'

Aileen was the daughter of one of Liverpool's most prominent families. Her father, Hamilton Stewart-Brown, was a stockbroker on the Liverpool exchange. Hamilton's father, Stewart Henry Brown, arrived in Liverpool in the 1850s from the United States to join the firm of Brown, Shipley & Co of which his uncle was senior partner. The bank prospered and Liverpool, with its own Stock Exchange and branch of the Bank of England, grew to be the second most important financial centre in Britain. Stewart Henry Brown became a successful businessman and Justice of the Peace; his five sons in turn were well known in financial and legal circles in the city and the family retained its connection with two private banking firms, Brown Shipley & Co. of Liverpool, and Brown Brothers & Co. of

New York. Aileen had grown up in the prosperous Aigburth district of the city, in a comfortable Victorian mansion, with three brothers and a sister. Her world was very much the upper social stratum of the city and did not include academics or people not from a well to do, commercial background. She was in her early twenties, and naturally marriage was high on her list of ambitions. The dark, sedate, don fascinated her, and she had the confidence and intelligence to encourage him. Chadwick was enchanted by Aileen, and the wedding was set for August 1925. Word spread through the drawing rooms of Aigburth that Aileen Stewart-Brown was going to marry a mad scientist fellow from Cambridge, who spent all his time working in a laboratory.

Since moving to Cambridge, Chadwick had had almost no contact with his family and does not seem to have invited them to the wedding. One can only surmise what might have led to this estrangement from his parents and brothers, but Chadwick rarely spoke about them for the remainder of his life. For different reasons, the wedding was also proving a testing time for the best man, Peter Kapitza. He wrote to his mother on the eve of the wedding:[35]

> Up to now I've been kept busy. I keep having to dress up, either in dinner jacket or morning dress, and to attend lunches and dinners . . . I've had no luck with this wedding. First, there's the outlay of money and then the loss of time—both very inconvenient . . . I don't know how it will all turn out but I hope I'll enjoy it. I have had to put up at the very smartest hotel and it's rather hard on my pocket. But, thank God, I did not have to buy a top hat—I borrowed one from Fowler whose head is the same size as mine.

The *Liverpool Post and Mercury* reported that the bride 'had a gown of palest pink, veiled with old family lace . . . The bridesmaids made a charming group in frocks of delphinium blue georgette over green, and wide green crinoline hats with posies of shaded delphiniums.'

Kapitza[36] provides a sharper account of the day, laced with delicious humour and a little conceit, in his next letter home.

> I returned safely from the Chadwick wedding. The English-style of wedding and my part in it were very interesting. I acted as the bridegroom's 'best man' and accompanied him all the time. In the church before the ceremony I sat in front on the right-hand side until the bride arrived. The bride enters the church and with the organ playing, she processes up the central aisle. She is preceded by the choir and followed by her entourage. She goes up to the altar where she waits for the clergyman. The groom stands beside the bride and the bride's father stands a little behind, facing the altar, on her side, while I stand on the groom's side.
>
> During all this time one must stay very serious, but I couldn't keep it up and started to grin. During the ceremony, when the clergyman asks: 'Who gives this woman to be married?' her father takes her by the hand and says: 'I do'. Then the clergyman asks the groom for the wedding ring. I take the ring from my pocket and give it to him. He asks some further questions, there is some singing and the organ plays. My duties are over and I sit down. There are some further ceremonies and the clergyman exhorts the man and wife about the purpose of marriage as if they themselves had no idea. When the ceremony is over we all go into the vestry

> behind the altar where the register is signed and I sign as a witness. There the bride and bridegroom receive congratulatons. According to custom, the best man has the right to kiss the bride and the bridesmaids. I claimed the former right but not the latter as the bridesmaids are not very well-favoured and it wasn't worthwhile.
>
> With the organ playing we all leave the church, I taking some elderly aunt in tow. In front, of course, go the bride and bridegroom. Photographs are taken as they leave the church. Then the whole company gets into cars and is driven to the bride's home for a large reception. There was a marquee in the garden for 200 people, refreshments, champagne, wedding cake and so on. After general congratulations, group photographs were taken and I shall send you my picture in top hat and tails as soon as I get it.

Kapitza also sent the photograph of himself and the bridegroom to Rutherford, who was in the middle of a long, triumphal tour of Australia and New Zealand. He was delighted by the picture of two of his 'boys' in formal dress, and replied:[37]

> I had a real good laugh when I saw the spick and span appearance of yourself and Chadwick at the wedding. I hope that you did your duties as best man to the satisfaction of all concerned. I understand from Fowler that his top hat completed your toilet. If you went to work in your laboratory in the same attire you would create a great sensation.

After a honeymoon in Scotland and Iceland, the Chadwicks returned to Cambridge, where James quickly found himself reimmersed in a scientific controversy that threatened to undermine his work with Rutherford. A serious challenge to the Cambridge school had been mounted by the Institut für Radiumforschung in Vienna, a laboratory headed by Rutherford's old friend, Stefan Meyer. For the first few years after the war, Rutherford and Chadwick had the field of artificial disintegration of elements to themselves, then in the summer of 1923 publications on this subject began to appear from Vienna under the names of Hans Pettersson, a Swede, and Gerhard Kirsch. They claimed to be able to break down atoms of silicon under intense bombardment with α-particles, whereas the Cavendish team had been unable to, and were therefore encouraged to re-examine other light elements 'found non-active in the experiments of Rutherford and Chadwick'.[38] They subsequently reported that beryllium, magnesium and lithium were also disintegrable. In a private letter to Meyer, dated 24 November 1923, Rutherford was moved to enquire whether Pettersson and Kirsch were working under his direction; describing their experiments 'as a valiant piece of work', he thought, nevertheless, on the basis of work carried out by two research students at the Cavendish, that 'one or two of the conclusions are doubtful'.

The differences of opinion became much more heated as a result of a meeting of the Physical Society of London in 1924, when a paper by Pettersson was read in which he attempted to discredit Rutherford's satellite model of the nucleus. Pettersson offered 'arguments for an alternative hypothesis which assumes that the α-particle communicates its energy to the nucleus as a whole, precipitating an explosion'. The presentation was too much for one old student of Rutherford's in the audience, E.N. da C. Andrade, who made the following caustic observation:[38]

> I have listened with great interest to Dr. Pettersson's paper, but I think it would have been of greater value if some of the experiments foreshadowed at the end had been performed before the theory was propounded . . . [The] distinction between the satellite and explosion theories is mainly a question of words at the present stage. . . . The only experimental evidence put forward [by Pettersson] . . . is not very striking. . . . The position is much as if a man having measured up a box and guessed from shaking it that it contained pieces of metal were to start speculating on the dates of the coins inside it.

The call to battle was unmistakable, and the first public response from Rutherford and Chadwick[39] came in a letter to *Nature* on 29 March 1924. In their previous work, they had concentrated on detecting protons ejected with a range more than 30 cm in air. The reports from Vienna referred to protons with ranges between 10–20 cm being measured in the new bombardment experiments. A new technique was adopted in Cambridge, based on the assumption 'that the particles of disintegration are emitted in all directions relative to the incident α-particles'. Rutherford and Chadwick now chose an angle of observation at 90° to the direction of the incident α-particles, and in this way were able to detect 'with certainty' disintegration protons with ranges of only 7 cm in air. In this way they were able to add seven more light elements, including magnesium and silicon, to their previous list of six. They detected a small effect with beryllium, which they thought might be due to a fluorine impurity in the metal, but 'the other light elements, hydrogen, helium, lithium, carbon and oxygen [gave] no detectable effect beyond 7 cm.' Pettersson and Kirsch were not inclined to seek common ground, and wrote a reply to *Nature* four days later in which they adopted a gloating tone. They were pleased that Rutherford and Chadwick had 'confirmed [their] results' for magnesium and silicon, and also commented that the new right-angle technique 'appears to be in many respects similar to one we have been using for some time'. They also reported important new results, especially 'that carbon, examined as paraffin, as very pure graphite and finally as diamond powder, gives off H-particles of about 6 cm range'.[38]

In a rebuttal which appeared in August 1924, Rutherford and Chadwick[40] stated that, in the case of several light elements, the results of Kirsch and Pettersson 'cannot be reconciled with ours'. In the case of carbon, they had looked for protons of range down to 2.6 cm in air and found nothing. 'This result is in complete disagreement with the experiments of Kirsch and Pettersson,' they wrote 'who found a very large number of H-particles of 6 cm range.' Rutherford and Chadwick restated their model of the nucleus, consisting of 'a main central nucleus with one or more hydrogen nuclei as satellites'. They assumed that there was a novel force of attraction between the massive, positively charged, central nucleus and the tightly bound proton satellites (also carrying a positive charge). 'The force at large distances [i.e.beyond the orbiting satellites] must become the usual repulsive force [between like charges] varying inversely as the square of the distance.

On this view there must, then, be a critical surface around the nucleus at which the force is zero. . . . The H-satellite is, of course, inside this surface and cannot escape from the nucleus with less final energy than corresponds to the potential at this critical surface.' After concluding that 'our picture of the nucleus has, at any rate, the great merit of simplicity', they hypothesized that 'an α-particle which penetrates within the surface of zero force will be attracted into the nucleus and will probably give rise to a disintegration particle'.

In his original 1919 paper on the disintegration of the nitrogen nucleus, Rutherford had tacitly assumed that following the collision with the nucleus, the α-particle would be unaltered and continue on its way. He and Chadwick now expressed a different view: 'The fate of the α-particle is a matter about which we have no information. . . . It is possible that the α-particle is in some way attached to the residual nucleus. Certainly it cannot be re-emitted with any considerable energy, or we should be able to observe it'.

In their April letter to *Nature*, Kirsch and Pettersson had made the additional claim that 'oxygen is also disintegrable and gives off α-particles of 9 cm range in the forward direction'. Rutherford and Chadwick[41] addressed this point in another paper submitted in July 1924, '*On the Origin and Nature of the Long-range Particles observed with sources of Radium C*'. They refuted this finding and instead confirmed the Cambridge view that the 9 cm range α-particles arose directly from the disintegration of the radioactive source radium C and not from the artificial disintegration of oxygen. They judged their evidence in this case to be 'overwhelming'.

During that same month of July, just as he was preparing to leave for Toronto, Rutherford received a personal letter from Pettersson enclosing a copy of a new letter to *Nature* from him and Kirsch. Rutherford reacted immediately, but more in sorrow than anger, and wrote to Pettersson suggesting that he should give further consideration before publishing. He praised Pettersson for his 'many good original ideas on this problem', but at the same time told him that he and Chadwick were of the 'definite opinion' that his 'work and conclusions' were 'demonstrably wrong'. He stated that the subject was 'full of pitfalls for the beginner and even for the veteran', and suggested that Pettersson should visit Cambridge during his absence to seek Chadwick's guidance. Rutherford was in no doubt that it was 'better to discuss these divergences of view in private than in print. Workers in this field are too few and too select to misunderstand one another.' On the same day, Rutherford wrote to Meyer, Pettersson's chief, in much the same vein, repeating his plea to stay publication of Kirsch and Pettersson's letter.[38]

Meyer stood firm behind his juniors whom he described to Rutherford as 'very capable trustworthy solid researchers'.[38] He did, however, agree that the disagreements between the two schools should be settled through 'personal communications'. Pettersson replied that he had asked the editor of *Nature* to defer publication as Rutherford had requested, but then vigorously defended its contents, by implication continuing his attack on the Cavendish results. Chadwick also received a copy of this letter and referred to it, when writing to Rutherford in Canada.

Chadwick[42] was on holiday in the Scottish Highlands, but his mind was clearly teeming with ideas.

> *Inveroran: Sept.[1924]* I think I ought to tell you the results of the last experiments but I don't know when you will get this letter as I very seldom see a post office.
>
> I counted the number of disintegrations particles from the aluminium under as definite conditions as were possible. As near as the experiments allow, the number agrees with that calculated on the assumption of an attractive inverse fourth and repulsive inverse square taking (1) zero force at 4×10^{-13} and (2) that the H particle appears when the alpha disappears. Of course the agreement cannot be very good on account of the counting errors and even in estimating the solid angles of the alphas used.
>
> Blackett has got 2 more photographs which are somewhat clearer than the others. They show the track of the H particles and the track of the recoil atom, but no track for the α. . . . If this is true it is a very fine addition to the evidence for the attractive field, and fits in very well with our expectations.
>
> I think we shall have to make a real search for the neutron. I believe I have a scheme which may just work but I must consult Aston first.
>
> I suppose you have received Pettersson's letter. He sent me a copy and also an additional short letter to which I made a very mild and conciliatory reply in the hope of inducing him to visit us. I don't think he will come, however.

Chadwick was right not to be optimistic about relations with Pettersson, and the next communications from Vienna were even more vexatious. In a crop of papers which appeared in several prominent German journals, Pettersson, and to a lesser extent Kirsch, redoubled their attack on the work of the Cavendish and again tried to supplant the satellite model of the nucleus. Pettersson claimed that his new microscope made scintillations appear much brighter than the optical system used by Rutherford and Chadwick—'a fact which no doubt explains why these authors find much smaller numbers of particles than we from the same elements and why they have so far failed to observe the H-particles from carbon'.[38]

Blackett's cloud chamber photographs, referred to by Chadwick in the letter quoted above, indicated that in the disintegration of nitrogen, the α-particle is captured by the nitrogen nucleus which at the same time loses one of its constituent protons. Although Chadwick and Rutherford could assimilate this into their model of the nucleus, and indeed had suggested it might be so in their most recently published paper, Rutherford became concerned that he had denied Pettersson priority in this discovery when he had persuaded him to withdraw the previous letter to *Nature* in July. An exchange of letters between the two men in February 1925 made it plain that not only did Pettersson believe this to be the case, but, worse, he also suspected Rutherford of passing on the idea to Blackett.[38] The situation was potentially calamitous: here was Sir Ernest Rutherford, recently awarded the Order of Merit by the King and about to begin office as President of the Royal Society, charged, albeit privately, with the suppression of ideas and scientific dishonesty. In a state of great agitation, he wrote to Pettersson again on 5 March to set the record straight. He pointed out that the French physicist Jean Perrin had suggested the possibility of α-particle capture in nuclear disintegration at a Solvay Conference in 1921, so that it was not a new idea. More importantly, he gave Pettersson an assurance that Blackett was 'quite ignorant of the contents of your original letter which, of course, I regarded as a private communication'.[38] To avoid any possibility being compromised in the future, Rutherford asked Pettersson not to send him any other papers before publication, although he would be glad to receive copies 'as soon as they have been printed'. As a public atonement, Rutherford[43] published a note in *Nature* on 4 April 1925 briefly reviewing the subject, and acknowledging the unpublished work of Pettersson and Kirsch.

The criticism from Vienna[38] continued relentlessly, and the stock of Kirsch and Pettersson grew steadily in the eyes of physicists outside the Cavendish. In two papers published in April, they concluded that 'probably all elements are disintegrable' and freely questionned the Cavendish research. Then in the July 1925 *Physikalische Zeitschrift*, Kirsch wrote a lengthy review supporting Pettersson's explosion hypothesis and discrediting the 'unnecesarily complicated and specialized satellite hypothesis' put forward by Rutherford and Chadwick. Chadwick read his copy of the journal on his return to the Cavendish after his honeymoon. Rutherford had withdrawn from the fray and was recharging his batteries in New Zealand. He learned of the latest developments in a letter from an indignant Chadwick[44] in October.

> Our friend Kirsch has now let himself loose in the Physikalische Zeitschrift. His tone is really impudent to put it very mildly. He takes our old experiments, the very first results, and proceeds to show what fools we are to talk of satellites and then tells what clever fellows he & Pettersson are to think of a nucleus which is simply chock full of satellites, rings and rings of them. Kirsch & Pettersson seem to be rather above themselves. A good kick from behind would do them a lot of good. The name on the paper is that of Kirsch but the voice is the familiar bleat of Pettersson. I don't know which is the boss but as Mr Johnson says there is no settling a point of precedence between a louse and a flea.

Chadwick also mentioned that he had 'found a letter from Arthur Compton waiting for me when I got back asking me to go to Chicago in October . . . they are reorganising and increasing the laboratory. But it was no use even thinking about it under the circumstances'. Compton had spent a year at the Cavendish after the war, and it would appear that he had been impressed by Chadwick's abilities as an organizer. It is not clear whether he just wanted Chadwick to visit and give advice, or was trying to lure him away from the Cavendish. Chadwick[45] clearly had no intention of going whatever the offer, and wrote again to Rutherford the following month, bringing him up to date with life at the Cavendish.

> I have admitted no new students. J.J. has taken two we refused. A Russian girl turned up from Berlin with a recommendation from Einstein. Newnham [College] have admitted her, but I have promised nothing except that her case will be considered after a couple of terms in Part II. This term is always a slow term and this year it seems slower than ever. It is partly due to the amount of demonstration done by some of the boys. In fact some of them seem to be so exhausted after a

> couple of hours' demonstration that they do very little the rest of the day. I started a class in working the lathe tools etc. by Crowe and Bert from 9 to half past every morning. I thought it was rather necessary after seeing some of the efforts in the workshop . . . I have done a few experiments . . . I can still find nothing from Be or C or O or gold.

Rutherford returned to Cambridge in the New Year of 1926, reinvigorated by his trip to the other side of the world. Although his laboratory was passing through a quiet phase, as Chadwick had indicated in his letter, theoretical physics was undergoing seismic changes, and the Bohr–Rutherford model of the atom was the major casualty. Ironically, Rutherford may have received the first inkling of the upheaval in a letter from Niels Bohr[46] conveying best wishes for the new year. Apart from general news about his institute in Copenhagen, Bohr also wrote 'Due to the last work of Heisenberg prospects have at a stroke been realized which although only vague[ly] grasped have for a long time been the centre of our wishes'. Heisenberg's revolutionary paper had appeared in *Zeitschrift für Physik* a few months earlier, and it set out 'to establish a basis for theoretical quantum mechanics founded exclusively upon relationships between quantities which are in principle observable'. Heisenberg was equally determined to overthrow the classical concept of orbiting electrons, and he firmly posited that 'even for the simplest quantum theoretical problems the validity of classical mechanics cannot be maintained'.[46] What Heisenberg did was to 'reinvent' an arcane branch of mathematics called matrix mechanics and use this to describe atomic phenomena in a non-physical way. He had sent a proof copy of his article to Cambridge in August 1925, not to Rutherford who would not have understood it nor sympathised with its spirit, but to Rutherford's son-in-law Ralph Fowler. He was the man from whom Kapitza had borrowed the top hat, and was regarded by Chadwick[6] as 'a mathematician of very broad ability, very wide' who 'gradually built up a small school of theoretical physics' in Cambridge. The lowly standing of the theoreticians was summed up by Nevill Mott,[47] a perspicacious undergraduate in the mid-1920s:

> At that time theoretical physics did not have much place in the Cavendish. Fowler had a room, but for students there was nowhere to sit, except the rather small and squalid library, which also served as a tea room. We were in the Faculty of Mathematics, and the tradition was that we should sit in our college rooms and think.

Fowler passed on Heisenberg's paper to 'the only man [in Cambridge] who could understand the new theories',[47] his research student Paul Dirac, who reworked and extended Heisenberg's ideas, providing a general mathematical link between the new quantum mechanics and classical mechanics.[48] Chadwick was, of course, an avid reader of the German scientific literature, and had probably read Heisenberg's paper and recognized its significance, even if the mathematics were beyond his comprehension. He was in the habit of discussing theoretical developments with Fowler, and may also have been aware of Dirac's work before its publication in November 1925. That same month, a long paper by Rutherford and Chadwick[49] appeared in the *Philosophical Magazine*; Chadwick[44] had corrected the proofs in Rutherford's absence and it seems likely that he had added this caveat to the discussion in the light of the recent proposals:

> While the ordinary laws of collision have sufficed as yet in accounting for the scattering of α- and β-particles, it is always well to bear in mind the possibility that in special cases scattering may in some way be governed by quantum conditions.[49]

The experiments continued the study of α-particle scattering in order to probe the 'variation of the law of force, either near to the nucleus or actually within its structure, and thus to gain information as to the actual way in which nuclei are built up of their component protons and electrons'. The most important and baffling finding in the new paper concerned the heavy, radioactive element, uranium. In its spontaneous disintegration, uranium emits an α-particle of range 2.7 cm. Rutherford and Chadwick made the apparently reasonable assumption that 'an α-particle from radium C of range 7cm, must penetrate deeply into the nuclear region. On these views, we should anticipate that the law of inverse squares should no longer hold for close collision of a swift α-particle with the uranium nucleus'. In fact they observed no deviation from the inverse square law; the high-energy α-particles seemed unable to penetrate the uranium nucleus and did not experience any attractive force, as they had expected it would. The tentative explanation put forward was that the satellites around the uranium nucleus were approximately neutral in charge, but Rutherford knew they had stumbled on another important paradox which, if it could be resolved, would yield a deeper level of understanding about nuclear structure.

Chadwick had continued to dissect the Cambridge–Vienna tangle in Rutherford's absence, and quickly brought him up to date. He had attempted a theoretical analysis equating the energy changes in artificial disintegration with the amount of mass lost. He was thwarted because 'the atomic masses of magnesium, aluminium, and silicon [had] not been determined with sufficient accuracy to afford any test of the mass relations'.[50] Thus he was unable to pronounce on theoretical grounds whether some of the elements in dispute, such as carbon and oxygen, could be disintegrated by α-particles emitted from radium C or not. Deprived of a solution based on relatively simple calculation and logic, Chadwick did not shrink from the less elegant alternative of painstaking experiment. One of the key claims made by the Vienna school was that with their superior optical system, they could discriminate qualitatively between the scintillation due to an α-particle and that due to a proton. Chadwick seized on this as a crucial difference to be examined rigorously. Rutherford[38] took up the point in his reply, dated 8 February 1926, to Bohr's New Year's letter.

> While I have been away Chadwick has been hard at work going over some of the results claimed by Kirsch and Pettersen [sic]. We are convinced that they are radically wrong for some reason. The idea that you can discriminate between a slow α particle and H particles by the intensity of the scintillation is probably the cause of their going wrong. Under the normal conditions of the experiments such a discrimination by eye is terribly dangerous.

The fruits of Chadwick's labours appeared in the *Philosophical Magazine* in

November 1926.[51] With characteristic clarity, he reviewed the main results obtained by Rutherford and himself, and then summarized the 'very different conclusions' reached by Kirsch and Pettersson. He made no attempt to soften the discrepancies, pointing out that 'even in the few cases where both our observations and those of the Vienna workers agree in showing a disintegration effect, the agreement is more apparent than real'. Before presenting the latest results of his own meticulous experiments, which confirmed the earlier observations made at the Cavendish 'in showing no evidence that the nuclei of lithium, beryllium, carbon, and oxygen are disintegrated by collision with an α-paticle', Chadwick presented a detailed comparison of the experimental arrangements used in the two centres under sections headed, *The Optical Systems used in Counting* and *The Efficiencies of Counters.* The latter reveals how systematic Chadwick had been in his nursery training course, and represents a fascinating attempt at controlling for human error in signal detection. Geiger and Werner had previously published a method for assessing the efficiency of an observer by comparing the records of two independent observers simultaneously watching the same scintillation screen. Chadwick reported that during the previous two years:

> About thirty students have been examined, the majority of whom had had no previous experience of counting. When the scintillations are produced by α-particles of a few centimetres range, even an untrained observer will count about 80 per cent. of the scintillations, and his efficiency generally rises in a short time to about 95 per cent. With scintillations due to α-particles of not more than a few millimetres range, the untrained observer may count only 60 to 70 per cent. at first, but in the course of time his efficiency will again rise to more than 90 per cent.

In a footnote aimed squarely at Vienna, Chadwick also commented that 'it sometimes happens, particularly when dealing with the weaker scintillations, that a scintillation which appears bright to one observer is weak to the other'. Just in case the point was missed, there was a majestic recapitulation of it later: 'we have never, in this laboratory, felt confident of our ability to pick out the scintillations of H-particles from those of a heterogeneous beam of α-particles, although we have had considerable experience in counting'. In conclusion, Chadwick found himself 'unable to suggest any explanation which will account satisfactorily for the differences between these results and those obtained in Vienna'.

There had been renewed communication between Rutherford and Stefan Meyer, at about the time Chadwick's critical review was published, trying to arrange for exchange visits. Rutherford[52] wrote again to Meyer on 23 December 1926:

> I was glad to hear from you about the prospects for an interchange of visits between this Laboratory and your Institute, to get to the bottom of the reasons of the differences in results obtained in the two Institutions. I thank you for suggesting that I should come over, but as you no doubt know, it is quite impossible for me to do so. My duties here as President of the Royal Society fully occupy my time. Chadwick has gone away for his holiday but he told me that he had received a letter from Pettersson. I have not had an opportunity to discuss the question with him but I am very doubtful whether he would be able to visit Vienna for the next

> few months. He is expecting an interesting domestic event in February and naturally cannot leave for some time. I will write you later when we have discussed the matter. I can quite understand that it might be best for Chadwick to visit your Institution and see for himself what is the cause of the divergences. I agree with you that it is highly important that this question should be amicably settled for I myself feel the whole subject of nuclear disintegration must remain in confusion pending a comparative investigation.

Rutherford was referring to the fact that Aileen Chadwick was pregnant, and the domestic event proved to be more interesting than anyone had foreseen with the arrival of twin daughters, Joanna and Judy, on 1 February 1927. The Chadwicks had just moved into a house, which they had built in Bentley Road. It had not been finished on time and they had spent the first eighteen months of their married life in rented, furnished accommodation. The pregnancy cannot have been easy for Aileen, and James, although he remained up to his ears in love, was certainly up to his eyes in work. Apart from being deeply immersed in his experiments, he had been commandeered, together with Charles Ellis, to help Rutherford prepare a new edition of his book on radioactivity. According to Chadwick,[6] 'Rutherford didn't have much to do with it really . . . He wrote the early part, a kind of introduction . . . about the disintegration theory which was not very different from what had appeared in the older edition. There may have been some new things in it, but I'm not sure that I didn't put them in'.

Chadwick began work on the book while he and Aileen were still in temporary digs, and with the east wind blowing from the Fens, there must have been times when he felt he was back at Ruhleben.

> It was a cold winter, and it was a very draughty house. There was a door between the drawing room . . . and the garage, but it didn't fit very well and the wind came through. There was no central heating. I was busy in the laboratory, and the only time I had to work was late at night, and it was so cold that I had to put a little table in front of the fire after my wife and the cook general had gone to bed; I put this little table in front of the fire with an overcoat on and sometimes with gloves, and wrote what I could before about 12 or 1 o'clock.[6]

Although Rutherford may not have contributed much new material to the book, he read the contributions of his co-authors quite closely. In 1927 he[53] published a theory of the structure of radioactive nuclei to explain the origin of α-particles in a way which would reconcile some of the conflicting results that he and Chadwick had found in their experiments. To Ellis[52] this paper represented 'perhaps the most characteristic example of Rutherford's genius during his Cambridge period'. Chadwick,[6] by contrast, had 'not much opinion of it'; he did not disguise his lack of enthusiasm, and this 'rather annoyed' Rutherford. Since his appointment as Assistant Director of Research, Chadwick had started to shed some of his insecurities, slowly at first, but with increasing speed after his marriage. He could now contemplate an argument with Rutherford over a scientific question or the organization of the laboratory, if not with complete equanimity, at least without raw fear. Aileen was his most devoted supporter and would shrink from no skirmish on his behalf. He was no longer to be at the Rutherfords' beck and call, and this led to some coolness between the two wives. It was perhaps natural that Chadwick should no longer be so closely involved with the Rutherfords socially, and Aileen introduced him to her family's friends in Liverpool, whom he found to be refreshing company. 1927 was an especially happy year, because in addition to becoming a father, he was elected Fellow of the Royal Society and now took his place amongst the country's scientific élite.

When it came to writing the book, it fell to Chadwick[6] to describe Rutherford's theory and to comment on it. He admitted to not being 'very complimentary' in his remarks. It was one of the chapters that Rutherford read at the proof stage, and he tackled Chadwick about it, saying, 'Well, you don't seem to think very much of my theory of the nuclear structure'. And Chadwick replied: 'No, I don't. It's artificial. It won't explain some of the observations'. After due deliberation, Rutherford said, 'Well, if that's what you think, we better leave it as it is'. Rutherford 'didn't alter a word', which shows remarkable forbearance and humility. That Chadwick spared him not at all can be judged from the following comments as they appeared in the book.

> This theory [of Rutherford's] gives no clue to the exponential law of transformation of the radioactive elements or to the Geiger–Nuttall relation, but is solely concerned with the energy relations of the nuclear changes. Apart from certain difficulties in connection with the assumptions involved, the theory is open to the objection that it does not explain why, in experiments where α particles of high energy are shot at the uranium nucleus, no evidence is obtained either of disintegration or of any departure from Coulomb forces.[54]

By the time the book, *Radiations from Radioactive Substances*, was published in 1930, the problem of α-decay, alluded to in the extract above and which had exercised Rutherford and Chadwick so greatly following their 1925 paper, had been solved. The conundrum was that two separate conditions seemed to apply in the case of the uranium nucleus, depending on whether an α-particle was leaving or attempting to enter the nucleus. The answer came almost simultaneously in 1928 from Gamow in Russia, and Condon and Gurney in the USA.[21] By applying the new quantum mechanics to the nucleus, they showed that an α-particle could tunnel its way out of the nucleus and emerge with relatively low energy: it did not have to surmount a fixed barrier as Rutherford and Chadwick had supposed from classical physics.

There would be no such revelatory end to the Cambridge–Vienna controversy. After Chadwick's authoritative 1926 review, Pettersson and Kirsch soon replied, in three separate publications, and in a vein that 'left no doubt in the mind of any reader that [they] regarded Chadwick's analysis as flawed'.[38] Much of their case again rested on the supposed superiority of their microscopes, and on the ability of the observers in Vienna to distinguish scintillations due to protons from those due to α-particles. In May 1927, Pettersson entered the lion's den and spent a few days in Cambridge, discussing the disintegration work in great detail with Rutherford and Chadwick. The atmosphere was friendly, but neither side was prepared to abandon their positions. Pettersson urged Chadwick to come to Vienna to see for himself the work that was being undertaken. In December, Chadwick was finally able to take up the offer, and he and Aileen travelled to Vienna, arriving on Wednesday 7 December. They stayed at the plush Hotel Regina and Chadwick immediately set to work at the nearby institute. He[55] wrote to Rutherford on 9 December that progress had been slow in part because 'Thursday was a holy day and no work could be done without danger to our future in the world to come. Today, however, we ended up with a fierce and very loud discussion'. Chadwick was pushing Pettersson to let him set up an experiment, but without success. Instead Pettersson carried out an experiment 'to his satisfaction', but not to Chadwick's, who regarded it as 'not conclusive' and 'unimportant'. Chadwick again suggested they should concentrate on the disintegration of carbon, a question which 'contains all the discrepancies in one result', but this 'precipitated a most fiery outburst, and all kinds of small matters were magnified to great importance'. He continued 'I am afraid this visit will not improve our relations. Stefan Meyer, Schweidler and all with no direct interest in the question are exceedingly pleasant and friendly but the younger ones stand around stifflegged and with bristling hair . . . I shall keep at this until the business is settled'.

Chadwick was not alone in finding the encounter tense. One of Pettersson's assistants, Elisabeth Rona formed an unfavourable impression of Chadwick: 'He seemed to us to be cold, unfriendly and completely lacking in a sense of humour. Probably he was just as uncomfortable in the role of judge as we were in that of the judged.'[38] The trial reached its climax on Monday 12 December. Rona, who was one of the scintillation counters, recalled the scene:

> All of us sat in a dark room for half an hour to adapt to the darkness. There was no conversation; the only noise was the rattling of Chadwick's keys. There was nothing in the situation to quiet our nerves or make us comfortable. Short spells of scintillation counting followed for each member of the group, and then the radiation source was exchanged with a blank, unknown to the persons who were doing the counting.

Since arriving in Vienna, Chadwick[56] had learned that 'their methods of counting are quite different from ours', and that 'not one of the men does any counting. It is all done by 3 young women. Pettersson says the men get too bored with routine work and finally cannot see anything, while women can go on for ever'. Chadwick gave his own version of events in a letter to Rutherford written on that Monday evening.

> Today, therefore, I arranged that the girls should count and that I should determine the order of the counts. I made no change whatever in the apparatus, but I ran them up and down the scale like a cat on a piano—but no more drastically than I would in our own experiments if I suspected any bias. The result was that there was no evidence of H particles [from carbon] . . . The results do not prove that there is nothing from carbon but I think they make it doubtful that there is much.

Chadwick also took the opportunity to test the counters with 'natural H-particles produced from paraffin. Counts were normal at the absorption they are accustomed to, irregular at a greater absorption'. Despite the fraught circumstances, he could see 'no reason why the counters should be off colour'. In fact the excuse was offered that the counters eyes were tired and so the experiment was repeated the next morning with the same negative outcome. In Chadwick's view[6] the 'young women were perfectly honest', there was 'no question of cheating'. He suspected that they were normally informed about the experimental arrangement in advance so that 'they were seeing what they expected to see'. What he did was to 'blind' them to the nature of the experiment and remove this source of observer bias.

The reason Chadwick eventually had such a free hand in a foreign laboratory was that Pettersson's family happened to be visiting from Sweden, and Chadwick only saw him for '5 minutes' that Monday. When Chadwick[6] eventually had the opportunity of explaining the findings which effectively destroyed his burgeoning reputation, Pettersson was 'very angry indeed', but agreed to meet Stefan Meyer, the institute's director, the next morning. In what was clearly a painful meeting for all three men, Meyer accepted Chadwick's evidence and offered to make a public retraction to set the record straight. Without consulting Rutherford, Chadwick knew that he would wish to spare Meyer any public embarrassment and immediately refused the offer. Instead Chadwick[6] recommended that they should simply 'drop the experiments and say nothing'. And so one of the most famous controversies of modern physics came to a private settlement that was honoured by all the participants. After four years, when exchanges in the scientific literature had become increasingly pointed, the matter was laid quietly to rest. The unbroken silence was the only sign to the outside world that hostilities had ceased, and eventually the affair was forgotten. Forty years later, Chadwick[6] was amused by an article in which Otto Frisch, who had been a student in Vienna at the time, admitted that he still did not know how Pettersson and Kirsch arrived at their erroneous results, but suspected subjective bias in the counting.

There were, of course, lessons for both sides; the dénouement forced the realization at the Cavendish that scintillation counting could no longer be relied upon to the extent that it had been. In the only brief reference to the controversy in the 1930 textbook, again in a chapter written by Chadwick,[57] 'any satisfactory explanation of these and other divergences' between results obtained in Cambridge and Vienna 'seems difficult to put forward', and the reader is told that 'this is not the place to attempt it'. Rather a fresh start is advocated using the 'Wilson expansion chamber' and 'an electrical method of detecting the particles of disintegration', instead of the scintillation method. The latter technique, in Chadwick's mind, now seemed to be consigned to the past as firmly as the 'philosopher's stone', which the alchemists of the Middle Ages sought in order to convert one form of matter into another. In a dramatic opening to the chapter,[57] he mused that despite the encouragement of rulers and princes:

> the successful transmutation of some common matter into gold was seldom reported. Even in these cases the transmutation could not be repeated; either the alchemist had vanished or his supply of the 'philosopher's stone' had been exhausted. The failures were many and the natural disappointment of the patron

usually vented itself on the person of the alchemist; the search sometimes ended on a gibbet gilt with tinsel.

One can picture Chadwick taking off his overcoat and gloves, with a smile of satisfaction, after writing this fine passage.

References

The letters of Peter Kapitza are all reproduced from *Kapitza in Cambridge and Moscow* by kind permission of his widow, Anna. It will be clear to the reader that I have drawn heavily on Roger H. Stuewer's meticulous account of the Cambridge–Vienna controversy.[38]

1 Kapitza, P. (6/7/22). Letter to O. Kapitza. In *Kapitza in Cambridge and Moscow* (ed. J.W. Boag, P.E. Rubinin and D. Shoenberg), p. 152. Elsevier Science Publishers B.V., Amsterdam.

2 Kapitza, P. (5/12/21). Letter to O. Kapitza. In *Kapitza in Cambridge and Moscow* (ed. J.W. Boag, P.E. Rubinin and D. Shoenberg), pp. 137–8. Elsevier Science Publishers B.V., Amsterdam.

3 Kapitza, P. (17/1/22). Letter to O. Kapitza. In *Kapitza in Cambridge and Moscow* (ed. J.W. Boag, P.E. Rubinin and D. Shoenberg), p. 140. Elsevier Science Publishers B.V., Amsterdam.

4 Thomson, J.J. (1910). Survey of the last twenty five years. In *A history of the Cavendish Laboratory, 1871–1910*, p. 96. Longmans Green, London.

5 Kapitza, P. (16/12/21). Letter to O. Kapitza. In *Kapitza in Cambridge and Moscow* (ed. J.W. Boag, P.E. Rubinin and D. Shoenberg), pp. 138–9. Elsevier Science Publishers B.V., Amsterdam.

6 Weiner, C. (1969). Sir James Chadwick, oral history. Niels Bohr Library, American Institute of Physics, College Park, Maryland.

7 Eve, A.S. (1939). *Rutherford*, p. 274, Macmillan, New York.

8 Wilson, D. (1983). Cambridge and the Cavendish. In *Rutherford: simple genius*, pp. 406–52, Hodder and Stoughton, London.

9 Marsden, E. and Lantsberry, W.C. (1915). The passage of alpha particles through hydrogen. *Philosophical Magazine*, 30, 824–30.

10 Rutherford, E. (19/8/20). Letter to B. Boltwood. In *Rutherford and Boltwood* (ed. L. Badash, 1969), p. 329. Yale University Press.

11 Rutherford, E. (28/2/21). Letter to B. Boltwood. In *Rutherford and Boltwood* (ed. L. Badash, 1969), pp. 341–4. Yale University Press.

12 Rutherford, E. and Chadwick, J. (1921). The disintegration of elements by α-particles. *Nature*, 107, 41.

13 Eve, A.S. (1939). *Rutherford*, p. 291. Macmillan, New York.

14 Oliphant, M.L. (1972). *Rutherford— recollections of the Cambridge days*. Elsevier, Amsterdam.

15 Rutherford, E. and Chadwick, J. (1921). The artificial disintegration of light elements. *Philosophical Magazine*, 42, 809–25.

16 Bieler, E. (1923). The law of force in the immediate neighbourhood of the atomic nucleus. Unpublished Ph.D. thesis. University of Cambridge.

17 Rutherford, E., Robinson, H. and Rawlinson, W.F. (1914). Spectrum of the β rays excited by the γ rays. *Philososphical Magazine*, 28, 281–6.

18 Ellis, C.D. (1921). Magnetic spectrum of the β-rays excited by γ-rays. *Proceedings of the Royal Society A*, 99, 261–71.

19 Meitner, L. (1922). Über die Enstehung der β-Strahl-Spektren radioaktiver Substanzen. *Zeitschrift für Physik*, 9, 131–44.

20 Frisch, O.R. (1970). Lise Meitner (1878–1968). *Biographical memoirs of fellows of the Royal Society*, 16, 405–15.

21 Pais, A. (1986). Nuclear physics: the age of paradox. In *Inward bound*, pp. 296–323. Oxford University Press.

22 Chadwick, J. and Ellis, C.D. (1922). A preliminary investigation of the intensity distribution in the β-ray spectra of radium B and C. *Proceedings of the Cambridge Philosophical Society*, 21, 274–80.

23 Ellis, C.D. and Wooster, W.A. (1927). The continuous spectrum of β-rays. *Nature*, 119, 563–4.

24 Chadwick, J. and Mercier, P.H. (1925). The scattering of beta-rays. *Philosophical Magazine*, 50, 208–24.

25 Kapitza, P. (17/8/22). Letter to O. Kapitza. In *Kapitza in Cambridge and Moscow* (ed. J.W. Boag, P.E. Rubinin and D. Shoenberg), p. 155. Elsevier Science Publishers B.V., Amsterdam.

26 Kapitza, P. (22/10/22). Letter to O. Kapitza. In *Kapitza in Cambridge and Moscow* (ed. J.W. Boag, P.E. Rubinin and D. Shoenberg), p. 158. Elsevier Science Publishers B.V., Amsterdam.

27 Kapitza, P. (18/3/23). Letter to O. Kapitza. In *Kapitza in Cambridge and Moscow* (ed. J.W. Boag, P.E. Rubinin and D. Shoenberg), p. 166. Elsevier Science Publishers B.V., Amsterdam.

28 Chadwick, J. (?1971). Undated notes on Oliphant's biography of Rutherford (ref. 14). CHAD II, 2/1, CAC.

29 Rutherford, E. (30/5/22). Letter to B. Boltwood. In *Rutherford and Boltwood* (ed. L. Badash, 1969), p. 351. Yale University Press.

30 Chadwick, J. Undated notebooks on the Cavendish Laboratory. CHAD II, 3/7, CAC.

31 Oliphant, M.L. and Penney, W.G. (1968). John Douglas Cockcroft, 1897–1967. *Biographical memoirs of fellows of the Royal Society*, 14, 139–88.

32 Cochran, W. and Devons, S. (1981). Norman Feather, 1904–78. *Biographical memoirs of fellows of the Royal Society*, 27, 255–83.

33 Oliphant, M.L. (1982). The beginning: Chadwick and the neutron. *Bulletin of the atomic scientists*, Dec, 14-18.

34 Kapitza, P. (26/6/25). Letter to O. Kapitza. In *Kapitza in Cambridge and Moscow* (ed. J.W. Boag, P.E. Rubinin and D. Shoenberg), p. 186. Elsevier Science Publishers B.V., Amsterdam.

35 Kapitza, P. (10/8/25). Letter to O. Kapitza. In *Kapitza in Cambridge and Moscow* (ed. J.W. Boag, P.E. Rubinin and D. Shoenberg), pp. 187–8. Elsevier Science Publishers B.V., Amsterdam.

36 Kapitza, P. (25/8/25). Letter to O. Kapitza. In *Kapitza in Cambridge and Moscow* (ed. J.W. Boag, P.E. Rubinin and D. Shoenberg), pp. 189–90. Elsevier Science Publishers B.V., Amsterdam.

37 Rutherford, E. (9/10/25). Letter to P. Kapitza. In *Kapitza in Cambridge and Moscow* (ed. J.W. Boag, P.E. Rubinin and D. Shoenberg), p. 258. Elsevier Science Publishers B.V., Amsterdam.

38 Stuewer, R.H. (1985). Artificial disintegration and the Cambridge–Vienna controversy. In *Observation, experiment, and hypothesis in modern physical science* (ed. P. Achinstein and O. Hannaway), pp. 239-307. The Massachusetts Institute of Technology, Cambridge, MA.

39 Rutherford, E. and Chadwick, J. (1924). The bombardment of elements by α-particles. *Nature*, 113, 457.

40 Rutherford, E. and Chadwick, J. (1924). Further experiments on the artificial disintegrations of elements. *Proceedings of the Physical Society*, 36, 417–22.

41 Rutherford, E. and Chadwick, J. (1924). On the origin and nature of the long-range particles observed with sources of radium C. *Philosophical Magazine*, 48, 509–26.

42 Chadwick, J. (1924). Letter to E. Rutherford. Rutherford papers, University of Cambridge Library

43 Rutherford, E. (1924). Disintegration of atomic nuclei. *Nature*, 115, 493–4.

44 Chadwick, J. (10/10/25). Letter to E. Rutherford. Rutherford's papers, University of Cambridge Library.

45 Chadwick, J. (21/11/25). Letter to E. Rutherford. Rutherford's papers, University of Cambridge Library.

46 Pais, A. (1991). The discovery of quantum mechanics. In *Niels Bohr's times*, pp. 267–94. Oxford University Press.

47 Mott, N.F. (1986). *A life in science*, pp. 20–2. Taylor and Francis, London.

48 Matthews, P.T. (1987). Dirac and the foundation of quantum mechanics. In *Reminiscences about a great physicist: Paul Adrien Maurice Dirac* (ed. B. Kursunoglu and E.G. Wigner), pp. 199–227. Cambridge University Press.

49 Rutherford, E. and Chadwick, J. (1925). Scattering of α-particles by atomic nuclei and the law of force. *Philosophical Magazine*, 50, 889–913.

50 Rutherford, E. and Chadwick, J. (1929). Energy relations in artificial disintegration. *Proceedings of the Cambridge Philosophical Society*, 25, 186–92.

51 Chadwick, J. (1926). Observations concerning the artificial disintegration of elements. *Philosophical Magazine*, 2, 1056–75.

52 Eve, A.S. (1939). *Rutherford*, pp. 318–20. Macmillan, New York.

53 Rutherford, E.(1927). Structure of the radioactive atom and origin of the α-rays. *Philosophical Magazine*, 4, 580–605.

54 Rutherford, E., Chadwick J. and Ellis C.D. (1930). *Radiations from radioactive substances*, p. 327. Cambridge University Press.

55 Chadwick, J. (9/12/27). Letter to E. Rutherford. Rutherford's papers, University of Cambridge Library.

56 Chadwick, J. (12/12/27). Letter to E. Rutherford. Rutherford's papers, University of Cambridge Library.

57 Chadwick, J. (1930). The artificial disintegration of the light elements. In *Radiations from radioactive substances* (ed. E. Rutherford, J. Chadwick and C.D. Ellis), pp. 281–316. Cambridge University Press.

6 Discovery of the neutron

The long series of scattering experiments conducted at the Cavendish came to an end with Chadwick's visit to Vienna; the results of the last joint investigation with Rutherford, *The scattering of α-particles by helium*,[1] were published in September 1927. The aims of this experiment and its methodology would have been immediately understood by any physicist who was familiar with the two papers that had flowed from Chadwick's doctoral thesis, even if he had read nothing else since 1921. Einstein once characterized the nature of these scattering experiments, where perhaps only 1 in 300 000 of the bombarding α-particles gave rise to a scintillation of note, as 'like shooting sparrows in the dark'.[2] Looking back, Chadwick[3] thought the work, though interesting and important, was 'work of consolidation rather than of discovery'. In his view, this and 'other work on the law of force close to the nucleus gave some insight into the problem of nuclear structure, but it fell short of what was needed'. At the end of the 1920s, 'the general conception of the nucleus was that it was built up from protons and electrons, with α-particles probably as a sub-unit of structure'. This was very much the view that Rutherford[4] put forward in his opening remarks at a discussion on the structure of atomic nuclei, held at the Royal Society in February 1929.

Since Bohr formulated his original quantum model of the atom in 1913, its representation had become progressively more elaborate in order to account for various properties of orbital electrons. In 1925 the concept of 'spin'[5] had been introduced to account for the fine structure of atomic spectra, where seemingly single lines could be resolved into doublets and, more particularly, spectral lines could be split by placing atoms in a magnetic field. This anomaly, known as the Zeeman effect, was explained by the electron having angular momentum or spin which was constrained to only two possible orientations in an external magnetic field—either parallel or antiparallel to the field. By the convention of quantum mechanics, this property is described by a number, s, which can only take the value of $+\frac{1}{2}$ or $-\frac{1}{2}$. The mathematical foundation for the concept of spin was provided by Dirac working in Cambridge in 1926. The new quantum mechanics had been concerned almost exclusively with orbital electrons, and now accounted for previously obscure phenomena, but attempts to extend the rules to electrons in the nucleus led to new confusion. One prominent example will suffice—*the nitrogen paradox*. Each nitrogen nucleus was thought, in 1928, to consist of 14 protons +

7 electrons, each of which possessed half-integer spin. The total spin of such a system containing an odd number of spin-½ particles should be half-integer, but spectroscopic data showed the spin of the nitrogen nucleus was integer (implying an even number of particles in the nucleus).

Discrepancies such as these caused major headaches for the men who devised the new theories, but could be more easily ignored by the experimenters at the Cavendish. Chadwick[6] said: 'it took quite a time to absorb the meaning of the new quantum mechanics. It was rather slow . . . there was no immediate application to the structure of the nucleus, which was what we were interested in'. Rutherford saw no reason to abandon the proton/electron/α-particle compound nucleus, and would maintain his scepticism of the new, complex mathematical systems. His way of conceptualizing, where the starting point would usually be some unexpected experimental observation, was far removed from Dirac's,[7] for example. Dirac thought it 'advisable to allow oneself to be led forward by the mathematical ideas [which] . . . can lead us in a direction we would not take if we only followed up physical ideas by themselves'. In practice, the types of interaction between atomic nuclei and relatively large, slow-moving α-particles that Rutherford and Chadwick had been studying could be satisfactorily described by simple mechanical theory; Dirac's elucidation of quantum field theory had little immediate impact on experimental physics at the Cavendish Laboratory. As Bohr[8] noted in a 1927 lecture, 'interpretation of the experimental material rests essentially on the classical concepts'. Bohr was speaking at the Volta Centennial Congress at Lake Como, and Rutherford was in the audience. Sir Ernest did not enjoy the congress and was a notable absentee from the Solvay Conference held in Brussels the following month. This gathering, which included Bohr, Born, de Broglie, Einstein, Heisenberg, Pauli, and Schrödinger, indelibly marked the epoch of quantum mechanics. Dirac and Fowler both attended, but the only experimental physicist from the Cavendish was C.T.R. Wilson.

In Chadwick's mind, the deficiencies in theoretical physics in Cambridge and other English universities had a deep-seated cause. He thought that the Cambridge mathematicians had been isolationist and were quite uninterested in any problems of a physical nature. 'It was quite different from anything you would find on the Continent. It was largely due to the type of mathematical course in Cambridge that there was so little theoretical physics in this country until Ralph Fowler developed a small school'.[6] This view is supported by one of Fowler's distinguished pupils, McCrea,[9] who imagined himself a Cambridge mathematician until 'in 1928 I went to Göttingen to work for a year in Max Born's group. I discovered that in Germany (a) "mathematician" meant one who did no physics, and (b) that Born and all his people were labelled "physicists'.' McCrea also found 'it hard to understand how Cambridge managed to maintain such a high reputation for mathematics when it taught mathematics so badly [unless] . . . Oxford did it even worse'! Things improved under Fowler, and a frequent exchange of personnel and information between Cambridge and centres such as Copenhagen and Göttingen soon developed. Niels Bohr, who had long been respected by Rutherford, became 'the international oracle to younger theorists'[10] and was always a welcome visitor. By 1928, the Cavendish research students were receiving lectures on 'Wave Mechanics' from D.R. Hartree, a member of Fowler's group who had spent time with Bohr in Copenhagen. Fowler and Chadwick[6] agreed on the need for the cross fertilization of ideas and set up 'a colloquium rather on the style of the Ruben's Colloquium . . . to bring together the theoretical physicists and the experimental physicists'.

Although the colloquium was intended to benefit the younger scientists, it soon bore unexpected fruit for Chadwick himself—he always much preferred small discussion groups to big meetings. In January 1929, Nevill Mott had returned to Cambridge after a term in Copenhagen where he had been thrilled and inspired by Bohr's teaching. Mott had set himself the challenge of demonstrating what changes Dirac's relativistic wave equation would make to Rutherford's law of scattering. With Bohr he had discussed studying, in a mathematical way, the collision of free electrons with atomic nuclei; now back in Cambridge, Fowler suggested to him that he should apply his ideas to the α-particle, which has no spin. Given that the α-particle is identical with the nucleus of the helium atom, Mott[11] predicted that 'if alpha particles traversed some helium gas, the number deflected through 45° would be twice as many as predicted by the famous Rutherford scattering formula'. When he announced this departure from classical theory at the Cavendish colloquium, Chadwick offered to see if the effect predicted was real. Rutherford and he would not have observed any such effect in their 1927 study of α-particle scattering by helium, because they were using high-energy α-particles in order to explore the forces in the immediate vicinity of the nucleus, where the inverse square law no longer held.

Mott had realized that after any collision the scattered α-particles and the projected helium nuclei would be identical in all respects. He then employed quantum mechanics to treat the two streams of particles as waves, and compared the outcome with that derived from Rutherford's classical law for various angles of scatter. Chadwick[12] employed 'α-particles of much lower velocity than those used in the previous experiments, and . . . found that the amount of scattering at 45° is twice that expected on classical theory, as predicted by Mott's calculations'. Chadwick carried out the work at Christmas time, when the laboratory was quiet, and referred to it, with evident satisfaction, as 'a very simple but pretty experiment'.[13] The paper was ready for publication in April 1930, and in the published discussion, he stated:

> The discrepancy from classical scattering cannot, therefore, be ascribed to a divergence from Coulomb [electrostatic] forces and it must be attributed to a failure of the classical theory itself. The experiments thus provide a striking verification of Mott's calculation and of the assumption on which it is based, that it is impossible to distinguish one helium nucleus from another of the same velocity. In other words, the helium nucleus has no spin or vector quantity associated with it, its field of force is perfectly spherical.[12]

As soon as he had verified the effect, Chadwick took Mott along to see Rutherford, who was impressed. He said to Mott: 'If you think of anything else like this, come and tell me'. To Mott[11] this reception gave 'complete confidence in my ability to make a career in theoretical physics, so that I could cease to worry about it'.

This was the second time that Rutherford and Chadwick had been convinced that wave mechanics had some merit as a tool to unlock the secrets of the nucleus. The first occasion had been the autumn visit of the Russian, George Gamow, to the Cavendish in 1928, when he had expounded his theory that α-particles could tunnel through a potential barrier around the nucleus and therefore emerge with less energy than would otherwise be expected.[14] This explained to Rutherford and Chadwick their puzzling, experimental findings with uranium, where high-energy α-particles were unable to penetrate the uranium nucleus, but low energy α-particles were able to escape from it, in natural radioactive decay. On the strength of this work, Gamow became a frequent and valued visitor to the Cavendish. While he was in residence early in 1930, working on nuclear theory, he and Chadwick collaborated on a letter to *Nature*.[15] The subject was the artificial disintegration of atomic nuclei by α-particles; their contention was that in addition to the accepted process whereby the α-particle penetrates the nucleus, is captured and a proton emitted, disintegrations could take place in which the α-particle is not captured, but a proton is still ejected. Using wave mechanics, they considered the probability of disintegration of both types and the energy profile of the emitted protons. This work was squeezed into the Rutherford, Chadwick and Ellis book, *Radiations from radioactive substances*, as an appendix, and the idea of nuclear resonance introduced.[16] Chadwick concluded that 'it is already evident that the phenomenon of artificial disintegration now promises to reveal the intimate structure of the nuclei of the lighter elements'.

Neither Chadwick nor Rutherford would have dreamed of making such an optimistic prediction just two years earlier because the Vienna-Cambridge controversy had not only discredited the work at the Institut für Radiumforschung, it had thrown their own programme into disarray. By uncovering the observer biases in Vienna, Chadwick had bankrupted scintillation counting as a scientific technique. However carefully the Cambridge researchers were trained in the Nursery, and no matter how scrupulous their recorded observations, Chadwick could not depend on the rest of the world to adhere to the same standards. The method, therefore, had lost its validity as a means of inter-laboratory comparisons and could only jeopardize future advance. It would still be employed on rare occasions because of its simplicity and familiarity: for example Chadwick used scintillation counting to verify Mott's prediction in the experiment described above.

Whereas the Vienna–Cambridge controversy mired α-particle scattering research, 1927 had at last seen the successful persuasion of Lise Meitner to the continuous spectrum of β-rays by Charles Ellis. Seeking to capitalize on this new found common ground and being rather dismayed about the state of their own α-particle research, Rutherford and Chadwick decided to hold an international conference in Cambridge on β- and γ-rays. Chadwick was the secretary to the conference and the date was set for July 1928. The timetable circulated for the meeting shows it to have been a leisurely affair with never more than three lectures being scheduled on any of the five days, and including one day of rest. In his short prospectus for the conference, Chadwick[17] wrote:

> The main object of this Conference is to provide an opportunity for a full and free discussion of the outstanding problems. It is hoped therefore that it will be possible to give the papers in a somewhat informal way.

Many of the leading experimental physicists from Europe attended, and there was an irreverent rumour circulating among the junior members of the laboratory that the main purpose in calling the conference was for Rutherford, Chadwick and Ellis to make sure they had not left anything important out of their forthcoming book. The French and German schools of radioactivity were well represented but not the Viennese, which made discreet conversations about the recent events in Stefan Meyer's laboratory permissible. Appropriately, perhaps, for a group of experimenters, the lasting effects of the conference came not from any philosophical trading in ideas, but from two gifts of equipment and material. Geiger[18] had written to Rutherford just before he travelled to Cambridge to tell him about the success he and his post-doctoral student, Walter Müller, had achieved in developing a new electrical counter.

> Experiments have made good progress in the last weeks. We have a counter now which counts β-rays over an area of 100 cm^2 and perhaps more. It is extremely sensitive . . . The counter must be screened from all sides with very thick iron plates to cut down the natural effect, which otherwise amounts to 100 throws [of the electrometer] per minute or more.

Apart from its great sensitivity, the Geiger–Müller counter was immune to the fatigue suffered by the human eye and allowed both continuous measurement and a wide range of counting frequency. The instrument had been devised for cosmic ray research, but could equally be used for disintegration studies. Both Rutherford and Chadwick had successfully used prototype electrical counters when working with Geiger before the war, but had not employed them at all since arriving at the Cavendish. A month after the conference, Geiger[18] sent 'one of the new counting tubes' to Rutherford with the maker's guarantee that 'there is not the slightest difficulty to get it going'. The major drawback to its use was that the standard α-particle source at the Cavendish, radium, emitted plentiful beta and gamma rays and these would trigger the counter just as effectively as α-particles or protons would.

The solution to this problem was to change source material and use polonium instead of radium. Polonium is a 'pure' α-emitter, although the particles are of lower energy than those from radium. Polonium was even more scarce than radium, but it could be recovered in tiny amounts from the spent radon tubes or 'seeds' used by pioneering radiotherapists to treat patients with cancer. Chadwick was spared the task of scouring the London hospitals in search of old seeds by Lise Meitner, who promised to send him some polonium from Germany after the conference. The small but precious package, containing perhaps 2 millicuries of polonium,[6] duly arrived in October 1928. In his letter of thanks Chadwick[17] wrote:

> I have been so busy starting our new research students that I have had no time to think of my own work. I hope I shall be able to do quite a number of experiments before the polonium decays much. I shall do as you suggest and have the range of the α-particles measured. But what is the range of the polonium α-particle? I have never been able to get as high a value as Geiger even with what appeared to be a clean source.

This letter was dated 22 October, and Chadwick's laboratory notebook[19] shows that he was making measurements on the new source the next day. He also enlisted the help of two fresh 1851 scholars, B.W. Sargent from Canada and H.C. Webster, an Australian, who were just starting the Nursery course. Chadwick called the two men into his office and gave them an hour's tutorial on scintillation counting. Since he was so busy, Sargent[20] was 'greatly impressed by his generosity in giving us so much of his time. Our project was to determine our efficiencies in counting scintillations of polonium α-particles . . . and to measure the extrapolated range of the α particles in air'. Chadwick would have been furious if he had been aware of the subversive thoughts that crossed the minds of his acolytes.

> Although Webster and I had been in the Cavendish Laboratory only a week or 10 days, we had gained the impression that the scintillation method was going out of fashion. Nevertheless, we feared that one or both of us might be assigned a research investigation involving the counting of scintillations for the Ph.D. degree. Knowing how slow and fatiguing the method was, we were determined to avoid it. Accordingly, we discussed whether we should cheat to keep our efficiencies below a level that Chadwick might regard as useful. When our efficiencies worked out at only 85% each, we decided that cheating was unnecessary.[20]

After a month in the Nursery, Chadwick assigned Webster his Ph.D. project. This was to extend an observation of Slater in 1921 that when heavy atoms such as gold are bombarded by α-particles, a small amount of high energy or hard γ-radiation is emitted, which was thought to orginate in the nucleus. Chadwick[6] had spoken to Lise Meitner and Walter Bothe, another friend from Berlin days, about this topic at the Cambridge conference. The first published observation of γ-rays being excited as a result of α-particle bombardment had been made by Chadwick[21] himself at Manchester in 1912. His original experiment had been limited by the use of radium C as a source and the intense primary γ-radiation it emitted, and Chadwick now wished to repeat the work with polonium to obviate the problem of γ-ray contamination. As Webster[22] explained:

> Interest lies in the possibility of exciting nuclear gamma radiation in light atoms where a swift α can get 'inside' the 'potential hump' of the nucleus (where the nucleus attracts rather than repels). An unstable 'stationary state' of the nucleus might then be produced followed by reversion to a stable state accompanied by radiation . . . Experiments (unpublished) by Dr. Chadwick . . . indicated that the effect to be investigated must be very small. It was therefore decided to use a Geiger–Müller tube counter as the most sensitive method of measurement.

The words may be Webster's, but there seems little doubt that the scientific direction was provided by Chadwick. The unpublished experiment referred to was carried out by Chadwick in 1925, while Rutherford was away in New Zealand; it would be summarized in their last joint publication, *Energy relations in artificial disintegration,*[23] in 1929, as follows:

> It appears from this evidence that the process of disintegration of an aluminium nucleus by an α particle of given energy is not exactly the same for each individual nucleus. We may explain this most simply by assuming that the energy change W is not the same in each disintegration. The variation in energy change must be due to variations in the internal energy either of the initial aluminium nucleus or of the final nucleus . . . If we adopt this view some aluminium nuclei must have a greater mass or energy content than others; the masses of the nuclei must be supposed to differ by as much as 0.006 mass units.

The work by Chadwick represented a valiant attempt to synthesize many of the different ideas about the nucleus that were floating around the Cavendish in the mid-1920s, although he would later dismiss his own interpretation of the data as 'just a bit of nonsense'. Ellis had first postulated the existence of different energy levels in the nucleus and suggested that these levels represented stationary states. Chadwick's own pet concept was the neutron, which he looked for, in vain, in a small adjunct to his main experiment. He also tried to support the idea, referred to in the quotation above, about discrepancies in the mass of different nuclei by comparison with Aston's mass spectrography results; but good though Aston's measurements were, they were not of sufficient accuracy to be of use in this context. Now one has some understanding of the cryptic line in his letter to Rutherford, written in September 1924, that 'I think we shall have to make a real search for the neutron. I believe I have a scheme which may just work but I must consult Aston first'.

The barest mention of 'neutrons' in the 1929 Rutherford and Chadwick paper was one of the few indications to the outside world that this was still a subject of interest at the Cavendish. According to Chadwick,[6] Rutherford by this time 'had lost interest to a certain degree' in the neutron as a result of his ever increasing public commitments and limited time in the laboratory. For his own part, Chadwick[6] 'just kept on pegging away', and looked for evidence of the neutron in any way he could devise. When Geiger sent him two of the new counters, his first thought was to try them out as neutron detectors, using 'all manner of tricks in the hope of finding some trace of the neutron'. In general, he followed two lines of attack throughout the 1920s. One was to attempt to synthesize the neutron from a proton and electron, which he and Rutherford believed to be its constituent parts; the other was to try to knock the neutron out of an atomic nucleus.[24] The approach was futile; Chadwick[25] later described some of the experiments as 'quite wildly absurd' and he probably discussed them only with Rutherford. Yet there was some value in these quick, speculative attacks on an apparently impregnable stucture.

> [Rutherford] would do some damn silly experiments at times, and we did some together. They were really damned silly. But if we'd got a positive result, they wouldn't have been silly. But he never hesitated . . . He would say things which

> put down on paper were stupid or would have been stupid. But when one thought about them, you began to see that those words were inadequate to express what was in his mind . . . There was always just the possibility of something turning up, and one shouldn't neglect doing, say, a few hours' work or even a few days' work to make quite sure.[6]

Chadwick was so busy with administration and supervision of research students that the only real opportunity to work at the laboratory bench himself for more than an hour or so came in the university vacations or by breaking Rutherford's house rule and staying in the laboratory after 6 pm. Another of his research students in the early thirties, W.E. Duncanson,[26] particularly remembered one occasion when Chadwick announced at 7.30 pm: 'I must go home as we have some people coming to dinner, but I shall be back later'. He returned at 9.30–10 pm and worked for several more hours; Aileen's views were not relayed.

In June 1929, Webster[22] sent a report of his progress to the Commissioners of the 1851 Exhibition and already concluded that 'hard gamma rays are excited when α-rays bombard matter'. He also mentioned that he was working on improving the Geiger–Müller counter by linking it to an automatic recording device so that kicks from a string electrometer were registered on moving paper strip by a magnetically operated pen [rather in the fashion of a modern electrocardiograph trace]. The technology of automatic particle counting became a preoccupation at the Cavendish in these years,[20] and Chadwick wholeheartedly supported it because he recognized that standardized, objective methods were necessary to make any further scientific headway. An amateur radio enthusiast (and 1851 Scholar), C.E. Wynn-Williams led a small group who applied their familiarity with 'wireless' valves to this problem. It had been shown by Greinacher working in Bern that the ionization due to a single fast particle (α-particle or proton) could be electronically amplified and thereby detected. Wynn-Williams and his team at the Cavendish improved on this basic technique and constructed a cascade of valves connected in series, a linear amplifier, which would detect the ionization caused by a single α-particle and record up to 500 α-particles per minute. Their equipment was exquisitively sensitive to electromagnetic or mechanical vibration, including Rutherford's booming voice, which led to the erection of the famous illuminated sign, 'TALK SOFTLY PLEASE'. This was more diplomatic than a prototype put up by another of Chadwick's research students, Jack Constable, which bluntly read 'Shut Up, Damn You'![26]

Constable and his partner, Ernest Pollard,[27] were charged by Chadwick to construct a linear amplifier like Wynn-Williams', but which could be used for counting protons rather than α-particles. This seemingly simple task took them six months 'with steady and patient pushing from Chadwick'. Pollard remembered it as a time of confusion, excitement and vigorous discussion, in which Chadwick participated with good humour—'rather of the quiet and canny kind, but with the occasional streak of mischief'. He was also 'splendid at shooting down a flight of enthusiasm, generally engendered by reading . . . some article. Chadwick had always read it and knew precisely the flaws'. Apart from Pollard and Constable, Chadwick had several other research students to supervise and would visit all two dozen or so workers in the laboratory every day to monitor progress and decide on what extra equipment needed to be ordered. Many of the research students would go on to lead distinguished careers themselves, and even the lesser known ones would become competent and accomplished scientists. In Pollard's opinion,[27] Chadwick played a vital role in all their development: 'he imparted a sense of excitement. Watching him operate you realised that if you made sure you knew what you were doing, absolutely sure, then what would result from your experiments would certainly reveal something new about Nature.'

In addition to overseeing research activity in the Laboratory, Chadwick watched the scientific literature like a hawk, partly to be armed with enough ammunition to keep the postgraduates on their toes and also to be able to brief Rutherford in their daily meetings so that the Cavendish would not overlook any promising avenue for further research. Like a true Rutherford disciple he always found time to think, however busy his day, and the intensity of his concentration brooked no interference. Duncanson[26] remembered:

> Often when I went into his room to ask a question or to discuss some matter he would completely ignore me, continuing at his little table doing a calculation or writing some notes. After five, ten or even fifteen minutes I would depart and there would still be no recognition that I had been in the room. On other occasions he would note my presence, in which case I would say my piece but get no response; again I would make myself scarce. Later—maybe several hours later—he would come out of his room and discuss the point I had raised without any recognition of a lapse of time.

Not surprisingly, Chadwick gained a reputation with some for being unapproachable and even rude, but according to Duncanson[26] 'the more one worked with Chadwick the more one appreciated his encouragement, friendliness and concern for those working under him'. Pollard[28] was another junior colleague who passed beyond the hard outer shell and found Chadwick to be a very sympathetic man: 'while he had the reputation of being dour and humourless, I knew a lot better'. During the long summer term, the dons at Gonville and Caius would challenge the research students to tennis and then give them dinner in the college. One research student of the period remembered Chadwick taking part in the tennis and then playing bridge after dinner, in a competition where the dons invariably trounced the students.[29]

A glimpse of Chadwick's way of handling junior research staff comes from a letter to Norman Feather, who having completed his Ph.D. in Cambridge, spent the academic year 1929–1930 at Johns Hopkins University in Baltimore. By the spring of 1930 he was sounding Chadwick out about a return to the Cavendish and received the following reply:[30] 'I shall have to find a small corner for you. Don't expect an office as well or a spittoon or a rack for chewing gum or any of the modern labour saving conveniences'.

Apart from the good natured banter, Chadwick made serious use of Feather's presence in the USA, and added a PS to his letter as follows:

> This is the sting. You have, I suppose, read the papers of Davis and Barnes [two Americans] in the Physics Review on the capture of electrons by alpha particles. The whole affair is most mysterious and to me incomprehensible.

He asked Feather to arrange a visit to Davis' laboratory to have a look at the experimental arrangement and 'see if he knows what he is doing'. Remembering his own sleuthing in Vienna and the unpleasant personal exchanges that followed, Chadwick ended apologetically: 'This is a nasty job to give you but it is really most important to know whether the experiments are right or wrong'. Two months later, Chadwick[31] wrote again with another request. 'It would be a marvellous stroke of business if you could collect a large quantity of radium D etc. The larger the better, of course'. Radium D etc. meant polonium. While the original present from Lise Meitner had been invaluable in helping the Cavendish modify its experimental paradigm, Chadwick could see by his involvement with the work of Webster and of Constable and Pollard that it was of insufficient activity to produce clearcut results in studies of the nucleus. Feather, anxious to please, exceeded Chadwick's wildest hopes. He[32] approached the authorities at the Kelly Hospital in Baltimore about their used radon seeds. He was aided by Dr West, a fellow Yorkshireman, who was in charge of radioactive materials at the hospital and came away at the end of his year with several hundred spent seeds containing as much polonium as any physicist outside Mme Curie's institute had ever had access to.

In the meantime, Webster soldiered on with his Ph.D. using the original weak polonium source prepared by Chadwick. He was also experiencing problems with equipment: 'A few experiments were made with counters very kindly sent over by Professor Geiger, but as these ceased to function satisfactorily after a time, new counters had to be constructed'.[33] Despite these difficulties, Chadwick set him the additional task of attempting to repeat the findings of Davis and Barnes, which he had found 'most mysterious' and 'incomprehensible'. It is not surprising that both parts of Webster's thesis, written in September 1930, (the hard γ-radiation emitted by elements bombarded by α-particles and the capture of electrons by α-particles) were inconclusive. His external assessor, O.W. Richardson*, commented 'I think this scholar has been very unfortunate . . . The reality of the radiation which was the objective of his first reasearch was disputed and he set about to make experiments to settle it. Unfortunately, the radioactive material at his command seems to have been inadequate to lead to a definite conclusion. However this may be, the point seems to have since been settled by some German experimenters'.[33]

The German experimenters were Walther Bothe and his student Herbert Becker. Bothe and Chadwick were the same age and met for the first time at the Reichsanstalt in 1914. At the outbreak of war, Bothe joined the German Army and was later taken prisoner by the Russians and sent to Siberia. Not able to conduct experiments (unlike Chadwick in Ruhleben) Bothe nevertheless studied mathematics and theoretical physics in the prison camp. According to Segrè,[34] 'he was somewhat difficult as a person, but his sterling intellectual and scientific integrity commanded

* Owen W. Richardson (1879–1959), Wheatstone Professor of Physics at King's College, London and 1929 Nobel laureate for work on thermionics.

great respect in the physics community'. He returned to Berlin in 1920 with his Russian wife and became involved in the development of electrical counting for the detection of subatomic particles.

Bothe's scientific work in the 1920s mirrored Chadwick's in many aspects, and he published important papers on the scattering of α- and β-particles. The two men met for the first time since the pre-war Reichsanstalt at the Cambridge conference of 1928. Bothe had just completed a study of the energy relations during artificial disintegration; no doubt Chadwick told him of his own work on this topic, carried out a few years earlier but not published until 1929 with Rutherford. Both men were fascinated by the possibility of disintegration protons occurring in groups of different ranges; Bothe thought he had evidence for nuclear disintegration without capture of the α-particle, whereas Chadwick with Constable and Pollard[35] would later find no evidence for it.

In 1928, Bothe was setting out on his project with Becker in which they essentially repeated Rutherford and Chadwick's earlier work on the disintegration of light elements by α-particles. Like Webster, they were looking for evidence of the concomitant emission of high-energy γ-rays during this process. Unlike Webster, they had a good polonium source which enabled them to make definite observations. Bothe and Becker found γ-radiation from the bombardment of aluminium, boron and magnesium (all elements which Rutherford and Chadwick had shown to be disintegrable with the emission of protons). More surprisingly, they found penetrating radiation from lithium and beryllium which did not, in the Cavendish at least, yield protons on bombardment with α-particles. The radiation from beryllium was of particularly high energy, higher perhaps than that of the bombarding α-particles, so that Bothe thought it must come from the disintegration of the beryllium nucleus despite the absence of detectable protons.

Chadwick[24] had been intrigued by beryllium for some years, believing it might be possible for it to be split into two α-particles and a neutron. He had bombarded it 'with alpha-particles, with beta-particles and with gamma-rays' without any noticeable effect. Now he prepared a much stronger polonium source and asked Webster, who had not previously looked at the case of beryllium, to repeat Bothe and Becker's experiment. Webster was able to replicate their findings and added a crucial observation of his own with a beryllium target. In Chadwick's words,[6] Webster 'found that the radiation emitted in the same direction as the alpha particles was harder [more penetrating] than the radiation that came backwards. And that, of course, was a point which excited me very much indeed, because I thought, "Here's the neutron" '.

The reason for Chadwick's excitement was that any emission of γ-ray energy from the beryllium nucleus should be the same in all directions, but if particles were being given out, their kinetic energy would be greater in the forward direction than backwards. Furthermore, they would have to be uncharged or neutral particles in order to show the great penetration through lead that Webster had measured. Although Chadwick[36] explained his thoughts and made 'some rough calculations', he did not think Webster 'realised how peculiar the results with beryllium were'. Chadwick thought that the neutron, despite being uncharged, might be very sparsely ionizing. He therefore suggested that Webster should pass the secondary radiation from beryllium through an expansion (Wilson cloud) chamber and look for very thin tracks.

Chadwick[6] could not be present while this was done because he had promised to take Aileen and the girls on holiday to Anglesey; the family summer holiday was always from the middle of June to the middle of July, and was a necessary break from his 'rather absorbing' duties at the Cavendish. In his absence, Webster teamed up with F.C. Champion, Blackett's postgraduate research student, who had been using a cloud chamber to photograph the thin tracks of fast β-particles. They set Webster's apparatus on top of Champion's expansion chamber and took 50 photographs with no sign of any particle tracks originating from the bombarded beryllium. Chadwick[36] was not impressed: 'When I got back the photos were lost or destroyed. I never saw them. I was extremely annoyed'. Webster many years later said that he and Champion suspected that the shock waves from the expansions in the cloud chamber caused a small displacement between the polonium source and the beryllium.[20] For some reason they never admitted this to Chadwick, who remained 'still peeved about it',[37] thirty years later.

The failure to find the elusive neutron was satirized at a show put on by younger members of the laboratory.[26] It was billed as a meeting of the 'Phoolosophical Society', and one research student delivered a paper on 'The Fewtron'. The fewtron had few if any properties, but if no tracks appeared on a photographic plate after exposure to a Wilson cloud chamber, then you knew that a fewtron had passed through the chamber!

These events took place in June 1931 and Webster left Cambridge that summer to take up a position as a lecturer in Bristol. Chadwick intended to obtain expansion chamber photographs himself—so sure was he that here was the neutron—but the opportunity did not present itself in the next six months. Webster wrote up his own work and Chadwick communicated the paper to the Royal Society in the middle of January 1932. Webster[38] mentioned the possibility that his observations were due to 'high-speed corpuscles consisting, e.g., of a proton and an electron in very close combination', but in view of the negative expansion chamber results held to the conclusion that the secondary radiations were electromagnetic in nature.

About a week after sending off Webster's paper, Chadwick received his copy of the French journal *Comptes rendus*. In it was a paper from Frédéric Joliot and his wife Irène Curie setting out their own experience with the mysterious penetrating radiation from beryllium. In the tradition of the Curie laboratory, both the Joliot-Curies were skilled radiochemists, and in 1931 had described their success in preparing a very strong source of polonium, amounting to approximately 200 millicuries.[39] Irène, Marie's daughter, had studied the properties of polonium for several years; she was adept at preparing sources for experiments—difficult and hazardous work, which she continued to do in late 1931 despite being pregnant. With their powerful polonium source, the Joliot-Curies were able to confirm the phenomenon of Bothe and Becker without any difficulty, and like Webster added a tantalizing observation of their own. When they passed the radiation emitted from beryllium through paraffin wax, it increased in intensity. The reason was that protons were being knocked out of the wax and entered the ionization chamber being used for measurement. They went on to determine the range of the protons and concluded that they had an energy of up to 4.5 million electron volts (MeV). By analogy with the Compton effect which describes the collision of X-rays with electrons, the Joliot-Curies postulated that in order to produce protons of such momentum, the γ-radiation from beryllium must have an energy of 55 MeV—much higher than any radiation previously encountered. It would need to be so high because the proton is about two thousand times heavier than an electron.

Chadwick had first met the Joliot-Curies at the 1928 Cambridge meeting; he and Aileen had entertained them to lunch at home. Frédéric was debonair, handsome, entertaining and smoked incessantly. Chadwick found him charming, but seems to have teased him about his name. Like all schoolboy jokes, this one lost something in translation: Joliot[40] was 'puzzled to find that Chadwick referred to me as "Curie-Joliot". He was the only one to do so, but I attributed this to a curious, not deliberate, attitude, which I could not really understand, on his part'. Irène was 'much more reserved. Indeed it was a little difficult to know her properly', according to Chadwick,[6] who admitted: 'She might have said the same thing about me. But on the whole we got along very well'. At the lunch, Irène caused consternation to the Chadwick's domestic servant, who was waiting at table, by her habit of breaking up bread, putting some in her mouth and throwing the rest over her shoulder.

The French research was published in *Comptes rendus* on 18 January 1932. Chadwick[24] vividly remembered the morning when his copy of the journal arrived and he read the paper. A few minutes after reading the paper,

> Feather came to my room to tell me about this report as astonished as I was. A little later that morning I told Rutherford. It was a custom of long standing that I should visit him about 11 am to tell him any news of interest and to discuss the work in progress in the laboratory. As I told him about the Curie-Joliot observations and their views on it, I saw his growing amazement; and finally he burst out 'I don't believe it'. Such an impatient remark was utterly out of character, and in my long association with him I recall no similar occasion.

Years later, Chadwick[41] told Oliphant privately that Rutherford had 'refused to believe the Curie-Joliot observations' at first, but this was a 'temporary aberration'. From the first, Chadwick had no doubt about the veracity of the observations nor that the Joliot-Curies were mistaken in their explanation of the events. To him it was obvious that the protons were being ejected as a result of neutrons hitting the wax target. The heavy neutrons would only need a kinetic energy slightly higher than that of the protons they set in motion. He needed to repeat the observations and to establish that the mysterious 'radiation' from beryllium also gave rise to elastic collisions with nuclei of other elements, not merely in hydrogenous materials. He also had to move quickly, if he was going to be the first to prove the existence of the elusive particle which had fascinated him for a decade. The Joliot-Curie paper had caused a major stir in the still small international world of nuclear physics. At least one other scientist, Ettore Majorana, working with Fermi in Rome, shared Chadwick's interpretation of the findings. 'What fools', he is reported as saying, 'They have discovered the neutral proton and they do not recognize it.'[34]

Chadwick[24] dropped all other responsibilities and started the most intensive research of his career:

> It so happened that I was just ready to begin experiment, for I had prepared a beautiful source of polonium from the Baltimore material. I started with an open mind, though naturally my thoughts were on the neutron. I was reasonably sure that the Curie-Joliot observations could not be ascribed to a kind of Compton effect, for I had looked for this more than once. I was convinced there was something quite new as well as strange.

The polonium source, deposited on a disc of silver, was mounted inside a small, matt black cylindrical tube which also contained the beryllium target. The apparatus resembled the boiler of a toy steam engine with an elongated chimney through which all air was evacuated. The stream of 'radiation' could then be passed through a thin target of wax or other material, and any resultant particles knocked forwards into an ionization chamber connected to the valve amplifier that Constable and Pollard had previously constructed. The output from the linear amplifier was detected by an oscillograph, which gave a series of traces which were recorded photographically. As Blackett[42] pointed out, Chadwick's experiment was 'truly in the Rutherford tradition and needed little apparatus but much inspiration and physical intuition'. With Rutherford's permission, Chadwick did much of the work at night to avoid interference to the notoriously sensitive amplifier. C.P. Snow,[43] a research student in the department at the time, recalled:

> He worked night and day, for about three weeks. The dialogue passed into Cavendish tradition:
>
> 'Tired, Chadwick?'
>
> 'Not too tired to work.'

At the end of this heroic effort, Chadwick[44] had shown that the radiation from beryllium 'ejects particles from hydrogen, helium, lithium, beryllium, carbon, air and argon'. The particles from hydrogen were obviously protons, whereas those from the other light elements tested were of shorter range but highly ionizing. Chadwick concluded that they were in fact recoil atoms of each element. In collaboration with Feather, he studied the tracks of these heavy recoil atoms in an expansion chamber. By 17 February, almost exactly a month after first reading the Joliot-Curie paper, Chadwick was ready to send a letter, *Possible Existence of a Neutron*, to *Nature**. The letter is not only 'a model of clear physical thinking',[42]

---

* See Appendix I.

but remarkable for its concision. After setting out his results, Chadwick pointed out that they 'are very difficult to explain on the assumption that the radiation from beryllium is a quantum radiation, if energy and momentum are to be conserved in the collisions'. The next sentence brings triumph: 'The difficulties disappear, however, if it be assumed that the radiation consists of particles of mass 1 and charge 0, or neutrons'.

There was a general air of excitement in the Cavendish laboratory in February 1932, and the first announcement of the discovery of the neutron came at a meeting of the Kapitza Club on the evening of the 23rd. The Club had been started by Kapitza in 1922 and membership was by invitation. It was a lively and informal group, which met in Kapitza's rooms in Trinity College once a week. No full record of the meetings was kept, but there was a small book in which the speaker signed and wrote out the title of his talk. The page for February 1932 is shown in Fig. 6.1. There were often several talks given on the same evening, and the list of previous speakers included many distinguished visitors, such as Heisenberg, Gamow and Millikan in addition to Cavendish physicists. With Kapitza in the chair, there was always a note of levity, reflected in the anonymous limerick composed in the minute book in 1925.

> In discussion, our chief, P. Kapitza,
> Goes off suddenly, like a howitzer,
> He's been known to say: 'No!'
> It is not at all so,
> It is true, as you're bound to admit, Sir!

Mark Oliphant[45] was a member of the expectant audience that night and described the occasion.

> Kapitza had taken him to dine in Trinity beforehand, and he was in a very relaxed mood. His talk was extremely lucid and convincing, and the ovation he received from the select audience was spontaneous and warm. All enjoyed the story of a long quest, carried through with persistence and vision, and they rejoiced in the success of a colleague. Chadwick's meticulous recognition of the parts played by others in pointing the way was a lesson to us all.

For once Chadwick spoke without inhibition, but the title he entered in the Kapitza Club book for his presentation, *Neutron?*, still bore his characteristic mark of caution.

Flushed with success and Trinity port, he asked at the end of his talk 'to be chloroformed and put to bed for a fortnight'.[43] He arose the next morning and wrote to Bohr, informing him of the discovery and sending a copy of his communication to *Nature.* In reference to the tentative title of the letter, *Possible Existence of a Neutron,* Chadwick[46] had this to say:

> I put this forward rather cautiously, but I think the evidence is really rather strong. Whatever the radiation from Be may be, it has most remarkable properties. I have made many experiments which I do not mention in the letter to 'Nature' and they can all be interpreted readily on the assumption that the particles are neutrons.

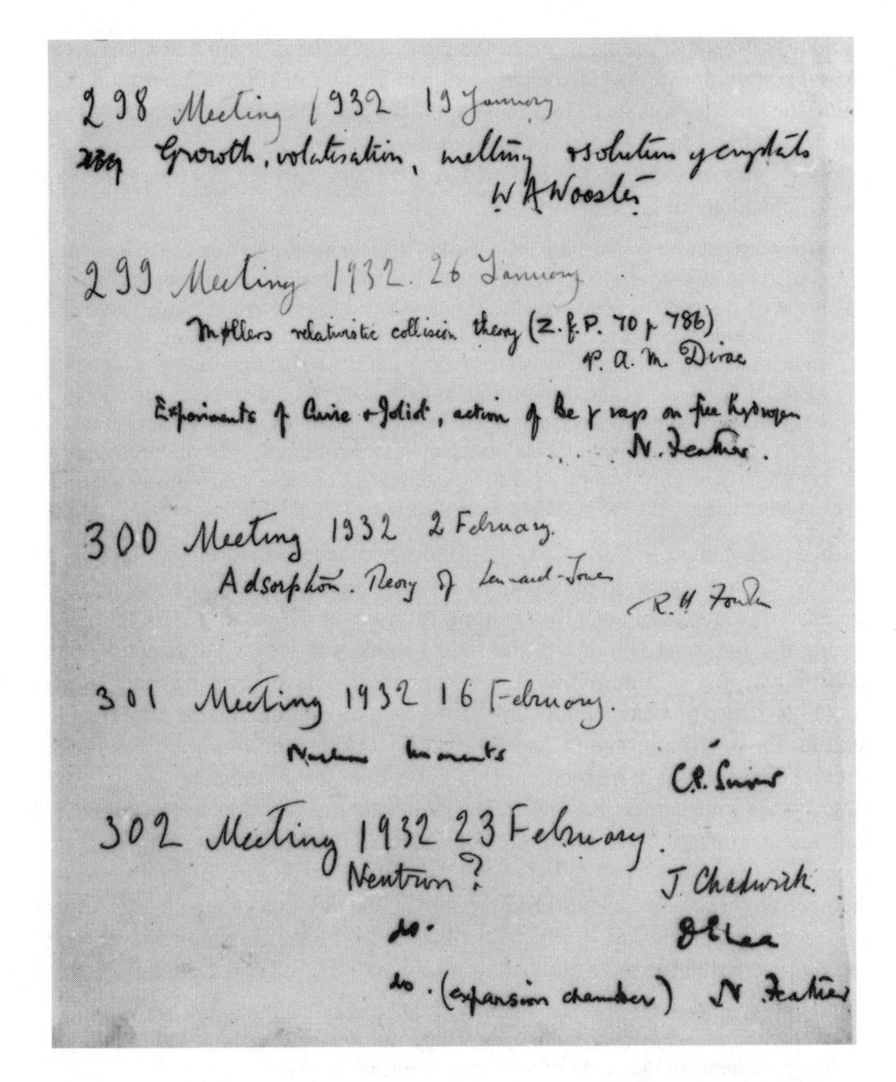

298 Meeting 1932 19 January
Growth, volatisation, melting & solution of crystals
W A Wooster

299 Meeting 1932. 26 January
Møllers relativistic collision theory (Z. f. P. 70 p 786)
P. A. M. Dirac

Experiments of Curie & Joliot, action of Be γ rays on free hydrogen
N. Feather.

300 Meeting 1932 2 February
Adsorption. Theory of Lennard-Jones
R. H. Fowler

301 Meeting 1932 16 February
Nuclear moments
C. P. Snow

302 Meeting 1932 23 February
Neutron?
J. Chadwick.
do.
D E Lea
do. (expansion chamber) N. Feather

Fig. 6.1 The page for February 1932 from the notebook recording the weekly meetings of the Kapitza Club.

> Feather has taken some pictures in the expansion chamber and we have already found about 20 cases of recoil atoms. About 4 of these show an abrupt bend or fork and it is almost certain that one arm of this fork represents a recoil atom and the other some other particle, probably an α-particle.

It is a commonplace observation that most great discoveries are made when their time is ripe: if Chadwick had not discovered the neutron in February 1932, the argument runs, someone else would have done so before long. That Bothe and Becker, Webster, and the Joliot-Curies had already obtained neutrons, without realizing it, seems to make the case. But none of them had Chadwick's mental preparedness. As Chadwick[47] wrote to Feather in 1968: 'The reason that I found the neutron was that I had looked, on and off, since about 1923 or 4. I was convinced that it must be a constituent of the nucleus'.

Joliot[40] was understandably annoyed at having missed the discovery, but was philosophical about it.

> The word neutron had already been used by the genius Rutherford in 1923, at a conference, to denote a hypothetical neutral particle which, together with protons, made up the nucleus. This hypothesis had escaped the attention of most physicists, including ourselves. But it was still present in the atmosphere of the Cavendish Laboratory where Chadwick worked and it was natural—and just—that the final discovery of the neutron should have been made there. Old laboratories with long traditions always have hidden riches. Ideas expressed in days gone by by our teachers, living or dead, taken up a score of times and then forgotten, consciously or unconsciously penetrate the thought of those who work in these old laboratories and, from time to time, they bear fruit: that is discovery.

Chadwick[36] cited the fact that 'the Joliots had never heard of the idea of the neutron' to correct the impression, which grew up later, that he was merely at the head of a queue of scientists waiting to discover the particle. 'As they were among the foremost workers in the field I think it is clear that the idea of the neutron was not so common as is now supposed to have been the case'. It was not in his nature to make an outward display of exhilaration, but the climax of the hunt for the neutron brought Chadwick intense excitement. For a few minutes or hours, he was the only man on earth to know one of the fundamental secrets of nature—a privilege rarely bestowed. The intellectual satisfaction was profound and long lasting; nevertheless, he 'could not help but feel that I ought to have arrived sooner . . .[and] that I had failed to think deeply enough about the properties of the neutron, especially about those properties which would most clearly furnish evidence of its existence'.[24] Other scientists judged his efforts more favourably, and Segrè[34] wrote the most eloquent testimonial.

> To his great credit, when the neutron was not present he did not detect it, and when it ultimately was there, he perceived it immediately, clearly, and convincingly. These are the marks of a great experimental physicist.

The discovery of the neutron was not only a revolutionary event in physics; it would in time change the course of world history.

References

The notebook recording the weekly meetings of the Kapitza Club is to be found in the Churchill Archives Centre's collection of Sir John Cockcroft's papers (CKRFT 7/2). The page for February 1932 is reproduced by kind permission of Christopher Cockcroft and the Master and Fellows of Churchill College, Cambridge. Chadwick's letters to Bohr are quoted with the permission of the Niels Bohr Archive in Copenhagen.

1 Ruther-
ford, E. and Chadwick, J. (1927). The scattering of α-particles by helium. *Philosophical Magazine*, 4, 605–20.

2 Frisch, O.R. (1979). *What little I remember*, p. 63. Cambridge University Press.

3 Chadwick, J. (1954). The Rutherford Memorial Lecture, 1953. *Proceedings of the Royal Society A*, 224, 435–47.

4 Rutherford, E. (1929). Discussion on the structure of atomic nuclei. *Proceedings of the Royal Society A*, 123, 373

5 Pais, A. (1986). First encounters with symmetry and invariance. In *Inward bound*, pp. 265–95. Oxford University Press.

6 Weiner, C. (1969). Sir James Chadwick, oral history. American Institute of Physics, College Park, Maryland .

7 Dirac, P.A.M. (1983). The origin of quantum field theory. In *The birth of particle physics* (ed. L.M. Brown and L. Hoddesdon), pp. 39–55. Cambridge University Press.

8 Pais, A. (1991). The spirit of Copenhagen. In *Niels Bohr's times*, pp. 295–323. Oxford University Press.

9 McCrea, W. (1987). Cambridge 1923–6: undergraduate mathematics. In *The making of physicists* (ed. R. Williamson), pp.53–65. Adam Hilger, Bristol.

10 McCrea, W. (1985). How quantum physics came to Cambridge. *New Scientist* (17 Oct), 58–60.

11 Mott, N.F. (1986). *A life in science*, pp 30–1. Taylor and Francis, London.

12 Chadwick, J. (1930). The scattering of α-particles in helium. *Proceedings of the Royal Society A*, 128, 114–22.

13 Chadwick, J. (22/4/30). Letter to N. Feather. Feather's papers, CAC.

14 Cockcroft, J. (1984). Some recollections of low energy nuclear physics. In *Cambridge physics in the thirties* (ed. J. Hendry), pp. 74–80. Adam Hilger, Bristol.

15 Chadwick, J. and Gamow, G. (1930). Artificial disintegration by α-particles. *Nature*, 126, 54–5.

16 Rutherford, E., Chadwick, J. and Ellis, C. D. (1930). *Radiations from radioactive substances*, p. 575. Cambridge University Press.

17 Chadwick, J. (1928). Correspondence with Lise Meitner. Meitner collection, 5/3, CAC.

18 Trenn, T.J. (1986). The Geiger–Müller counter of 1928. *Annals of Science*, 43, 111–35.

19 Chadwick, J. (1928). Unpublished laboratory notebook. CHAD III, 1/4, CAC

20 Sargent, B.W. (1985). On the fiftieth anniversary of major discoveries at the Cavendish Laboratory. *American Journal of Physics*, 53, 208–20.

21 Chadwick, J. and Russell, A.S. (1912). Excitation of gamma rays by alpha rays. *Nature*, 90, 463.

22 Webster, H.C. (1929). Unpublished report to Royal Commisioners to the 1851 Exhibition. In Webster's unpublished notes, 1851 Archives.

23 Rutherford, E. and Chadwick, J. (1929). Energy relations in artificial disintegration. *Proceedings of the Cambridge Philosophical Society*, 25, 186–92.

24 Chadwick, J. Undated typed notes. CHAD IV, 11/51. CAC.

25 Chadwick, J. (1964). Some personal notes on the search for the neutron. *Proceedings tenth international congress on the history of science, Ithaca 1962*, 1, 159–62.

26 Duncanson, W.E. (1984). Reminiscences 1930–4. In *Cambridge physics in the thirties* (ed. J. Hendry), pp. 90–4. Adam Hilger, Bristol.

27 Pollard, E. (1991). Neutron pioneer. *Physics World*, Oct., 31–33.

28 Pollard, E. (1990). Recollections of Chadwick. In *Science and people*. Leaflet, obtainable from Woodburn Press, PO Box 348, Lemont, PA 16851.

29 Lewis, W.B. (1979). Letter to E.C. Pollard, lent to the author.

30 Chadwick, J. (22/4/30). Letter to N. Feather. Feather's papers, CAC.

31 Chadwick, J. (9/6/30). Letter to N. Feather. Feather's papers, CAC

32 Feather, N. (1984). The experimental discovery of the neutron. In *Cambridge physics in the thirties* (ed. J. Hendry), pp. 31–41. Adam Hilger, Bristol.

33 Webster, H.C. (1930). Unpublished report to Royal Commisioners to the 1851 Exhibition. In Webster's unpublished papers, 1851 Archives.

34 Segrè, E. (1980). The wonder year 1932. In *From x-rays to quarks*, pp. 177–97. W.H.Freeman, New York.

35 Chadwick, J., Constable, J.E.R. and Pollard, E.C. (1931). Artificial disintegration by α-particles. *Proceedings of the Royal Society A*, 130, 463–89.

36 Chadwick, J. (22/10/59). Letter to N. Feather. Feather's papers, CAC.

37 Chadwick, J. (11/8/62). Letter to N. Feather. Feather's papers, CAC.

38 Webster, H.C. (1932). The artificial production of nuclear γ-radiation. *Proceedings of the Royal Society A*, 136, 428–53.

39 Blackett, P.M.S. (1960). Jean Frédéric Joliot, 1900–1958. *Biographical memoirs of fellows of the Royal Society*, 6, 87–101.

40 Goldsmith, M (1976). *Frédéric Joliot-Curie*. Lawrence and Wishart, London.

41 Chadwick, J. (?1971). Undated notes on Oliphant's biography of Rutherford. CHAD II, 2/1, CAC.

42 Blackett, P.M.S. (1962). Memories of Rutherford. In *Rutherford at Manchester* (ed. J. Birks), pp. 102–13, Heywood & Co., London .

43 Snow C.P. (1982). *The physicists*, p. 85. Papermac, London.

44 Chadwick, J. (1932). Possible existence of a neutron. *Nature*, 129, 312.

45 Oliphant, M.L. (1982). The beginning: Chadwick and the neutron. *Bulletin of the Atomic Scientists*, 14–18, Dec.

46 Chadwick, J. (24/2/32). Letter to N. Bohr. Niels Bohr Archive, Copenhagen.

47 Chadwick, J. (25/7/68). Letter to N. Feather. Feather's papers, CAC.

7 International renown

While the advent of the neutron did not bring its discoverer instant fame, it did make front page news—in the United States at least. The leading story in Great Britain was the setting of the new land speed record by Sir Malcolm Campbell; in terms of university news, more space was given in *The Times* to the Cambridge athletics meeting and preparation for the University Boat Race than to Chadwick's find. The most informative account in the British press appeared, under the general heading of 'Our London Correspondence', in the *Manchester Guardian* on Saturday 27 February 1932. It was written, anonymously, by their science correspondent J.G. Crowther, who was friendly with Chadwick and had been invited to the meeting of the Kapitza Club on the previous Tuesday evening. Crowther told his readers that other physicists regarded it as 'one of the fundamental discoveries in physics and ranks with the discovery of the electron, the proton, and the X-rays. It is hailed as the most important achievement in experimental physics since Lord Rutherford* demonstrated the nuclei *(sic)* structure of the atom in 1911'. In Crowther's judgement, Chadwick 'now joins Niels Bohr and Moseley as the most brilliant of Lord Rutherford's Manchester pupils'. After pointing out to 'the impatient' that there was no apparent direct application for neutrons in medicine yet, Crowther thought their practical use would doubtless be discovered before long. He ended his well written account with the following comments which accurately reflected Chadwick's view that the neutron was not a pure fundamental particle, but an amalgam of proton and electron.

> The neutron represents the first step in the evolution of matter, the first step in the building up of the common materials of everyday life out of the primeval electrons and protons, the first stage in the growth of order out of what Milton called 'chaos and old night.'

Chadwick gave several interviews to the press over the weekend and on Monday 29 February the front page of *The New York Times* carried the headline 'Chadwick Calls Neutron "Difficult Catch".' In his interview with their correspondent, Chadwick had been at his cautious best, refusing 'to make any sweeping claims' and preferring 'to understate his own case in the interest of scientific accuracy'. Chadwick described his experiments as convincing, but planned further

---

* He had become Baron Rutherford of Nelson in the New Years Honours List for 1931.

research at the Cavendish Laboratory 'until the chain of evidence becomes unbreakable'. One prominent American who was not at all convinced was Robert A. Millikan—President of the California Institute of Technology, the 1923 Nobel laureate in physics, and the man who first proposed the term 'cosmic rays'. Millikan, speaking from Pasadena, referred to the work of Bothe and still favoured the German scientist's original interpretation that the very penetrating radiation obtained from the bombardment of beryllium with α-particles consisted of γ-rays rather than neutrons. In principle he thought Bothe's explanation was identical with his own interpretations of cosmic ray phenomena, and naturally preferred it. Although a man of strong convictions, Millikan did allow that 'Chadwick is one of the foremost of our living experimental physicists and his interpretation should therefore be given much weight'.

Lord Rutherford was quick to endorse Chadwick's discovery. Speaking at the Royal Institution on 18 March,[1] he traced the history of the neutron back to his 1920 Bakerian Lecture and then described the confirmatory studies going on at the Cavendish under Chadwick's supervision. At the end of March, Chadwick received a delayed reply from Bohr[2] thanking him for his letter about the neutron and the *Nature* article, and inviting him to a small informal meeting to start in Copenhagen on 7 April so that he could give a first hand account of his experiments. Chadwick[3] had arranged to take a short holiday with Aileen, so declined the invitation, but the neutron was a major topic of discussion at the conference anyway. In his address at the meeting, Bohr[4] was at pains to point out that the current theories of the nucleus were inadequate and could not advance understanding in their present form.

> Just as little as it is possible at the present stage of atomic mechanics to account in detail for the stability of ordinary nuclei, it is impossible at present to offer a detailed explanation of the constitution of the neutron. Of course its mass and charge suggest that a neutron is formed by a combination of a proton and an electron, but we cannot explain why these particles combine in such a way as little as we can explain why 4 protons and 2 electrons should combine to form a helium nucleus or α-particle.

Ralph Fowler, an inveterate traveller, had attended the Copenhagen meeting and briefed his father-in-law about it on his return to Cambridge. Rutherford[5] immediately wrote to Bohr concerning the neutron and other momentous events at the Cavendish.

> April 21st 1932
>
> My dear Bohr,
>
> I was very glad to hear about you all from Fowler when he returned to Cambridge and to know what an excellent meeting of old friends you had. I was interested to hear about your theory of the Neutron. I saw it described very nicely by the scientific correspondent of the Manchester Guardian, who is quite intelligent in these matters. I am very pleased to hear that you regard the Neutron with favour. I think the evidence in its support, obtained by Chadwick and others, is now complete in the main essentials. It is still a moot point how much ionization is, or should be, produced to account for the absorption, disregarding the collisions with nuclei.
>
> It never rains but it pours, and I have another interesting development to tell you

about of which a short account should appear in Nature next week. You know that we have a High Tension Laboratory where steady D.C. voltages can be readily obtained up to 600,000 volts or more. They have recently been examining the effects of a bombardment of light elements by protons. The protons fall on a surface of the material inclined at 45 degrees to the axis of the tube and the effects produced were observed by the scintillation method, the zinc sulphide screen being covered with sufficient mica to stop the protons. In the case of lithium brilliant scintillations are observed, beginning at about 125,000 volts and mounting up very rapidly with voltage when many hundreds can be obtained with a protonic current of a few milliamperes. The α-particles apparently had a definite range, practically independent of voltage, of 8 cms, in air. The simplest assumption to make is that the lithium 7 captures a proton breaking up with the emission of two ordinary α-particles. On this view the total energy liberated is about 16 million volts and this is of the right order for the changes in mass involved, assuming the Conservation of Energy.

'They', as Bohr well knew, were Cockcroft and Walton, and their achievement was to split the lithium nucleus using a stream of protons produced by a machine—the first example of artificial disintegration of the atom. The reaction can be represented as:

$$^{1}p + {}^{7}Li \rightarrow 2\alpha(2^{4}He)$$

Chadwick[6] had been urging Rutherford to explore the possibility of building a high-voltage accelerator at the Cavendish since the early twenties, but Rutherford, probably wisely, resisted for a number of years. He judged that the small scale apparatus that he and Chadwick employed was still useful, and there were formidable technical problems to be overcome in getting high-energy machines to work. A major difficulty was to produce the high voltage in a highly exhausted vacuum tube so that accelerated particles would not be scattered by air. By 1927, Rutherford[7] decided that the necessary technology was at hand and in his annual address to the Royal Society called for such machines that would provide a copious stream of high-speed electrons and atoms 'which have an individual energy far transcending that of the α- and β-particles from radioactive bodies'. In fact the Cockcroft–Walton experiment was accomplished with protons accelerated to quite modest energies. Cockcroft[8] had realized this would be possible after hearing Gamow lecture at the Cavendish in 1928 about the escape of relatively low-energy α-particles through the potential barrier around the nucleus. Cockcroft reasoned that if quantum mechanics provided tunnels for particles emerging from the nucleus, then they should be able to enter in the other direction—and so it proved. He and Walton laboured for about two years building their apparatus, sealing the joints by applying Plasticine* with their thumbs to ensure the necessary vacuum. Cockcroft[9] recalled the moment of fruition:

* Plasticine had replaced red Bank of England sealing wax, which had been widely used as a vacuum cement until a few years before. Blackett in a 1933 essay on *The craft of experimental physics* (In *University studies*, {ed. H. Wright}, Ivor Watson & Nelson.) commented on this change, saying, 'it is hardly possible to exaggerate the controlling importance of such simple technical matters, however trivial they may seem to be'.

> At last we obtained a high energy proton beam and brought it out into the air through a window to check on its energy and range in air. After wasting a certain amount of time in this way, until prodded by Chadwick and Rutherford, we directed it on to a lithium target and at once observed, with a zinc sulphide screen, the bright scintillations which were obviously due to particle emission from lithium.

Rutherford had been showing his usual impatience with the time it took to get apparatus working, and his pressure probably led to the 'time wasting' which horrified Chadwick. He[10] discovered on his daily tour of the laboratory that Cockcroft and Walton had obtained a beam of protons and were measuring their range in air rather than bombarding a light element, 'the purpose for which all their work was a preparation'. Cockcroft and Walton replied defensively that 'they wanted to justify their work by getting some immediate definite results'. Chadwick told them 'perhaps rather angrily that they were not getting on with their real job and that I should go at once to tell Rutherford. This I did and of course Rutherford was incensed'.* The rebuke from Chadwick seems to have led to some lingering resentment on Walton's part. Rutherford's wrath evaporated as soon as meaningful results started to appear. It is interesting to note that the time-honoured scintillation method was used to detect the α-particles, and Rutherford was brought in to observe them. With particular glee, and mindful of the Cambridge–Vienna controversy, he pronounced: 'Those scintillations look mighty like alpha-particle ones. I should know an alpha-particle when I see one for I was in at the birth of the alpha-particle and I've been observing them ever since'[11]. As a measure of how important the Cockcroft–Walton experiment was, Rutherford swore them to secrecy until more data were acquired.

In the meantime Chadwick had continued his neutron work and important ancillary studies were undertaken by Feather and Philip Dee, a research student who had been working with C.T.R. Wilson. Both Feather and Dee used cloud chamber techniques in order to photograph tracks of charged particles that had been set in motion by neutrons (themselves not directly visible in a cloud chamber). Feather filled his chamber with nitrogen and obtained photographs of recoil atoms which had been struck by the much smaller neutrons. He also found double tracks which represented disintegration of nitrogen nuclei with the expulsion of an α-particle. Dee[12] had approached Chadwick at the end of his lecture to the Kapitza Club because it had occured to him during the course of Chadwick's talk that the interaction of neutrons with electrons should be very different from proton–electron interactions, and this could reinforce Chadwick's interpretation of events. Chadwick and Feather were already extensively involved in their own work and Chadwick told Dee he could have polonium/beryllium source for overnight experiments. The furious rate of scientific discovery at the Cavendish in the early part of 1932 seems to have led to a temporary suspension of Rutherford's rule that the laboratory must close at 6 pm.

* The story went round the Cavendish that Rutherford called for Cockcroft and Walton and said: "If you don't put up a scintillation screen by tomorrow, I'll sack both of you'.

Dee could not find the tracks he was looking for, but did discover that there were some γ-rays emitted from the polonium source. His old supervisor C.T.R. Wilson became caught up in the excitement and brought all his gold medals, including the Nobel Prize, to the laboratory for Dee to use to filter out the γ-rays. Dee recalled that Rutherford's initial reaction to his continued inability to find the electron tracks was scathing (but later became a favourite subject for teasing him); Chadwick was unperturbed—he had calculated that the neutron–electron collision cross section (the target presented by an electron to a neutron) was very small, and encouraged Dee to try to set an upper limit to its magnitude, which he did. In Dee's words 'during March and April the neutron became accepted in the Cavendish as a definite, almost familiar, entity'.

Chadwick[13] sent his paper *The existence of a neutron* off to the Royal Society in May with those of Feather and Dee, and the three subsequently appeared together in the *Proceedings*. After giving a full account of his experiments, Chadwick made estimates of the mass of the neutron because he believed that this was the best way to define its nature. His best estimate was 1.0067 atomic mass units. The masses of the proton and electron were accurately known and if added together came to 1.0078 mass units, a slightly larger number. At face value, one might therefore deduce that a neutron cannot consist of a proton and electron combined because it has less mass. But this would be erroneous because it ignores the *mass defect*, a phenomenon which can be explained by Einstein's famous equation, $E = mc^2$, which links mass and energy. When any two subatomic particles, such as a proton and electron, combine, there is a resultant tiny decrease in mass, the mass defect. This quantity of mass becomes the *binding energy* of the nucleus formed and is equivalent, by Einstein's equation, to the energy that would have to be expended in pulling the two particles apart again. Chadwick therefore wrote:

> Such a value for the mass of the neutron is to be expected if the neutron consists of a proton and an electron, and it lends strong support to this view. Since the sum of the masses of the proton and electron is 1.0078, the binding energy, or mass defect, of the neutron is about 1 or 2 million electron volts. This is quite a reasonable value. We may suppose that the proton and electron form a small dipole, or we make take the more attractive picture of a proton embedded in an electron.

Although Chadwick concluded that the neutron was a complex particle, he did not entirely discount that it could be an elementary particle, but thought 'this view has little to recommend it at present, except the possibility of explaining the statistics of such nuclei as ^{14}N'. Here Chadwick was referring to the *nitrogen paradox* mentioned in Chapter 6, p. 92: if the nitrogen nucleus consisted of 7 protons and 7 complex neutrons, there would still be a total of 14 protons and 7 electrons present—an odd number so the paradox remained. Bohr[4] had considered this point in his Copenhagen talk, and reluctantly swept the nuclear electrons under the carpet. He had come to the conclusion that 'the statistics of all nuclei are determined solely by the number of protons contained, and . . . we may say that the electrons in nuclei have lost their individuality as separate mechanical units'. Bohr was so troubled by numerous complexities and seeming inconsistencies of nuclear processes that he was prepared to set aside the laws of energy and momentum conservation in their consideration: these were precisely the laws that Chadwick relied on to make his case for the neutron's existence.

At this turbulent time in the history of nuclear physics, Niels Bohr became the central figure. He was both the fountain-head and conduit for new ideas. Just as Chadwick and Rutherford immediately wrote to him about breakthroughs at the Cavendish, he was in regular correspondence with most of the other leading European physicists about their work. A particularly close colleague was Werner Heisenberg, who, like Bohr, had become very interested in the neutron and was also unsettled by the continuing presence of electrons within the nucleus. In a letter to Bohr[14] dated 20 June 1932, he addressed the very problem that Bohr had identified in his April lecture and his broad solution was 'to shove all difficulties of principle onto the neutron and then apply quantum mechanics to the nucleus'. The two men continued to debate the neutron question in a series of letters during that summer, and Heisenberg completed a triptych of papers in which he sketched a revolutionary theory of nuclear forces. He stated that the nucleus was composed of protons and neutrons, and then explored how quantum mechanics could be applied to such a system. Heisenberg did not entirely reject the idea that the neutron comprised a proton plus an electron, but assigned it properties (e.g. spin) as an elementary particle. He expounded these ideas at a summer physics meeting held in the University of Michigan during August 1932.

> So intimate is this union [between proton and electron] that the neutron takes on characteristics of its own, which cannot be predicted from the properties of the original proton and electron. The statistics it obeys cannot be derived from those of the proton and electron, it weighs less than the two together, and this 'mass defect' does not give a measure of the neutron's stability when affected by external forces . . .
>
> Professor Heisenberg's theory now holds that there are no electrons in the nucleus except as parts of neutrons. The nucleus is then composed of protons and neutrons, and by considering the latter to be independent units of nature, with their own properties, these properties can be considered as the cause of happenings inside the nucleus which could not be explained by the old theories, which held that the nucleus was made up of protons and free electrons. With the neutrons unique and independent units, they may be interpreted in terms of the usual quantum theory.[15]

When Heisenberg's theories reached the Cavendish, they seem to have been regarded with some initial disdain. According to Chadwick,[6] Fowler was annoyed with himself when he read Heisenberg's paper on nuclear forces and said: 'Well, I could have published that months ago'. The first public airing of the collective English response came at the British Association for the Advancement of Science meeting in York early in September. Chadwick held to the notion that the neutron was a composite particle and was supported by O.W. Richardson, who said:

> I am glad that Dr Chadwick has stuck to the view that it is a combination of a proton and electron. Some people have said it was a new kind of ultimate particle.

> It was really too much to believe—that a new ultimate particle should exist with its mass so conveniently close to that of the proton and electron combined. It was nothing but a bad joke played on its creator and on the rest of us.[16]

Referring to the intellectual torment of the European theoreticians like Niels Bohr and Heisenberg, Professor Charles Darwin seemed to draw comfort from the South Seas voyages of his illustrious grandfather when offering the following comments.

> The misconduct of the nucleus, I think, is very much exaggerated. The nucleus is really very well behaved, and in time we will find all fitting in very nicely. If we dwell too much on their misconduct we are taking the attitude of the missionary in New Guinea, who deplores the cannibalism of a few natives and forgets all about their many virtues.[16]

By the end of the year, Chadwick was beginning to change his opinion of the fundamental nature of the neutron, and by the time of his own Bakerian Lecture[17] to the Royal Society in May 1933, his mind was completely open. He reconsidered the mass of the neutron and came to the conclusion

> there can be no doubt that [it] is distinctly less than that of a hydrogen atom . . . This argument from the mass is certainly in favour of the complex nature of the neutron but it is by no means conclusive. The most direct proof would be the splitting of the neutron into a proton and an electron in a nuclear collision, but both calculation and experiment show that this must be a very rare event.

Chadwick then presented his audience with a number of counter-arguments which could be adduced to show the neutron is an elementary particle. First among these was the restriction imposed by Bohr's quantum theory so that the hydrogen atom was the only allowable combination of an electron and a proton. Chadwick also raised a more general question.

> If the neutron is a proton and electron why does not the hydrogen atom transform into a neutron with a release of energy? There is ample evidence to show that such a transformation must be exceedingly rare. This consideration seems to me to argue strongly for the elementary nature of the neutron.

Tongue in cheek, he then posed a perplexing alternative: perhaps 'the proton is the complex particle and [one might] regard it as a neutron plus a positive electron'. In that case, a neutron could not consist of a proton plus a negative electron—it becomes a chicken and egg conundrum. The positively charged electron, or positron, had finally been discovered in August 1932 by Carl Anderson, a research fellow working with Millikan on cosmic ray research at the California Institute of Technology. 'Finally' because positrons had been overlooked by as many scientists as had misidentified the neutron prior to Chadwick. The Joliot-Curies had again come within a hair's breadth of the discovery, and Chadwick himself had spotted an important lead as early as 1928 at the Cambridge conference on β-and γ ray problems. One of the invited speakers to that conference was Dmitry Skobeltzyn[18] from Leningrad, who demonstrated a collection of cosmic ray tracks taken in a Wilson cloud chamber, which, as he said, 'produced some impression on the audience'. On one photograph, there seemed to Chadwick to be the track of an electron 'going the wrong way'. He wrote to Skobeltzyn trying to clarify what this pattern meant, but no conclusion was reached and Chadwick was too busy with other matters to pursue it further.

This was perhaps the first among multiple observations made by various physicists[19] over the next few years of electrons 'moving backwards', 'curving the wrong way' or 'coming up from the floor' before Anderson recognized that they were due to positrons. Millikan himself showed more clearcut examples in photographs taken by Anderson, when talking to the Kapitza Club in November 1931, but it would be another nine months before their true significance was appreciated. Whereas Skobeltzyn, and Millikan and Anderson used cloud chambers that were very inefficient and only recorded cosmic ray tracks sporadically, Patrick Blackett with his new assistant at the Cavendish, Guiseppe Occhialini, devised an ingenious arrangement whereby their cloud chamber was triggered by the passage of a cosmic ray itself. Occhialini vividly described the excitement 'that Saturday morning when we first ran the chamber, [Blackett] bursting out of the dark room with four dripping photographic plates held high, and shouting for all the Cavendish to hear "one on each, Beppe, one on each."[20] The new counter-controlled cloud chamber provided a much higher yield of useful photographs, and Blackett and Occhialini were soon able to confirm and extend Anderson's findings regarding the positive electron. Some months earlier, Chadwick had assigned a research project to a student which had been thwarted mainly by an inability to take good photographs of tracks in a cloud chamber. Now he approached Occhialini about attempting to produce positrons artificially and using the new coincidence recording technique to photograph them. With Blackett's help they were successful and wrote a letter to *Nature* on 27 March 1933 titled 'New evidence for the positive electron';[21] the same effect was described independently by Lise Meitner and Philipp, and by the Joliot-Curies almost at the same time.

Chadwick's reasons for wishing to create positrons in the laboratory were firstly to study how they were created in the interaction of radiation with matter and secondly to measure their properties more easily. Curie and Joliot had proved that γ-rays rather than neutrons from the beryllium were responsible for the production of positrons, and the Cambridge workers quickly accepted that fact. Early in 1934 Chadwick, Blackett and Occhialini[22] published the results of a year's work on the positron and decided that pairs of positive and negative electrons were 'probably created in the electric field outside, rather than inside, the nucleus'; they also made the most accurate estimate of the mass of the positron available at that time. They did, however, miss one critical observation, and this time the brilliant Joliot-Curies were *sans pareil.* The Cambridge group in one experiment had looked at positron production using α-particles from polonium to bombard boron. The Joliot-Curies were doing much the same with their polonium in Paris, and in July 1933 reported two kinds of reaction with the production either of protons or of neutrons and positrons. A few months later they speculated, perhaps having read Chadwick's Bakerian lecture, that the second type of yield might be due to the disintegration of a proton into a neutron and positron. By the end of the year, they had established the true explanation which was due to artificial radioactivity. Alert to every detail, Joliot noticed that these reactions were not instantaneous, but that the emission of positrons continued for some minutes when the polonium source was removed. He and Irène correctly conceived of a sequence of events whereby an intermediate, unstable isotope was being created by the initial α-particle bombardment and this then decayed with the further emission of positrons. The proof was irrefutable when they were able to detect the new, unstable isotopes chemically.

The happy spirit of the times was summarized by Chadwick,[6] thus:

> In those days there were many fewer workers. One, you might say, knew them all, and one was not competing with them. You were competing with nature. And in a sense, experimental physics particularly I regarded as a kind of sport. It was contending with nature. There were other things in life besides physics.

In his opinion, if the Cambridge workers missed 'the phenomenon of artificial radioactivity . . . it was entirely our own fault'.[6] The remarkable string of breakthroughs that began with Chadwick's find of the neutron made the Cavendish quite simply the leading laboratory for nuclear physics research in the world during the early 1930s. Rutherford's bold strategy to concentrate on the nucleus to the exclusion of much else had been vindicated beyond even his dreams. Apart from supplying crucial research work, Chadwick's contribution to the whole programme had been immense, and it was fitting that he should now emerge a little from Rutherford's broad shadow to join the world's scientific elite in his own right. The occasion marking Chadwick's arrival on the international stage was the seventh Solvay Conference on Physics held in Brussels in October 1933. The general theme of the meeting was the 'Stucture and Properties of Atomic Nuclei'[23] and many of the discoveries already described were major subjects for discussion. Madame Curie and Lord Rutherford represented the old guard amongst the invited participants: Chadwick, the Joliot-Curies, Bothe, Fermi, Blackett, Cockcroft and Ellis were amongst the next generation. Bohr and Einstein were on the organizing scientific committee and had ensured that the leading theoreticians, Dirac, Gamow, Pauli, Heisenberg and Schrödinger came.

John Cockcroft gave the first talk, the *Disintegration of elements by accelerating protons*, in which he referred to Rutherford and Chadwick's scattering experiments as a background to his project with Walton. Chadwick followed him and gave a three part presentation. He too started by considering the scattering of α-particles by atomic nuclei, and referred to Bieler's (his late student) theory about a very strong attractive force in extremely close proximity to the nucleus. Next he discussed energy states within the nucleus and related them to experimental findings. Finally came the neutron where he discoursed over its detection, mass, and potential as a tool for further exploring nuclear structure. Nevill Mott added a theoretical contribution before Chadwick's work was thoroughly analysed and praised by many of the distinguished physicists present. All of the succeeding talks (by the Joliot-Curies, Dirac, Gamow and Heisenberg) concerned subjects in which Chadwick had been directly involved during the previous three years. The last talk by Heisenberg brought up an unresolved issue, the continuous spectrum of β-emission, which had been the most vexing question in the whole of nuclear physics since about 1929 and had even led Bohr to suggest that the laws of conservation of energy and momentum did not apply. The problem had become so overwhelming that no one was concerned any longer with its origins; many had probably forgotten that the spectrum was first described correctly in 1914 by a young postgraduate student named James Chadwick.

The challenge posed by the continuous spectrum was that the original radioactive nucleus prior to the emission of a β-particle and the resultant daughter nucleus have discrete energy levels, according to quantum theory. A number of identical radioactive nuclei decaying to a number of daughters with lower but identical energy levels should emit β-particles each associated with an energy equal to the difference between the initial and final nuclear states. This was the case with α-particles, but β-particles emerged with a whole range of energies. After Heisenberg had touched on the difficulty in his talk, Wolfgang Pauli from Zurich tackled it head-on in discussion. He began by dismissing Bohr's explanation that the laws of conservation of energy and momentum do not hold when nuclear processes involve β-particles as not 'either satisfying or even plausible'. Pauli then repeated a proposal he had first made at a conference in 1931:

> The conservation laws [of energy and momentum] hold, the emission of beta particles occurring together with emission of a very penetrating radiation of neutral particles, which has not been observed yet. The sum of the energies of the beta particle and the neutral particle . . . emitted by the nucleus in one process will be equal to the energy which corresponds to the upper limit of the beta spectrum.[24]

Pauli was saying that his neutral particles acted as makeweights in the process of β-decay: one neutral particle would accompany every β-particle emitted and together the two would have a constant energy. His neutral particle was not Chadwick's neutron because it could not have a mass greater than an electron's, and Enrico Fermi ended any possible confusion by calling them *neutrinos (It.* small neutral one) in contrast to Chadwick's *neutrone* (the big neutral one). After Pauli's historic contribution at the Solvay Conference, Chadwick[23] joined the discussion to report that he had already made some attempts to detect neutrinos. Although the experiment had been unsuccessful, the mere fact that he had tried shows how assiduously Chadwick followed every lead in the rapidly unfolding drama of nuclear physics, and more significantly was always trying to provide the next clue. He had given the neutrino problem as a project to Douglas Lea, then a physics student from Trinity College, and their results were presented formally to the Cambridge Philosophical Society in December 1933. Chadwick and Lea[25] did not come close to finding 'a neutral particle of small mass', but the last sentence of their paper 'such particles would be exceedingly difficult to detect' was an accurate prediction—the neutrino remained elusive for another twenty years.

The glory reaped by the Cavendish laboratory served to insulate the institution and its staff to some extent from the chill winds of the great depression which were howling through the streets of commerce and industry. Several Cavendish physicists, most notably Blackett and Bernal, the crystallographer, were openly left-wing, but unlike other parts of the university, the Cavendish never became a breeding ground for communists. The one exception was Alan Nunn May, a research student under Chadwick's mantle in the early 1930s, who was described by one contemporary as 'a retiring, dreamy-looking character . . . who managed to convey the distant impression of always being spiritually in his laboratory'.[26] The rise of Europe's two evil dictators did have direct effects on Rutherford's laboratory. The major loss was Kapitza who was not allowed to leave Russia after making his annual summer visit in 1934. Rutherford was outraged by the brutality and injustice of both Hitler's and Stalin's regimes, particularly as they were directed against fellow scientists, many of whom he knew, and for the first time in his life felt compelled to take a public political stand. With his encouragement, many refugee German scientists began to appear in Cambridge, some to take up long term residence, others en route for other English universities or the United States.

One of the earliest arrivals, in the spring of 1933, was a gentle, friendly theoretician, Rudolf Peierls,[27] who had previously been Pauli's assistant in Zurich and was, at the age of 26, obviously destined for academic distinction. He had planned to spend six months of a Rockefeller Fellowship in Cambridge, but had already decided that Germany was heading for disaster and had no intention of returning with Hitler in power. An even younger and completely unknown graduate student, Maurice Goldhaber,[28] was moved to write to Rutherford after he had witnessed young Nazis rioting at the University of Berlin. Rutherford had replied promptly, offering him a research position at the Cavendish. Unlike Peierls, Goldhaber had no fellowship to support him and was concerned to find the cheapest way of joining the University. He paid a initial visit in August; the first person he spoke to at the Cavendish was Chadwick, who advised him to spend some extra money that he did not have to join a college because 'they do things for you'. Another research student, David Shoenberg, took Goldhaber onto Trinity Street and indicated three colleges in a row:

> At Trinity, the Senior Tutor said, 'Sorry, we are full up.'
> At St. John's, the answer was: 'We'll let you know in six weeks.'
> At Magdalene, the Senior Tutor, V.S.Vernon-Jones, said immediately, 'Ah, you're a refugee—I suppose we ought to have one.' Then he added, 'I suppose you have no money. We better give you a hundred pounds.'

This was enough to live on for about six months, and afterwards Goldhaber always listened carefully to Chadwick's advice, no matter how cryptic. Goldhaber's first work was a theoretical analysis of the lithium reactions which Cockcroft and Walton had obtained a year before. He was nominally under Fowler's wing, but worked independently. His ebullience soon made its mark on the Cavendish. According to the more sedate Peierls:[27]

> He was exceptionally bright but a little naive, and had not yet acquired a very polished manner. He caused raised eyebrows by telling everybody, including Rutherford and Chadwick, what experiments they ought to be doing. It was particularly aggravating that he was usually right.

When Goldhaber came to write his paper on the lithium reactions in April 1934, he needed to know the exact masses of the various isotopes. He talked to Nevill Mott who told him to go to see Chadwick, 'who knew all about masses'.[28] After obtaining the information he sought, Goldhaber decided to mention to Chadwick an idea he had been ruminating on for the past year. When the Nazi riots had stopped all classes in Berlin, Goldhaber had used the time to catch up on his reading. One of the things he read about was 'heavy' water, which instead of ordinary hydrogen atoms (each with one proton) contained a heavier isotope of hydrogen (each nucleus comprising one proton and one neutron). Goldhaber wondered whether this heavier nucleus could be split in a novel way, by using γ-radiation rather than a bombarding particle. Having found the courage to mention this idea, Goldhaber recalled the reaction: 'Chadwick, gentleman that he was, listened politely, but seemed to catch fire only when the point was brought out that the mass of the neutron could be determined rather accurately in this way'.

We have seen how the mass of the neutron was a crucial key to understanding whether it was an elementary particle or a combination of a proton and electron. Chadwick's best previous estimate of the mass was 1.0067 which, as explained in his Bakerian Lecture, was evidence for a composite particle because it was less than the sum of the masses of the proton and the electron. By contrast, the Joliot-Curies had suggested a mass as high as 1.012, pointing to an elementary particle (because it exceeded the summed masses and would be contrary to the concept of mass defect).

Goldhaber realized that his idea was 'a bit of a shot in the dark' and thought little more of it until Chadwick brought it up again some weeks later. He asked, 'Were you the one who suggested the photodisintegration of the diplon[4*] to me?' When Goldhaber confirmed that he was, Chadwick said: 'Well, it works—for the first time last night. Would you like to work on it with me?' Chadwick's deadpan manner did not diminish the young man's excitement and he soon obtained Fowler's permission to switch to experimental work. Chadwick put him in the capable hands of H. Nutt, his personal laboratory assistant.

For the γ-ray source they used a radioactive isotope known as thorium C', that emits a γ-ray of 2.62 MeV—the highest energy then available[**]. Chadwick and Goldhaber[29] filled an ionization chamber with deuterium and observed the protons produced in the following photodisintegration:

$$ {}^{2}_{1}\mathrm{H(deuteron)} + \gamma(h\upsilon) \rightarrow {}^{1}_{1}\mathrm{H(proton)} + {}^{1}_{0}\mathrm{n(neutron)} $$

---

* *Diplon* was the name favoured by Rutherford at the time, but the standard term soon became the *deuteron*.

** This isotope discovered by Charles Ellis is now recognized to be an isotope of thallium, ^{208}Tl, a member of the thorium radioactive series.

When the chamber was filled with ordinary hydrogen instead of deuterium, no effect was seen. A greatly reduced effect was observed with deuterium if radium, a source of lower energy γ-rays was used instead of the thorium. They therefore concluded that the binding energy of deuterium was between 1.8–2.6 MeV. Since the masses of the proton and neutron produced in the reaction are approximately identical, the available energy from the incident radiation (once the binding energy of the deuteron has been exceeded) can be assumed to be divided equally between the two particles. Chadwick and Goldhaber measured the kinetic energy of the protons produced which was found to be almost 1.05 MeV. The mass of the neutron could now be calculated from the energetics of the reaction:

$$m_D + E_\gamma = m_H + m_n + 2E_p$$

The only unknown in this equation is now the mass of the neutron, m_n, since the masses of the deuteron and the proton were already accurately established. The calculation in Chadwick's neat handwriting took just one page of his small laboratory notebook (Fig. 7.1), and established that the mass of the neutron must be greater than 1.0077 but less than 1.0086.

The established mass of a free proton was 1.0078 so that the neutron was heavier and it had to be an elementary particle. This result effectively ended any notion that it consisted of a proton–electron composite, and the electron was finally banished from the nucleus. Heisenberg's theoretical approach gained good experimental support, and the nitrogen paradox was solved because the nucleus contained an even number of particles (7 protons + 7 neutrons) with spin-½ as its atomic spectrum suggested it should.

The experiment had taken just six weeks of intensive work and Goldhaber, a very gifted greenhorn, found Chadwick a pleasure to work with. It seems that Chadwick's time at home was limited during this period and one morning Aileen decided to bring the girls in to see their father. Goldhaber remembers the door of the research room opening and: 'in walked first his wife, but he did not look up; then in succession his two twin daughters came in and again he did not look up. Then came the family dog and Chadwick turned around and said 'Hello boy!'.[30]

Now that he and Chadwick had measured the photodisintegration cross section, Goldhaber tried to calculate the value theoretically as a check. When he failed to come up with an answer within about a week, Chadwick approached Rudolf Peierls,[27] who by then had taken a job in Manchester but happened to be visiting the Cavendish with another German refugee, Hans Bethe. Without telling them that he already knew the answer from the experiment, Chadwick issued the simple challenge to Peierls and Bethe: 'I bet you could not calculate the cross section for the photodisintegration of the diplon'. Peierls and Bethe[31] completed the initial calculation that evening on their way back to Manchester on the train, and later published a very successful quantum theory of the diplon.

The reason for Chadwick's impatience was that he and Goldhaber[32] had sensed an interesting discrepancy. Some months earlier, Lea (who had joined Chadwick

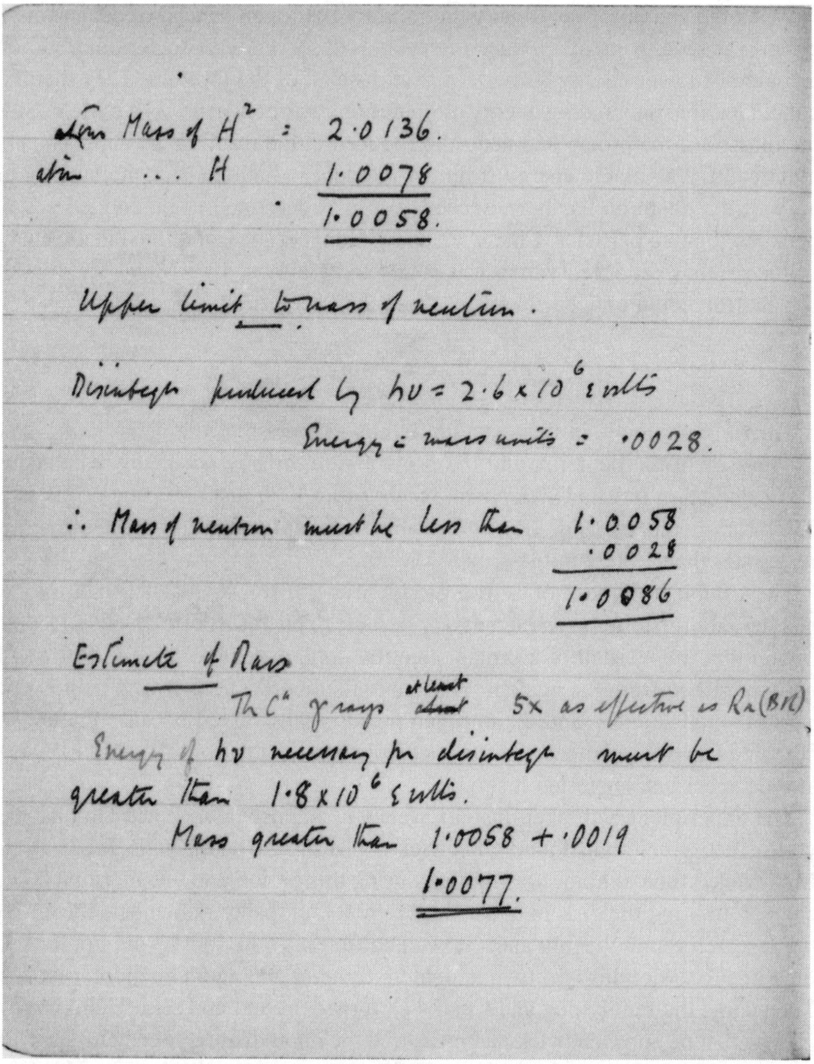

atomic Mass of H^2 : 2·0136.
atomic .. .. H 1·0078
1·0058.

Upper limit to mass of neutron.

Disintegn produced by $h\nu = 2{\cdot}6 \times 10^6$ e volts
Energy in mass units = ·0028.

∴ Mass of neutron must be less than 1·0058
·0028
1·0086

Estimate of Mass
ThC″ γ rays at least 5× as effective as Ra(B+C)
Energy of $h\nu$ necessary for disintegn must be greater than $1{\cdot}8 \times 10^6$ e volts.
Mass greater than 1·0058 + ·0019
1·0077.

Fig. 7.1 The page from Chadwick's laboratory notebook for 10 July 1934.

in the unsuccessful hunt for the neutrino) had reported an experiment where he had detected γ-rays after exposing paraffin to neutron bombardment. He explained this in terms of a reaction that was the exact reverse of the Chadwick and Goldhaber process, thus:

$$^1p + {}^1n \rightarrow {}^2d + \gamma$$

Goldhaber calculated from his experimental results with Chadwick what should be expected for the reverse reaction. He found that Lea's cross section for his neutrons was perhaps several hundred times larger than it should have been, and did not believe that the production of γ-rays could have resulted from the *primary* neutrons from Lea's source. Instead Goldhaber supposed that the primary neutrons were first slowed down in the paraffin and were then absorbed by carbon atoms which might have a much larger cross section for slow as opposed to fast neutrons. He and Chadwick were writing their short paper for *Nature*[29] outlining their results, and in early drafts included speculation about slow neutrons. But, as Goldhaber knew, 'speculation was somewhat frowned upon at the Cavendish', and a few days later Chadwick suggested that they omit this passage from their report; the final paper appeared with much more cautious, non-committal, comments. Goldhaber believes the change of heart probably came after Chadwick had discussed their interpretation with Rutherford. One can see that Chadwick might have wished not to jeopardize a short experimental paper that already contained results of great significance by adding further material which might have been erroneous. But Goldhaber's hypothesis was sound and a few months later Fermi announced from Rome the discovery of slow neutrons. Rutherford was remorseful and came to see Goldhaber to say, 'He'll give you credit, won't he?'.[28]

Fermi and his colleagues had been mining another rich seam of research since the spring of 1934. Following on the Joliot-Curie's demonstration of artificial radioactivity, they had been systematically irradiating elements with fast neutrons. From the first sixty elements studied, about forty produced new radioactive isotopes by various nuclear reactions. Their results were of great interest to all nuclear physicists, and Rutherford wrote to Fermi congratulating him on his 'successful escape from the sphere of theoretical physics'.[33] Almost no effort was made to repeat this work in Cambridge—partly, no doubt, because there was little attraction in using valuable resources to duplicate another scientist's results and also according to Chadwick[6] because 'that was Fermi's business . . . One didn't barge into another man's subject unless by arrangement with him'.

While not disputing the primacy of Fermi's group in Rome, Chadwick seems to have felt that the Cavendish had some proprietary claim to slow neutrons, and he and Goldhaber set out on a series of experiments using them. They first reported[34] that slow neutrons would disintegrate lithium, boron and nitrogen amongst the light elements. They then extended their studies[35] to some heavier elements, including uranium, and reported the results to the Cambridge Philosophical Society in the late summer of 1935. Uranium is a naturally radioactive substance and, in order to stop its relatively low-energy α-particles, Chadwick and Goldhaber wrapped it in aluminium foil. They were looking for particles of greater range resulting from the slow neutron bombardment, but found nothing. What they could not have foreseen was that the foil concealed an even more important process produced by slow neutrons in uranium: as Goldhaber later remarked, their cleverness with the aluminium foil earned them 'the dubious honor of being the first to miss the fission fragments later found by Otto Frisch'.[28] Even Chadwick[6] was not too hard on himself for this oversight: 'it was, I suppose, one of those stupid things one does do'.

The experiments with Goldhaber set the seal on Chadwick's great neutron work. Without question he was amongst the first rank of experimental physicists during an extraordinary era—indeed his discovery of the neutron is often cited as the birth of nuclear physics proper. His reserved and diffident manner sometimes served to obscure his genius, even with other scientists. When the President of the Royal Society, Sir Robert Robinson,[36] awarded him the Copley Medal in 1950 'for his pioneering researches on the constitution of matter', he ventured the opinion that Chadwick was 'a brilliant experimenter and a sound, but not a highly speculative philosopher'. This would seem to be fair comment, but was interpreted by Chadwick as implying that he lacked imagination. The remark irritated and amused him at the same time because:

> I knew that I wasn't lacking in imagination. What I did fall down on was in telling people about the things I was thinking. I very seldom indeed told even Rutherford, to whom I would naturally speak. But my difficulty was not so much that I had a lack of imagination but that I could generally find two or three explanations of the same phenomenon. I was always unwilling to speak about them until I had done the experiments, until I had evidence on the way things were going. And, of course, that depended upon being able to do the experiments, and I was restricted, not only by facilities but by time. There's a limit to what one can do. But I had thought about a great many things.[6]

The Cavendish style started by Clerk Maxwell and perpetuated by Rutherford and Chadwick was an intimacy between the physicist and his apparatus. The creative process involved original thought and depended for its expression on the fashioning of equipment that would illuminate just that hidden corner of the physical world that the scientist wished to investigate. There were now younger men in the Cavendish, like Kapitza, Cockcroft, Walton and Allibone who still needed imaginative intellects to generate hypotheses, but who approached practical problems like engineers rather than as hand craftsmen. Chadwick recognized that theirs was the way of the future.

References

The page from Chadwick's laboratory notebook for 10 July 1934 is reproduced with permission of the Churchill Archives Centre and can be found in CHAD III, 1/6.

1 Rutherford, E. (1932). Origin of the gamma rays. *Nature*, **129**, 457–8.

2 Bohr, N. (25/3/32). Letter to J. Chadwick. Niels Bohr Archive, Copenhagen.

3 Chadwick, J. (30/3/32). Letter to N. Bohr. Niels Bohr Archive, Copenhagen.

4 Bohr, N. (1932). On the properties of the neutron. In *Niels Bohr: collected works*, Vol. 9. (ed. R. Peierls, 1986), pp. 117–18. North-Holland, Amsterdam.

5 Bohr, N. (1962). Reminiscences of the founder of nuclear science and some of the developments based on his work. In *Rutherford at Manchester* (ed. J.B. Birks), pp. 159–61. Heywood, London.

6 Weiner, C. (1969). Sir James Chadwick, oral history. American Institute of Physics, College Park, Maryland .

7 Rutherford, E. (1928). Presidential address to the Royal Society for 1927. *Proceedings of the Royal Society A*, **117**, 300

8 Cockcroft, J. (1984). Some recollections of low energy nuclear physics. In *Cambridge physics in the thirties* (ed. J. Hendry), pp. 74–80. Adam Hilger, Bristol.

9 Oliphant, M.L. and Penney, W.G. (1968). John Douglas Cockcroft, 1897–1967. *Biographical memoirs of fellows of the Royal Society*, **14**, 139.

10 Chadwick, J. (?1971). Undated notes on Oliphant's biography of Rutherford. CHAD II, 2/1, CAC.

11 Oliphant, M. (1972). *Rutherford – recollections of the Cambridge days*, p. 86. Elsevier, Amsterdam.

12 Dee, P. (1984). Some reminiscences of the discovery of the neutron. In *Cambridge physics in the thirties* (ed. J. Hendry), pp. 46–8. Adam Hilger, Bristol.

13 Chadwick, J. (1932). The existence of a neutron. *Proceedings of the Royal Society A*, **136**, 692–708.

14 Pais, A. (1991). Looking into the atomic nucleus. In *Niels Bohr's times*, pp. 324–45. Oxford University Press.

15 Heisenberg, W. (1932). Quoted in *New York Times*, 17/8/32.

16 British Association for the Advancement of Science (1932). Meeting report in *New York Times*, 6/9/32.

17 Chadwick, J. (1933). The neutron (Bakerian lecture). *Proceedings of the Royal Society A*, **142**, 1–25.

18 Skolbeltzyn, D. (1983). The early stage of cosmic-ray particle research. In *The birth of particle physics* (ed. L.M. Brown and L. Hoddesdon), pp. 111–19. Cambridge University Press.

19 Hanson, N.R. (1963). *The concept of the positron*, p. 138. Cambridge University Press.

20 Lovell, B. (1975). Patrick Maynard Stuart Blackett, Baron Blackett, of Chelsea, 1897–1974. *Biographical memoirs of fellows of the Royal Society*, **21**, 1–115.

21 Chadwick, J., Blackett, P.M.S. and Occhialini, G.P.S. (1933). New evidence for the positive electron. *Nature*, **131**, 473.

22 Chadwick, J., Blackett, P.M.S. and Occhialini, G.P.S. (1934). Some experiments on the production of positive electrons. *Proceedings of the Royal Society A*, **144**, 235–49.

23 Mehra, J. (1975). The structure and properties of atomic nuclei. In *The Solvay conferences on physics*, pp. 209–26. Reidel, Boston.

24 Brown, L.M. (1978). The idea of the neutrino. *Physics Today*, (Sept.), 23–8.

25 Chadwick, J., and Lea, D.E. (1934). An attempt to detect a neutral particle of small mass. *Proceedings of the Cambridge Philosophical Society*, **30**, 59–61.

26 Boyle, A. (1979). *The fourth man*, p. 65. The Dial Press, New York.

27 Peierls, R. (1985). *Bird of passage*. Princeton University Press.

28 Goldhaber, M. (1993). Reminiscences from the Cavendish Laboratory in the 1930's. *Annual Review of Nuclear and Particle Science*, **43**, 1–25.

29 Chadwick, J. and Goldhaber, M. (1934). A nuclear photo-effect: disintegration of the diplon by γ-rays. *Nature*, **134**, 237–8.

30 Goldhaber, M. (4/3/92). Letter to the author.

31 Bethe, H. (17/8/95). Letter to the author.

32 Goldhaber, M. (1971). Remarks on the prehistory of the discovery of slow neutrons. *Proceedings of the Royal Society of Edinburgh A*, **70**, 191–5.

33 Amaldi, E. (1977). Personal notes on neutron work in Rome in the 30s and post-war European collaboration in high-energy physics. In *Story of twentieth century physics* (ed. C. Weiner), pp. 290–345. Academic Press, New York.

34 Chadwick, J. and Goldhaber, M. (1935). Disintegration by slow neutrons. *Nature*, **135**, 65.

35 Chadwick, J. and Goldhaber, M. (1935). Disintegration by slow neutrons. *Proceedings of the Cambridge Philosophical Society*, **31**, 612–16.

36 Robinson, R. (1951). Presidential Address to the Royal Society, 1950. *Proceedings of the Royal Society A*, **205**, 1.

8 Liverpool and the Nobel Prize

The only American physicist to attend the 1933 Solvay Conference was Ernest Lawrence from the University of California at Berkeley. He had arrived at Berkeley five years earlier and decided then to pursue nuclear research because 'the pioneer work of Rutherford and his school had clearly indicated that the next great frontier for the experimental physicist was surely the atomic nucleus'.[1] In 1929 he had designed an instrument that would accelerate subatomic particles in a circular racetrack to extremely high energies. The apparatus became known as the *cyclotron*, and Lawrence and his team devoted their efforts over many years to building bigger and ever more powerful models. Whereas work at the Cavendish had been undertaken by individuals or small groups of scientists with the ad hoc technical support of a laboratory assistant like George Crowe or Mr Nutt, John Cockcroft[2] found on a visit to Berkeley in the summer of 1933 that the approach was more like a factory.

> The experimenters were divided into shifts: maintenance shifts and experimenters. When a leak or fault developed in the cyclotron the maintenance crew rushed forward to plug the leaks by melting the numerous wax joints and fixed the fault when the operating shifts rushed in again.

Cockcroft, who shared a common interest in electrical engineering with Lawrence, came away deeply impressed by his enthusiasm and organization. Ernest Lawrence was an entrepreneur amongst physicists, with a knack for raising funds in a country which encouraged brash ambition. By the time of the Solvay meeting in October 1933, *The New York Times* reported that he had succeded in accelerating deuterons in 'the whirligig atom-gun'[3] to energies of three million electron volts, which far exceded any other system, man-made or natural. Lawrence met Rutherford and Chadwick for the first time in Brussels. Chadwick[4] remembered him 'bubbling over with enthusiasm' and liked him very much. In particular, he was struck by Lawrence's openness; Rutherford too was favourably impressed, and even told Chadwick, 'Lawrence reminds me of my young days. He's just like I was at his age'. There were indeed some close parallels.[5] Lawrence was then thirty-two, and two years earlier had become the youngest professor ever appointed at the University of California; Rutherford had been just twenty-seven when he took up his first chair at McGill University. Both men were the grandsons of immigrants and had grown up in frontier, agricultural communities—Lawrence in

South Dakota, Rutherford on the South Island of New Zealand. They both kept a staightforward, plain speaking approach to life and were at ease in any company. Their energy, generosity of spirit and assertive character made them outstanding leaders, each inspiring younger scientists and thereby founding a great school in physics. There was though a marked difference in their science. Lawrence knew one big thing, the cyclotron, and strove to make it ever more powerful, leaving its application largely to others. Rutherford by contrast had no interest in equipment for its own sake, only what he could do with it.

It has often been said that the year 1932, with the discovery of the neutron and the other momentous developments in the field, marks the birth of nuclear physics proper. Before 1932 nuclear physics was not an academically glamorous specialty and attracted only a small band of workers. With one or two notable exceptions like Ernest Lawrence, it was almost exclusively a European pursuit. In the leading American journal, *Physical Review*, papers related to nuclear physics comprised a mere 8% of the total published in 1932; this increased to 18% the following year and nearly half the increase was accounted for by publications about the neutron.[6] The strong rising trend in the number of nuclear physics papers continued and was accompanied, in the United States at least, by an even more remarkable expansion in funding for this research. Under Lawrence's driving leadership, the Berkeley school emerged as the leading American centre during the early 1930s, and began to challenge the premier position of the Cavendish.

The challenge did not go unnoticed in Cambridge and was the source of considerable frustration among Rutherford's more senior staff. The difference of opinion which Chadwick and Rutherford had experienced a decade earlier over the introduction of high-voltage equipment into the laboratory now threatened to resurface over the cyclotron. Chadwick[4] felt

> . . . it was becoming very difficult to push on without some new equipment . . . I was at an end really with the equipment which I had or could see myself getting, and it was quite clear to me, as it was no doubt to others, that we needed a means of accelerating protons or other particles . . . at high energies. But that meant more space, particularly more money, and particularly engineering. It meant complicated equipment, and Rutherford had a horror of complicated equipment.

Cockcroft[2] had first 'tried to persuade Rutherford to allow us to build a cyclotron' on his return from Berkeley in the summer of 1933, but Rutherford believed that it was unnecessary to go to higher energies, especially in view of Cockroft and Walton's success with their accelerator. Even without a cyclotron, the laboratory's appearance was being transformed in Rutherford's eyes to 'a great hall [containing] massive and elaborate machinery rising tier on tier, to give a steady potential of about two million volts. Nearby is the tall accelerating column with a power station on top, protected by great corona shields—reminding one of a photograph in the film of Wells', *The shape of things to come*'.[7] Although Cockcroft and Chadwick formed a significant alliance, Rutherford still had the last word on laboratory policy and that was '*No*'. There were good reasons for him to resist the cyclotron in 1933 or '34. Despite the unprecedented successes of the Laboratory during the previous few years, carping criticism of the Cavendish continued to be heard in high places, and the national economy was still contracting. In 1930 Lord Rutherford[8] had succeeded Sir William McCormick as Chairman of the Advisory Council of the Department of Scientific and Industrial Research (DSIR), a government body charged with applying scientific knowledge to industry. The financial state of many industry-related research associations was parlous as a result of the Depression, and Rutherford continued to be attacked for not applying the work of the Cavendish more towards the industrial needs of the nation. The mundane nature of much of the applied research covered by the mantle of the DSIR was reflected in the agenda of the first meeting with Rutherford in the chair; topics for discussion included Lenses in Railway Signals, Dry Rot and the Deathwatch Beetle, and Atmospheric Pollution. With the disastrous economic deflation, the Treasury issued the DSIR with some forthright guidelines:

> it would be wise to proceed with researches which are well established . . . rather than commit ourselves to new researches of which the value and period of cost of development is largely problematical.[9]

The Treasury, though, shrunk from telling the DSIR which projects should be funded and which refused, on the grounds that 'The Advisory Council [of the DSIR] can decide much better than we can which of their kittens is to be drowned'.[9] Oliphant,[10] who had become Rutherford's confidant, described him returning a little dejected from a London meeting at a time he thought that any misgivings about nuclear physics research should have been laid to rest. Despite Chadwick's discovery of the neutron and the Cockcroft and Walton's demonstration, Rutherford complained: 'They have been at me again, implying that I am misusing gifted young men in the Cavendish to transform them into scientists chasing useless knowledge.'

There was, therefore, a powerful political reason for Rutherford to proceed cautiously on any further expansion of the Cavendish, and with Oliphant he soon provided scientific evidence to buttress his conservative position. Lawrence had observed that when using a beam of deuterons in his cyclotron, any target bombarded seemed to emit an enormous number of protons. He interpreted this as the deuteron breaking up into a proton and a neutron. Oliphant[10] did not believe this explanation, although he did observe the same copious production of singly charged particles when studying deuterons in his low energy accelerator in the Cavendish. With Rutherford's encouragement, he devised a way of bombarding deuterons with other accelerated deuterons in order to explore the question as directly as possible. One reaction they identified was two deuterons combining and then breaking up into two unequal components—a proton and a previously unknown isotope of hydrogen with atomic mass 3, which became known as tritium. Rutherford also predicted, correctly, that two diplons (deuterons) could unite and then disband into a neutron and a helium isotope of atomic mass 3*. They had

* Rutherford informed Oliphant of this hypothesis by phone at 3 am. When asked for his reasons, Rutherford roared, "Reasons! Reasons! I feel it in my water!"

found two new isotopes, measured their masses and understood the remarkable deuteron reactions without recourse to a cyclotron.

While Chadwick understood Rutherford's attitude, he thought that the Cavendish, which had been the world centre of nuclear physics, was beginning to fall behind, and that without equipment such as a cyclotron 'the fall seemed bound to continue'.[11] He was not prepared to quarrel with Rutherford, not wishing to repeat the period of coolness which they had passed through in the mid-1920s, and he resolved to leave the Cavendish. Although Chadwick always maintained that this was his primary reason for deciding to move, he had reached his forties and wanted to run his own show. He had begun to find life in Cambridge somewhat narrow and wanted to meet people from outside the academic sphere. The social strata in Cambridge were inclined to be rigid, with a wife's position depending on her husband's standing in the University. Aileen never found friendships easy to make there, partly because she had been brought up to mix with her own rank in the equally closed world of Liverpool's upper middle classes.

The perfect opportunity presented itself early in 1935, when Chadwick was approached by Liverpool University about the Lyon Jones Chair of Physics. Professor Lionel Wilberforce,[12] the incumbent since the turn of the century, had finally agreed to retire at the age of 74 years. He was a grandson of the great philanthropist and slave abolitionist, and nephew of Bishop Wilberforce.[13] The latter had achieved fame as Bishop of Oxford, when he clashed with Thomas Huxley at the British Association Meeting of 1860 over Darwin's theory of evolution. Despite the Bishop's aversion to scientific ideas in that debate, he had taken a first in mathematics when a student at Oxford University. He was later nicknamed 'Soapy Sam', because of his smooth manner and perceived slipperiness over the Church of Rome, but retorted that the sobriquet showed that 'though often in hot water, he always came out with clean hands'. Lionel Wilberforce seems to have shared not only the Bishop's mathematical ability, but also his humour—Chadwick found him one of the wittiest men he ever met.[14]

Wilberforce had been a demonstrator at the Cavendish when Rutherford first arrived in 1895. He had been more welcoming than some of his colleagues, and Rutherford described him to his fiancée as 'a very decent fellow, and a married man to boot'.[15] Wilberforce was one of the first to report the shortage of equipment as the number of students at the Cavendish increased in the closing years of the nineteenth century.

> Competition among the research students and between them and the advanced classes for instruments, retort-stands and rubber tubing waxed very keen, and it was occasionally conducted by means of raids, which forced one victim to describe himself as pursuing his investigations with his apparatus in one hand and a drawn sword in the other.[16]

Such unruly behaviour may have swayed him against laboratory research, and after he became the second Lyon Jones Professor of Physics at Liverpool in 1900,

Wilberforce much preferred teaching by prepared demonstration or illustrative experiments. His first achievement was to supervise the building of the new George Holt Laboratory, with funds from a local shipping family. In its early years, this was the site of some important research by Charles Barkla,[17] who had graduated from Liverpool and then spent three years on an 1851 Scholarship at the Cavendish, as a research student under J.J. Thomson. At Cambridge, he was also the bass soloist in the King's College Chapel Choir, and despite being offered a valuable choral scholarship to stay on, returned to Liverpool as Oliver Lodge Fellow in 1902. He continued investigations he had begun in the Cavendish into the properties of X-rays. In 1905 Barkla was the first to demonstrate the polarization of X-rays showing they were similar to light waves; this followed a discussion with Wilberforce 'who first suggested to me the the idea of producing a plane polarized beam by a secondary radiator, and of testing the polarization by a tertiary radiator.' Barkla also laid the foundation of X-ray spectroscopy by discovering the emission of characteristic X-ray lines from different elements; the two main series he christened the K and L series*—he first called them the B and A series,[14] but then, realizing there might be others as yet undiscovered, he chose two other adjacent letters from his name.

Wilberforce fully supported this work, but when Barkla left in 1909 to become Wheatstone Professor of Physics at King's College, London, laboratory research in Liverpool withered. In 1921–1922 there were only 12 physics students, compared with 224 studying chemistry, and a combined total of over 800 taking engineering or medicine (who would receive some grounding in physics).[18] By the late 1920s, the department was languishing and tended to attract local students who could not afford to travel or were not good enough to get into a more distinguished school of physics. The majority of undergraduates settled for an ordinary degree, and often only two or three were chosen to take the third year honours course. Unusually for a small English department, there was a theoretical physicist, James Rice,[19] who had been one of the first in the country to understand and appreciate the special theory of relativity. In 1923 he had published '*Relativity; a systematic treatment of Einstein's theory*', which had become the most popular university textbook on the subject in the English language. It was through him that Einstein visited the department with Lorentz during the 1920s, but such international visitors were rare. There was a brief revival of research activity in the early 1930s when two young post-graduates, Beevers and Lipson, made a name for themselves in the field of X-ray crystallography. Their success owed more to the advice and encouragement from Professor W.L. Bragg, 40 miles away in Manchester, than it did to Wilberforce, whose only contribution seems to have been epigrammatic, describing one of their findings as 'the spectre in the spectrum'.[20]

The university authorities must have been relieved when Wilberforce agreed to step down, and lost no time in finding a replacement. A subcommittee of the Faculty of Science[21] was charged with generating a list of suitable candidates and

---

* Charles Glover Barkla (1877-1944) won the Nobel Prize for Physics in 1917.

came up with twenty five names. This was progressively shortened to three candidates, including Chadwick, and testimonials were obtained. Blackett's opinion was candid:

> I think that there can be no question that you would be well advised to get Dr.Chadwick if you can. He has, in fact, directed large parts of the Cavendish Laboratory for many years, and he would certainly get a fine school of research going in Liverpool very quickly. I think it would be excellent for Liverpool to get Dr.Chadwick.

Rutherford also wrote a strong letter of recommendation, informing the University of Liverpool they would be 'fortunate if they could acquire the services of such a distinguished physicist as Chadwick.' 'I am sure', he continued, 'that under his direction the Laboratory would acquire prestige as an important centre of teaching and research. Personally, I shall be exceedingly sorry to lose the services of Dr.Chadwick, but I feel that it is important that some of our best young men should be available for posts in other Universities.' On 11 March 1935, the Faculty of Science voted unanimously to invite Chadwick to the Lyon Jones Chair from 1 October.

The only person perhaps more pleased than Chadwick himself was Aileen, who would be returning to her native city and family, and as a professor's wife. While she was always devoted to her husband and never seems to have resented his preoccupation with his work, the promise of a well-established position in familiar surroundings was very appealing. The Chadwicks had spent many weekends in Liverpool since their marriage, and James had been introduced to the large circle of the Stewart-Brown's acquaintances. He found their company very relaxing in a way that high table conversation in Cambridge was not, and had formed one or two close friendships with men such as Duncan Norman, who was the Chairman of Owen Owen Ltd (a well-known Liverpool department store). As cities, Cambridge and Liverpool were complete opposites. Cambridge was quiet, cold, aloof and inward-looking; Liverpool vibrant, teeming with raucous dockers and sailors, England's 'threshold to the ends of the earth'. Cambridge traded in ideas: Liverpool in cotton, textiles, sugar, tobacco, fruit and timber. The university had been founded in 1881 largely as a result of municipal pride, backed by donations from wealthy merchants, soap manufacturers, shipowners, brewers, chemical manufacturers and Lord Derby. The origins of the university were perpetuated in its 'named buildings and professorships' which 'read like a social register, a Who's Who' of nineteenth-century Liverpool.[22]

By 1935, the prolonged slump in world trade had brought endemic unemployment and widespread hardship to the city. The fine eighteenth-century mansions which had been built for successful merchants had degenerated into slum properties, each housing five or six large families. The brick and terracota university buildings were situated on Brownlow Hill, and until 1930 they looked out onto the Workhouse which had been the shelter of Liverpool's destitute, the majority of Irish Catholic descent, for 150 years. The Workhouse was finally closed in 1928 and it was decided to construct a new Catholic Cathedral on the site. The Roman Catholic Archbishop of Liverpool, Archbishop Downey, was an ambitious man and Sir Edwin Lutyens was commissioned to design the Cathedral. *The Illustrated London News* published an article, '*Liverpool of the future*', in October 1934, with an artist's impression of Lutyens' magnificent domed design superimposed on an aerial photograph of the city. Work began on the crypt, but Archbishop Downey neglected to obtain approval for the project from Rome.[23] This omission would have an unforseeable consequence for the Physics Department of the university.

Although he had got to know the city reasonably well, Chadwick had not had cause to visit the university or the dingy environs of Brownlow Hill. His first contact was with the Vice-Chancellor, Hector Hetherington, who had been instrumental in Chadwick's acceptance of the Chair, promising generous funding to support expansion of the department and laboratory research. This was much needed as Chadwick[4] discovered when he paid his preliminary visit to the George Holt Laboratory.

> When I went there, I found a laboratory which was much behind the times . . . I was quite shocked. Little oil lamps were used to illuminate the galvanometer mirrors. You can imagine the state of the laboratory.
>
> Now, that didn't put me off, because it meant that even if I couldn't do very much, at least I could put the laboratory straight, and I knew that I would get as much help from the University authorities as they could possibly give me. That Hector Hetherington made quite clear. But, you see, some of the chief people of the University Council, the business people, were friends of my wife's family, perhaps even connected by marriage.

The laboratory had not changed much since it had first opened 30 years previously. The electricity supply was still direct current and the few machines in the workshop were operated by belts driven from an overhead shaft. But there was plenty of unused space, and in his mind's eye, Chadwick could already see a cyclotron standing in the basement. He felt that his decision to leave the Cavendish was inevitable and had not discussed it much with Rutherford for fear of upsetting him. When the announcement was made, Chadwick was delighted to receive the following note from J.J. Thomson,[24] recognizing his new status.

> Cavendish Lab.
> Cambridge
> 25 March 1935
>
> My dear Professor,
>
> I did not hear of your appointment to Liverpool until last evening as I had missed it in the Times. My first feeling was—what a loss this will be for Cambridge. You have done magnificent work here, and we owe you a big debt of gratitude. I think that for your sake your decision is a wise one—you ought to have a laboratory under your own control and be able to turn research into the directions you think most promising. I think the people who have most to say in University matters in Liverpool are very open minded and have over and over again shown great generosity. I hope that you and Mrs. Chadwick will find Liverpool everything you could wish.

Yours most sincerely,
J.J. Thomson

As he was preparing to leave Cambridge at the end of the summer term, Chadwick asked Goldhaber to prepare a draft of a summary paper on their work on the nuclear photo-effect. Goldhaber[25] came to his house, and surrounded by packing cases, Chadwick sat down to write a final manuscript in his immaculate hand, as Goldhaber dictated. At the end of the session, Chadwick handed Goldhaber a piece of paraffin that he had employed to knock out protons in the discovery of the neutron, saying: 'I don't need it; do you want it?' Goldhaber gladly accepted the trophy, which is now in the Science Museum, South Kensington. Their hand-written paper was received by the Royal Society at the beginning of August and subsequently published in the *Proceedings.*[26] They reviewed their experiments with the deuteron, (the term *diplon* had apparently been dropped by then), and compared their findings with the theoretical predictions made by Bethe and Peierls, finding satisfactory agreement. Again they concluded that the neutron was heavier than the proton, making it an elementary particle. There was one new twist: Goldhaber[25] had shocked himself with the realization that a free neutron (*i.e.* not one bound in a nucleus) might undergo radioactive decay. At the end of their discussion, Chadwick and Goldhaber quietly introduced the extraordinary idea that an elementary particle might have a lifetime:*

> If the neutron is definitely heavier than the hydrogen atom, then one must conclude that a free neutron is unstable, i.e., it can change spontaneously into a proton + electron + neutrino, unless the neutrino has a mass of the order of the mass of the electron. These speculations must, however, depend on more exact measurements of the masses of hydrogen and deuterium.

Chadwick had started planning his new department almost as soon as he had accepted the appointment, and in May 1935 the Leverhulme Trustees agreed to fund a new lectureship in physics, the first holder to be Dr Norman Feather. Rutherford does not seem to have minded Chadwick's poaching, but he managed to lure Feather back to Cambridge within the year by arranging a fellowship for him at Trinity College. When the time for Chadwick's parting came, Rutherford was 'most kind and understanding' and showed him 'no reproach, but hope for the future, more than I deserved'.[11] He wrote Chadwick a farewell letter in August 1935 recording his heartfelt gratitude:[27]

> Before you finally leave us, I would like to express to you on behalf of the Staff and the Research Students in the Cavendish Laboratory, our deep appreciation of the the services you have rendered to the laboratory for so many years.
>
> In particular, I would like to tell you how much I have personally valued your cooperation in the development of the research activities of the laboratory, and the

---

* Goldhaber (1993) says that such a concept 'now appears trivial'. He calculated in 1935 that the half-life for a free neutron should be 'about half an hour'. It is now known to have a half-life of about a quarter of an hour, and the decay process is understood in terms of a constituent down quark changing to an up quark with the emission of a virtual W boson.

way in which you have thrown yourself wholeheartedly into the life of the Cavendish. During your time of office there has been an unprecedented development in the research activities here, and much of the credit of this is due to your help and guidance. The rapid changes in technical methods during this time have made the work of close direction of researches much more difficult and important, and the weight of this has largely fallen on your shoulders. We are grateful too for the strong personal interest you have taken in helping men to obtain suitable posts in the Universities and in Industry.

We shall all miss your presence and help in the laboratory, but we hope that in your new post you will find a still wider field for your abilities. I wish you all success and happiness at Liverpool, and I hope that you will still keep a keen interest in the laboratory with which you have been so long connected.

Chadwick had not seen his own parents more than once or twice since his marriage, but he did inform them of his move. Ann Chadwick,[28] showing a true mother's love and a Dickensian turn of phrase, wrote to Rutherford and sent him a piece of embroidered lace.

Dear Lord,

I am sending you a small present in appreciation for the kindness you have given to our son during the time he has been under your supervision both at Manchester and Cambridge universities.

We are proud that our son as had such a distingwished a gentleman to help him.

I hope you will accept this small token as it is a piece of my work.

Yours gratefully,
A.M.Chadwick

The arrival of the new professor with his zeal for research and personal determination soon revolutionized the Liverpool Physics Department. The first to experience the new broom was Mr Welch, who 'had been used to acting rather like a gentleman's valet to Wilberforce, for example meeting him when he arrived in the laboratory each morning and taking his hat and coat, but his attempts to provide the same service for the new head were brushed aside unceremoniously'.[14]

Wilberforce's one useful legacy to the department was the stock of apparatus for lecture demonstrations, and Chadwick made good use of this. He now had to give lectures to undergraduates and needed to fill in some gaps in his own knowledge of classical physics that he had missed when an undergraduate in Manchester. He gave a course on optics, and John Holt[14] attended the first lecture.

I remember his entrance, a tall bird-like figure, followed by the chief technician Welch, whose job it was to assist with the slide lanterns and demonstrations. After glaring at us for a few moments he looked down in thought and then began a concise presentation of the wave optics of lenses, using diagrams which he drew on the blackboard using board compasses. It was a beautifully prepared course which progressed through physical optics to electro-magnetic waves . . . He liked to make these demonstrations quantitative whenever possible, and would carry out the measurements, commenting on the likely errors, and finally work out the results on the board. His spontaneous grin when the

> results agreed with the expectations conveyed something of his pleasure in experimental physics, and indicated that it had not all been rehearsed beforehand.

Holt had been taught at school that all matter is composed of protons and electrons, and during his first year at Liverpool had heard nothing to the contrary. He 'had seen reference in the popular press to a neutral proton or neutron, but had dismissed these as science fiction. The arrival of Chadwick, the discoverer of the neutron, whose name I then heard for the first time, altered all that and opened up the prospect of new unexplored territory.' Surprisingly, Chadwick only ever gave the course on optics, and never taught the undergraduates nuclear physics, for example. He never overcame his dislike of lecturing, but would insist at the start of every year that he be put down for the optics course since he felt it was an obligation. The teaching schedule was organized by Roberts, the senior lecturer, who had been appointed by Wilberforce. He always had to be ready to follow Chadwick's notes for the course because he would often get a message from the professor at the last minute saying he had got backache or had to attend a university meeting and Roberts would have to substitute for him.

The new professor quickly established himself as a force in the general affairs of the university. His technique was usually to listen carefully the first time a subject was presented, often sucking or fiddling with his pipe. By the time of the next meeting, Chadwick would be ready to address the issue incisively in measured, deep tones. The clarity and unmistakeable quality of his judgements more than compensated for his nervous presentation, when he would furiously rotate his pipe between both hands and appear to be addressing himself to an unseen audience beneath the committee table. He made a great impression on James Mountford,[29] the Professor of Latin and later Vice-Chancellor, who was struck that he showed 'no backward glance of regret for Cambridge'. Mountford felt 'here was a man whose patently disinterested devotion to excellence in any academic sphere inevitably demanded and gained attention, respect and eventual gratitude. In Senate and its committees, as in Faculty, he stripped sloppy thinking of its pretensions and though invariably kind to individuals and sensitive to their difficulties, he was never deflected from his own high standards of judgement.'

Naturally, Chadwick's standing both in the university and the Royal Society was boosted by his winning the 1935 Nobel Prize for Physics.[30] The prize had not been awarded in 1934, in part because the Nobel Committee could not decide between Chadwick and the Joliot-Curies as recipients. Heisenberg had been minded to support a joint award in 1934, but mislaid the Nobel Committee's letter under a heap of old papers on his desk for six months, so missing the deadline. In 1935, he nominated the Joliot-Curies again, and Chadwick received support from scientists in England, France, Germany, Japan, Sweden and America for his discovery of the neutron. Naturally, his most influential backer was Lord Rutherford, who wrote a four page letter recapping Chadwick's contributions since the war, beginning with his demonstration that the charge on the nucleus is numerically equal to its atomic number. In his concluding paragraph Rutherford wrote:

All the work of Chadwick is characterised by originality of method, accuracy of measurement, and judgement in interpretation of results. He has played a notable part in the pioneer attack on the properties of nuclei and on the transformation of the elements, an attack which, after a period of slow advance to reconnoitre the ground, is now being pushed home rapidly with the aid of new and powerful technical methods.

It was Rutherford[31] too who suggested a way out of the impasse by suggesting the Joliot-Curies for the Chemistry Prize. When Chadwick learned of the award in November 1935, he[32] wrote to Rutherford thanking him for his support. He had already exchanged congratulatory telegrams with the Joliot-Curies and described his emotions to Rutherford with typical brevity: 'In a sense I was not terribly surprised but I was completely staggered when it happened. One of the nicest things about it is the letters and telegrams from all parts.' There were letters from the Masters of two Cambridge colleges, redolent of the life he had just left there.

The Lodge
Gonville and Caius College
Cambridge
14 November 1935

Dear Chadwick,
Stammers has just rung me up with the news—not of the first General Election results—but of your Nobel Prize, as a spot item on the wireless 'Late News'. It is splendid. I only wish you were here so that we might drink your health in the port over which you exercised a watchful care.

The wireless man (so I am told) described you as a Fellow of Gonville and Caius College. I hope the day may come when that description will once more be true.

Yours sincerely,
J. Cameron[33]

Trinity Lodge
Cambridge
17 November 1935

My dear Chadwick,
Very hearty congratulations on the Nobel Prize. With that and C.T.R.'s Copley Medal it has been a great week for the Cavendish.

Yours most sincerely,
J J Thomson[34]

He also heard from Niels Bohr,[35] who thought the prize was 'so well deserved an acknowledgement of your splendid work', and invited him and Aileen to stay in Copenhagen on their way back from Stockholm. Ernest Lawrence[36] sent an effusive letter from California, telling Chadwick that he should have received the award 'long ago'. Lawrence had no doubt that everyone at the Cavendish was missing Chadwick, but took comfort from the fact that his move to Liverpool 'means that there is to be another great center of nuclear physics in England'.

In his replies, Chadwick was self-effacing, telling both men that he was having difficulty in accepting that he deserved the honour. The prize was to be given in Stockholm in the middle of December which left little time for preparation. The family were in a state of high excitement and Aileen went out to buy a number of evening gowns so that she would not have to wear the same one twice. Chadwick was preoccupied with writing his prize lecture, 'rather a trying ordeal',[32] on *The neutron and its properties*.

Their stay in Stockholm was memorable for a number of small gaffes which would have caused temporary embarrassment to any distinguished visitor not wishing to transgress the elaborate protocol of such a grand occasion, and which absolutely mortified Chadwick.[4] First he did not know there would be a reception committee awaiting their arrival at Stockholm station and got straight into a taxi to the Grand Hotel without seeing them. The committee thought the prizewinner had got lost! Next he was assigned a guide, 'a Count von Tolstoy, some relative of the famous Tolstoy,' who was supposed 'to look after me generally and to see that I got to the places at the right moment, to tell me what was about to happen . . . But he wasn't very good'. Chadwick did get to the presentation ceremony in the Stockholm Concert House on time and correctly dressed to receive his prize from King Gustav. The chairman of the Nobel Committee for Physics, Professor H. Pleijel, made a lengthy speech in Swedish and English about the significance of Chadwick's work, and the events that followed were described in the *Liverpool Daily Post* on 11 December.

> Two sailor trumpeters blew a blast, and Professor Chadwick was presented by King Gustav with the certificates, the souvenir, and the cheque.* Returning to his place, Professor Chadwick dropped the cheque. Prince Wilhelm retrieved it and handed it to him. When the Professor returned to his place on the platform, the orchestra played the 'Pomp and Circumstance' march.

Chadwick[37] delivered his Nobel Lecture without a hitch two days later, and was followed by the Joliot-Curies who each gave a short talk. In a reversal of their usual roles in the laboratory, Irène gave a short account of the physical processes, while Frédéric described the chemistry. His presentation contained one strikingly prophetic passage.

> If, turning towards the past, we cast a glance at the progress achieved by science at an ever-increasing pace, we are entitled to think that scientists, building up or shattering elements at will, will be able to bring about transmutations of an explosive type, true chemical chain reactions. If such transformations do succeed in spreading in matter, the enormous liberation of usuable energy can be imagined.[38]

Chadwick[4] was impressed by Frédéric Joliot's lecture, not so much for its scientific content nor the concluding remarks quoted above, which probably struck him as fanciful, but for the panache of its delivery. 'He was a great actor. He liked that kind of thing, and he did it very well.'

---

* The Nobel Prize for 1935 had an approximate monetary value of 8,350 pounds.

Chadwick could not hope to match Joliot's poise and he was not helped by his guide, Tolstoy, who would suddenly announce a move to a new venue so that Chadwick[4] was never quite sure what would be coming next or what was to be expected of him. There were a number of large and formal banquets which may have suited Aileen, but not James. Rutherford[39] warned him in a letter, which unfortunately arrived after Chadwick left for Stockholm, that 'you will no doubt be expected to make a speech at the dinner, but with a glass of champagne inside you that ought not to present any difficulties'. So Chadwick's first intimation that he was expected to speak at the grand dinner, given by Prince Wilhelm, was when the prizewinner in physiology stood up and made a well-crafted speech from prepared notes. He was seized by panic.

> Well, I was in a dreadful pickle. I was extremely nervous by that time, having gathered that speeches were going to be made, and not the faintest idea what to say. I mumbled a few words and left it at that.[4]

The following day, the Chadwicks[40] visited Uppsala with their friends the Siegbahns and saw Professor Svedberg, who had been one of Chadwick's supporters for the Nobel Prize. They left Sweden on Saturday 14 December and arrived back in Liverpool via London on the Monday night. On Tuesday morning at 9 am, Chadwick began examining students.

He finished a momentous year laying the groundwork for his ambitious plan to install a cyclotron in the department. When Lawrence[36] wrote to him to send congratulations on the Nobel Prize, he mentioned that A.P.M. Fleming, the Research Director of the British electrical engineering firm Metropolitan-Vickers, had visited Berkeley and 'discussed the possibility of building a magnetic resonance accelerator [cyclotron] for you'. Fleming had first got to know Rutherford through their common involvement in scientific research applied to military needs in the Great War, and the relationship continued later through the DSIR. 'Metro-Vick', largely because of Fleming, became a great supporter of the Cavendish after the war. They supplied talented research workers like Cockcroft and Allibone as well as much of the heavy equipment for Kapitza. Through this connection, Fleming had got to know Chadwick well and was fully aware of his plans for Liverpool. Chadwick[41] replied to Lawrence on 29 December:

> I am exceedingly pleased to hear from you that Fleming of Metropolitan Vickers is thinking of helping me to build your magnetic resonance accelerator, which ranks with the expansion [cloud] chamber as the most beautiful piece of apparatus I know. As a matter of fact I have been turning such a project over in my mind for some time, but I had to see what the laboratory really needs before thinking of my own wishes . . . It is very good of you to offer to help with the drawings and specifications, and, I hope, with your advice. Although the method is so beautifully simple in principle I am sure there must be many troubles before one can get it to work smoothly.

Chadwick was also mindful of recruiting some staff with experience of running a cyclotron, and the obvious candidate was Bernard Kinsey, a Cambridge graduate, who had been Oliphant's research student at the Cavendish. According to

Oliphant,[10] Kinsey was an excellent scientist, and 'in voice and manner, was a character straight from Wodehouse'. After working at the Cavendish, he had spent three years at Berkeley, on a Commonwealth Fellowship, learning the practicalities of building and operating cyclotrons. Chadwick continued:

> In connection with the accelerator there was a further matter I wanted to mention. Kinsey is obliged to return to England at the end of the year. I was thinking it might be possible to offer him a temporary post here to help with the accelerator. As things are at present I cannot ask the University for money to create a post. I shall probably have a small studentship or fellowship to offer and the rest I may have to provide myself. Do you think Kinsey has sufficient technique? He was only a beginner when he came to you but he seems to have developed well, and I like him personally. I should be very glad to have your opinion on Kinsey.

Finally, Chadwick posed the hard questions concerning the capital costs and running expenses for such a project. He thought such considerations might seem mundane to Lawrence:

> It might sound rather absurd to enquire about running costs but you would be surprised to know what this laboratory has been running on in the last few years—less than some men spend on tobacco.

As a practical step, Chadwick completely re-equipped the laboratory workshop so that many of the components of the cyclotron could be made on the spot. The renovation cost about seven hundred pounds, but he knew that some of the larger electrical parts could only be purchased from Metropolitan-Vickers and their cost would exceed any amount that the University could afford.

References

I am indebted to the University Archives of the University of Liverpool for giving me access to material relating to Chadwick's appointment to the Lyon Jones Chair of Physics. Dr David Edwards and Professor John Holt, both physicists from the University of Liverpool, kindly gave me unpublished manuscripts of lectures they had given about Chadwick. I would like to acknowledge the generosity of the Nobel Committees for Physics and Chemistry of the Royal Swedish Academy of Sciences in allowing me to see correspondence relating to Chadwick's nomination. The letters between Chadwick and Lawrence are quoted through the courtesy of the Bancroft Library, University of California, Berkeley, owners of the Ernest O. Lawrence papers, and with the permission of Sir James Chadwick's daughters.

1 Lawrence, E.O. (1965). The evolution of the cyclotron. In *Nobel lectures, physics 1922–1941*, pp. 425–43. Elsevier, Amsterdam.

2 Cockcroft, J. (1984). Some recollections of low energy nuclear physics. In *Cambridge physics in the thirties* (ed. J. Hendry), pp. 74–80. Adam Hilger, Bristol.

3 *New York Times* (28/10/33).

4 Weiner, C. (1969). Sir James Chadwick, oral history. American Institute of Physics, College Park, Maryland.

Plate I Staff and research students, University of Manchester Physics Department, 1913. [Names from left to right. Front row: H. Robinson, D.C.H. Florance, Miss M. White, J.N. Pring, Prof. E. Rutherford, W. Makower, E.J. Evans, C.G. Darwin. Second row: A.B. Wood, E. Green, R.H. Wilson, S. Oba, E. Marsden. H. Gerrard, J. Chadwick, F.W. Whaley, H.G.J. Moseley. Third row: H. Richardson, J.M. Nuttall, B Williams, W. Kzy. Back row: T.S. Taylor, A.S. Russell.]

Plate 2 Civilian prisoners at Ruhleben, 1918, assembled in front of one of the stable blocks.

Plate 3 Staff and research students at the Cavendish Laboratory, Cambridge, 1923. [Names from left to right. Front row: J. Chadwick, G. Stead, F.W. Aston, Prof. Sir J.J. Thomson, Prof. Sir E. Rutherford, J.A. Crowther, Miss B. Trevelyan, G.I. Taylor. Second row: P. Kapitza, H. de W. Smyth, T. Alty, J.E. Crackston, H. Robinson, L.F. Curtiss, E.S. Bieler, A.G.D. West, P. Mercier. Back row: P.M.S. Blackett, R.E. Clay, H.W.B. Skinner, H.D. Griffith, A.W. Barton, L.F. Bates, J.S. Rogers, K.G. Emeleus.]

Plate 4 Aileen Stewart Brown photographed in 1925, the year of her marriage to James Chadwick.

Plate 5 The photograph which gave Rutherford 'a real good laugh'. Peter Kapitza (right), in borrowed top hat, is James Chadwick's best man.

Plate 6 The room which Rutherford and Chadwick used for their scattering experiments in the 1920s. The work was carried out in the dark, often to the accompaniment of Rutherford singing 'Onward Christian Soldiers'.

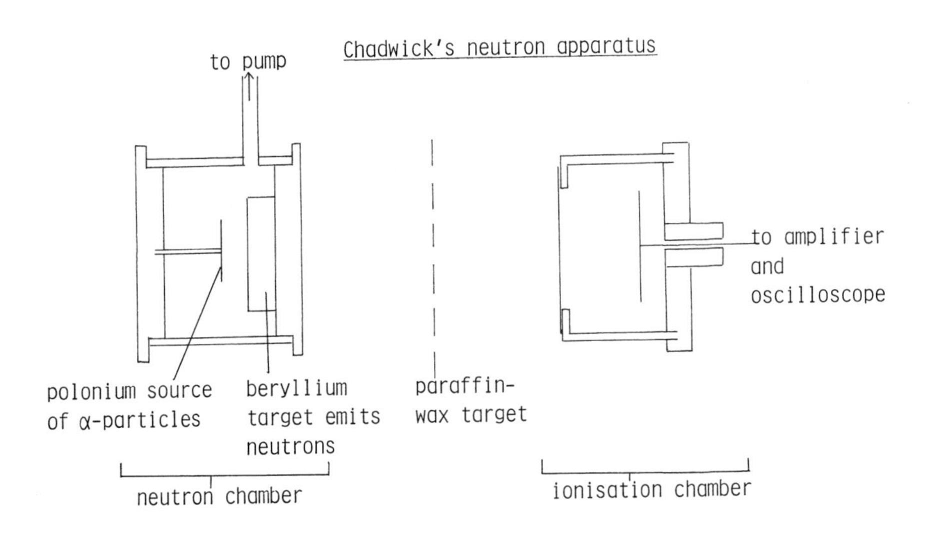

Plate 7 Chadwick's neutron chamber in which α-particles from a polonium source, at the right-hand end, bombarded a beryllium source, at the left-hand end. From there, neutrons emerged to pass through a paraffin wax target, releasing enough protons to register on an oscilloscope. A vacuum pump was attached to the finger-like chimney.

Plate 8 Chadwick's laboratory with its vibration-sensitive equipment was through the open door. Rutherford is talking to J.A. Ratcliffe.

Plate 9 Staff and research student of the Cavendish Laboratory, 1933. The photograph was taken in front of the new Mond Laboratory with Eric Gill's carving of the crocodile commissioned by Kapitza to honour Rutherford.[Names from left to right. Front row: Miss Sparshott, J.A. Ratcliffe, G. Stead, J. Chadwick, G.F.C. Searle, Prof. Sir J.J Thomson, Prof. Lord Rutherford, Prof. C.T.R. Wilson, C.D. Ellis, Prof. Kapitza, P.M.S. Blackett, Miss Davies. Second row: J.K. Roberts, P. Harteck, R.C. Evans, E.C. Childs, R.A. Smith, G.T.P. Tarrant, L.H. Gray, J.P. Gott, M.L. Oliphant, P.I. Dee, J.L. Pawsey, C.E. Wynn-Williams. Third row: B.B. Kinsey, F.W. Nicholl, G. Occhialini, E.C. Allberry, B.M. Crowther, B.V. Bowden, W.B. Lewis, P.C. Ho, E.T.S. Walton, P.W. Burbidge, F. Bitter. Fourth row: C.B.O. Mohr, N. Feather. C.W. Gilbert, D. Shoenberg, D.E. Lea, R. Witty, Halliday, H.S.W. Massey, E.S. Shire. Back row: W.J. Henderson, W.E. Duncanson, P. Wright, G.E. Pringle, H. Miller.]

Plate 10 The Seventh Solvay Council, Brussels, 1933, which was devoted to the atomic nucleus. [Names from left to right. Seated: Schrödinger, Irène Joliot-Curie, Bohr, Joffé, Marie Curie, Richardson, Langevin, Rutherford, De Donder, Maurice de Broglie, Louis de Broglie, Meitner and Chadwick. Standing: Henriot, Perrin, Frédéric Joliot-Curie, Heisenberg, Kramers, Stahel, Fermi, Walton, Dirac, Debye, Mott, Cabrera, Gamow, Bothe, Blackett, Rosenblum, Erra, Bauer, Pauli, Verschaffelt in front of Cosyns, Herzen, Cockcroft, Ellis, Peierls, Piccard, Lawrence and Rosenfeld.]

Plate 11 Dr Chadwick, sucking on his pipe, makes one of his daily tours of the Cavendish Laboratory to review progress. Maurice Goldhaber, with Norman Feather standing behind him, is studying a recording of kicks in an ionization chamber. The figure at Chadwick's side is Mr Lincoln, a technician who joined the Laboratory 1892.

Plate 12 Chadwick at the time of his appointment to the Lyon Jones Chair at Liverpool University in 1935.

Plate 13 Receving the 1935 Nobel Price from King Gustav in Stockholm. Moments later Chadwick would drop the cheque!

Plate 14 Delivering the Nobel Lecture.

Plate 15 The 37-inch Liverpool cyclotron, completed in 1939.

Plate 16 General Leslie Groves and Sir James Chadwick, Washington DC.

Plate 17 At a meeting in Cambridge in 1947 with Rudolf Peierls (left) and G.I. Taylor (right).

Plate 18 Master of Gonville and Caius College.

5 Oliphant, M.L. (1966). The two Ernests, I & II. In *History of physics: readings from Physics Today* (ed. S.R. Weart and M. Phillips, 1985), pp. 173–93. American Institute of Physics, New York.

6 Weiner, C. (1972). 1932—Moving into the new physics. *Physics Today*, (May), 40–49.

7 Wood, A. (1946). *The Cavendish Laboratory*, p. 48. Cambridge University Press.

8 Wilson, D. (1983). Politics and power. In *Rutherford: simple genius*, pp. 453–95. Hodder and Stoughton, London.

9 Moseley, R. (1980). Government science and the Royal Society. *Notes and Records of the Royal Society*, **35**, (2), 186.

10 Oliphant, M. (1972). *Rutherford—recollections of the Cambridge days*, Elsevier, Amsterdam.

11 Chadwick, J. (?1971). Undated notes on Oliphant's biography of Rutherford. CHAD II, 2/1, CAC.

12 University Archives. *Chadwick file*, D 97/1/1, (undated), University of Liverpool.

13 Edwards, D. (1991). *Sir James Chadwick (1891–1974)*. Lecture to the British Association, Plymouth.

14 Holt, J. (1988). *Seminar on James Chadwick*. University of Liverpool. See also Holt, J. (1994). Reminiscences and discoveries: James Chadwick at Liverpool. *Notes and records of the Royal Society*, **48**, (2), 299–308.

15 Eve, A.S. (1939). *Rutherford*, p. 31. Macmillan, New York.

16 Wilson, D. (1983). The wide, wide world. In *Rutherford: simple genius*, pp. 63–86. Hodder and Stoughton, London.

17 Allen, H.S. (1947). Charles Glover Barkla, 1877–1944. *Obituary notices of fellows of the Royal Society*, **5**, 341–66.

18 Williamson, R (1987). *The making of physicists*, pp. 180–1. Adam Hilger, Bristol.

19 Donnan, F.G. (1936). Prof. James Rice, obituary. *Nature*, **137**, 807–8.

20 Lipson, H. (1987). Physics in a minor department: 1927–36. In *The making of physicists* (ed. R. Williamson), pp. 94–100. Adam Hilger, Bristol.

21 Thompson, J. (1935). *First report on the chair of physics*. University Archives, S.3039, University of Liverpool.

22 Lane, T. (1987). *Liverpool: gateway of empire*. Lawrence and Wishart, London.

23 King, C.D. (1993). The design and construction of the Liverpool 156" synchrocyclotron. Unpublished M.Sc. thesis, University of Liverpool.

24 Thomson, J.J. (1935). Letter to J. Chadwick. CHAD II, 1/17, CAC.

25 Goldhaber, M. (1993). Reminiscences from the Cavendish Laboratory in the 1930's. *Annual Review of Nuclear and Particle Science*, **43**, 1–25.

26 Chadwick, J. and Goldhaber, M. (1935). The nuclear photoelectric effect. *Proceedings of the Royal Society A*, **151**, 479–93.

27 Rutherford, E. (19/8/35). Letter to J. Chadwick. Rutherford's papers, University of Cambridge Library.

28 Chadwick, A.M. (1935). Undated letter to Lord Rutherford. Rutherford papers, University of Cambridge Library.

29 Mountford, J. (1974). *University of Liverpool Recorder*, **66**, 15.

30 Chadwick, J., nominations for the Nobel Prize. Center for the History of Science, Nobel Committees for Physics and Chemistry, The Royal Swedish Academy of Sciences.

31 Rutherford, E. (26/11/35). Letter to J. Chadwick. Rutherford's papers, University of Cambridge Library.

32 Chadwick, J. (20/11/35). Letter to E. Rutherford, Rutherford's papers, University of Cambridge Library.

33 Cameron, J. (14/11/35). Letter to J. Chadwick. CHAD II, 1/17, CAC.

34 Thomson, J.J. (17/11/35). Letter to J. Chadwick. CHAD II, 1/17, CAC.

35 Bohr, N. (3/12/35). Letter to J. Chadwick. Niels Bohr Archive, Copenhagen.

36 Lawrence, E.O. (27/11/35). Letter to J. Chadwick. Lawrence's papers, Bancroft Library, University of California, Berkeley.

37 Chadwick, J. (1935). The neutron and its properties. In *Nobel lectures: physics, 1922–1941*, pp. 339–48. Elsevier, Amsterdam, 1965.

38 Joliot, F. (1935). Chemical evidence of the transmutation of elements. In *Nobel lectures: chemistry, 1922–1941*, pp. 369–73. Elsevier, Amsterdam, 1965.

39 Rutherford, E. (7/12/35). Letter to J. Chadwick. Rutherford's papers, University of Cambridge Library.

40 Chadwick, J. (12/12/35). Letter to N. Bohr. Niels Bohr Archives, Copenhagen.

41 Chadwick, J. (29/12/35). Letter to E.O. Lawrence. Lawrence's papers, Bancroft Library, University of California, Berkeley.

9 His own man—and Rutherford's

When Chadwick took his leave of the Cavendish, it was the second of three notable exits from the laboratory, each of which came as a surprise to its director, Lord Rutherford. The first and most dramatic departure was that of Peter Kapitza, who was not allowed to return from the Soviet Union after making an annual summer visit there in 1934 to see his mother. Rutherford first learned of Kapitza's detention from his wife, Anna, when she returned to Cambridge alone in October 1934. This was a keen personal loss for Rutherford, who had come to delight in the vitality and outlandishness of Kapitza, providing, as he did, an amusing contrast to the English reserve of many of the Cavendish scientists. He had actively supported Kapitza's career at every turn, so that the young Russian had scaled the bastions of the English scientific establishment, becoming in turn a Fellow of Trinity College, a Fellow of the Royal Society and then a Royal Society Messel Professor. Kapitza had always managed to secure larger funding for his work on intense magnetic fields and low-temperature physics than any other Cavendish researcher, and ultimately in 1933 had obtained his own purpose built institute, the Mond Laboratory. Rutherford, who maintained a deep-seated distrust of communism, seems to have sensed that Kapitza was placing himself in jeopardy by his visits home, and would warn him, only half in jest, 'They'll getcha'.[1] Ironically, Kapitza's conspicuous preferment in Rutherford's domains may have led to the unwelcome attention of the Soviet government and also to a level of resentment amongst some English colleagues.

Rutherford suggested in a letter that

> Kapitza in one of his expansive moods in Russia told the Soviet engineers that he himself would be able to alter the whole face of electrical engineering in his lifetime. I believe the Soviets now recognise that these chances are very remote. This seems to be a very probable explanation of their action and is due to our friend's love of the limelight.[2]

The Soviets were trying to expand their economy by means of a series of Five Year Plans which placed heavy demands on scientific and technical manpower. Kapitza, regardless of whether they listened to his boasting or not, was too valuable to pass up, and the Soviet Ambassador in London, Ivan Maisky, told Lord Rutherford so quite bluntly. After Rutherford had written to him privately 'to hasten the return of Professor Kapitza', Maisky replied that the USSR now 'needs all the scientific workers who have hitherto been working abroad'.[3] For about six months, Rutherford kept trying to pressure the Soviets in private, for fear that publicity would only harden their position. He mobilized all his political and international scientific connections, but with no success. In March 1935, the Soviets made their first overture to Rutherford about purchasing all Kapitza's special equipment from the Mond Laboratory in order that he could continue his work in Russia. The following month any chance of a private settlement evaporated when the British newspapers latched on to the story and the whole affair became public knowledge. Rutherford wrote a thoughtful letter to *The Times* asking that Kapitza be allowed to return, at least for a short period to complete his work in hand at the Mond Laboratory. This plea was effectively demolished by a rejoinder from H.E. Armstrong, a chemist and senior Fellow of the Royal Society, who disapproved of the presence of foreign scientists in senior positions at English universities, and suggested:

> Instead of leading a lotus life at Cambridge, he, too, may well be doing national work [in Russia] of a far higher importance than even that involved in magnetizing atoms to destruction.[2]

While Kapitza's colleagues and friends at the Cavendish were genuinely sorry about his plight, there was the feeling in several quarters that he had been riding for a fall. Dirac remarked that: 'He had trodden on too many toes. He was always impatient with important people who wanted to see his lab. He wasn't too polite.'[1] Blackett,[4] with whom his relationship was always a little strained, felt that Rutherford had given a disproportionate share of support and funds to Kapitza. Chadwick shared this opinion and had objections to the Mond Laboratory because he thought some of the costly engineering equipment was unnecessary. He confronted Kapitza with his views, but not Rutherford 'who was unapproachable on this whole business'.[5] This was particularly galling to Chadwick since Rutherford would not spend money on particle accelerators like the cyclotron, which he had come to regard as absolutely necessary for the development of nuclear physics; and nuclear physics, not magnetism nor low-temperature physics, was the centrepiece of Cavendish research.

At the stage when Lord Rutherford was attempting to obtain Kapitza's release by his own considerable personal influence, there was nothing that Chadwick could add to the campaign. When the affair became public in the spring of 1935, Chadwick still did not enter the fray on Kapitza's behalf, nor even write to him, which is harder to understand. The reason for his reticence was not any personal rift, but perhaps derived from his own experience as a prisoner in the Great War. He gave the following explanation for his inactivity many years later.

> From his entry to the Cavendish, Peter Kapitza was a valued friend. He was my best man at my wedding. Our differences in some matters did not affect our personal friendship. I did not write to him when he was detained in Russia for fear that this might do him harm; it was an unfortunate error on my part.[5]

Rutherford was still preoccupied with the Kapitza affair when Chadwick announced that he would be going to Liverpool. His departure had less immediate impact on Rutherford, but it was a break that he would come to feel deeply over the next year. Chadwick had been a fixture in his scientific life for a quarter of a century.

Of all Rutherford's scientific collaborators—Soddy, Geiger, Oliphant—Chadwick had the by far the longest association, and most resembled him in his style of working. Rutherford had a high regard for Chadwick's intellect and his gifts as an experimenter. As the years had gone by, he had come to rely more and more heavily on his judgement and ability to get things done. Chadwick became his prime source of intelligence both for happenings within the Cavendish, and, as importantly, for the scientific literature, so that the Cambridge school were always positioned to follow up any interesting findings from other laboratories. Indeed, it was long Chadwick's[6] idea to keep one or two rooms and equipment earmarked for this purpose so that a man could be put to work straightaway, but Rutherford could never agree to run a laboratory at less than what he perceived as full capacity—and this often meant fighting for equipment and space! Oliphant,[7] who succeeded Chadwick as Assistant Director of Research, wrote that 'no-one else bore anything like the burden carried by Chadwick, and no-one played so vital a part in the initiation and direction of the research work'.

Rutherford had come to think of Chadwick as a permanent number two: he had never displayed any ambition to move on and, for all his administrative skills, his quiet, austere character did not seem to fit him for leadership. For his part Chadwick always held Rutherford in awe—he had never known any other chief—and he could not imagine himself on an equal footing. Other more self-assured juniors were less tolerant of Rutherford's occasional tantrums and were more inclined to set their own course. For example, the young Harry Moseley, writing to his mother in May 1910 even before he had arrived in Manchester, was mapping out his future and could see Rutherford as 'an excellent stepping stone'.[8] At the Cavendish, Patrick Blackett was seen to emerge one day 'from Rutherford's office white faced with rage at being instructed by Rutherford to take a larger share of the lecturing burden [and said] . . . If physics laboratories have to be run dictatorially . . . I would rather be my own dictator'.[9] Chadwick was by temperament the perfect foil and would never have entertained, let alone expressed, such opinions.

While there are strong similarities in the way they thought about physics and unlocked its secrets at the laboratory bench, the two were different in most respects, with Chadwick the chalk and Rutherford the cheese. Rutherford was a larger than life, shambling figure, with the superficial demeanour of a New Zealand farmer. His booming voice and unrestrained laughter tended to dominate any social gathering. Chadwick's favourite description of him was Sir Hugh Anderson's, the Master of Gonville and Caius, who saw him as 'a great, big, lovable bear'. Chadwick was always tidy in appearance, comfortable in a buttoned three-piece suit and stiff collar, one of life's bank managers rather than a farm boy. The contrast was even present in their pipe-smoking: Rutherford's pipe was a volcanic affair spitting sparks and ash into the air, constantly in need of more tobacco and relighting; Chadwick's smoking was unobtrusive, contemplative, and the pipe was often unlit.

Over the years the two men had shared numerous conversations about topics of interest such as the neutron or nuclear forces, often in a darkened room while they were preparing for a session of scintillation counting. The flavour of these conversations is preserved in an exchange of letters early in 1936, following a visit of Niels Bohr to England. Bohr[10] was propounding a new theory of the nucleus, designed to account for some of the new observations that had been made using neutrons to bombard different elements. The central feature of Bohr's new theory was the 'compound nucleus' which supplanted the idea that the atomic nucleus was a rigid single body that exerted a fixed overall force on an impinging particle. Rather Bohr suggested that nuclear reactions should be considered as a more complex many body problem, in which the incoming neutron (or other subatomic particle) could temporarily coalesce with the bombarded nucleus to form a compound nucleus. This new entity, while not stable, would have a discrete set of energy levels (like a stable nucleus), governed by quantum mechanics. Bohr published his initial thoughts on this in 1936 and continued to refine them over the next eighteen months culminating in his famous liquid drop theory of the nucleus. Rutherford[11] wrote to Chadwick in Liverpool on 17 February 1936 to give his impressions.

> We have just had a visit from Bohr, who is full of his new qualitative theory of the nucleus. He considers that Heisenberg and Gamow & Co have all got hold of the wrong end of the stick, and that it is impossible to deal with the movements of an actual particle within the nucleus. He therefore considers that the space inside the potential barrier is a sort of 'mush' of which the vibration can be qualitively worked out on wave mechanics. One result of his theory is that there should be a number of very sharp [energy] levels each with very small energy differences which serve to account for the peculiarities of neutron absorption. He also thinks he can explain in a sensible way why some elements have such a big target area. He is publishing a note on his theory in 'Nature' shortly. I myself am inclined rather to like his point of view, and it certainly seems to help us in understanding some of the special properties of neutrons. He thinks he can explain in a natural way the levels in the radioactive bodies and he also considers that a nucleus can be excited to almost any extent without breaking up at once.

Rutherford's undiminished enthusiasm for new ideas is plain, and Chadwick was obviously at a disadvantage, not having heard Bohr firsthand. He[12] replied a week later.

> I was very interested to hear about Bohr's new idea about nuclear structure. It seems to me to be coming back to our earlier ideas when we considered the nucleus as a whole and talked about the energy state of a nucleus without putting a special particle into a special place. Hulme was here a few days ago told me a little about it, but I did not understand completely how the quantitative predictions appear. I shall have to wait until the note comes out in Nature.

Chadwick mentioned in a postscript that he had just given a lecture in the old laboratory in Manchester and 'it is still contaminated [with radioactivity] in places'. Attitudes towards radiation safety were still cavalier in the 1930s, although with the premature deaths of radiologists and a few notable physicists like Marie Curie, the risks were becoming more apparent. All the Rutherford school of physicists received radiation doses that would be regarded as dangerously high by present day standards, but as a group they were curiously lucky. The laboratory stewards who handled the sources the most were not so fortunate, and Rutherford[13] writing to Chadwick in May 1936 informed him that 'Crowe is in London undergoing a special grafting treatment on his fingers'. It was George Crowe who in 1919 had put Rutherford's precious radium from Manchester into solution, and he would regularly pump off the radon gas from the solution. His manual dexterity was crucial in preparing the materials for Rutherford and Chadwick's scattering experiments through the 1920s; he would usually ignore the laboratory rule and not wear the cumbersome protective gloves provided so that his fingers had shown signs of radiation dermatitis from 1926 onwards*.

Spurred on by Lawrence, Chadwick was now resolved to press ahead with the cyclotron for Liverpool. Lawrence had sent a characteristically enthusiastic reply to Chadwick's enquiries about the costs involved.[14] He wrote:

> Perhaps a satisfactory conception of the expense involved in a cyclotron {slang for the magnetic resonance accelerator} may be obtained from the cost of the cyclotron being installed in the laboratory at Princeton . . . For the complete installation, including various accessories, motor generator, etc.–in other words including all equipment needed for embarking on a program of nuclear investigations, they expect to spend a total of about $12,000.

He had no doubt that

> Kinsey would be an awfully good man for you. He is a very hard worker . . . very eager to have an opportunity to help with the construction and use of [a cyclotron] in England . . . said that nothing would please him more than an opportunity to assist you in such an undertaking. I feel confident that with Kinsey on your staff a cyclotron could be constructed and put into satisfactory operation in well within a year's time.

In case Chadwick had any lingering worries, Lawrence provided a reasssuring anecdote to counter the 'general impression that the cyclotron is a very tricky and difficult apparatus to operate'. Their most recent machine designed by a relative novice had produced 'a beam of 4.5 MV deuterons . . . immediately upon turning the switches'. His infectious salesmanship found a ready taker in Liverpool, and Chadwick[15] was now proceeding full-steam ahead. He met Fleming at a dinner and told him that he was 'intensely interested' in the cyclotron, and when Fleming's estimate of the cost was double what Princeton expected to pay, it did not mean that he had 'given up hope—far from it'. Chadwick wrote back to Lawrence saying he was 'looking around to see if I can pick up a little more money' so that he could provide a post for Kinsey.

Chadwick's original plan for fund raising was to tap some of his Liverpool business friends for donations, but then he discovered an obscure fund administered by the Royal Society which had been endowed by 'a man who had made his money in Liverpool'.[6] His application for a £2000 grant 'to build a magnetic resonance accelerator of the type devised by Professor E.O. Lawrence of Berkeley, California',[16] was made in March 1936. Chadwick pointed out that several such cyclotrons were under construction in the United States, and he thought it was 'highly

---

* George Crowe returned to his position after plastic surgery and gave skilled service to the Cavendish Laboratory until his retirement in 1959—a career of over 50 years.

desirable that at least one should be built in this country'. He outlined an extensive programme of research which would include basic physics, using the cyclotron as a means of producing a copious supply of neutrons. It could also be a source of new radioactive isotopes and some of these such as radioactive carbon and nitrogen could be used as indicators to unravel biochemical processes, which were of increasing interest in medicine. Finally he suggested that some of the direct biological effects of neutrons and radioactive isotopes could now be studied, again with possible applications in cancer treatment. This was an ambitious, but well conceived, multi-disciplinary programme for Liverpool University, and Chadwick presented it to the Royal Society with Hetherington, the Vice-Chancellor and a philosopher, in support. Hetherington was able to tell them that the University had already promised £2000 towards what Chadwick thought would be a total capital cost of £4000–£5000. The Royal Society committee was used to making grants of £100 or less, but they were persuaded of the plan's merits, and Chadwick and Hetherington returned to Liverpool with the money.

Within a few days, Chadwick received another enthusiastic letter from Lawrence,[17] who told him that he had gained the 'distinct impression' from Fleming that Metropolitan-Vickers 'would be willing to stand a large share of the cost' and 'they should be glad to undertake the construction at once'. He continued with some exciting news.

> Fleming's interest in the cyclotron is primarily connected with the biological applications, particularly the treatment of cancer, and since he was here there has been a very important development, which I think you should pass on to him. It is that we have recently carried out experiments on the comparative effects of neutrons and x-rays on a certain convenient malignant tumor called 'mouse sarcoma 180', and have obtained very impressive evidence that the neutrons are relatively much more lethal on these malignant cells than on healthy tissue. Briefly, we have measured the dose of neutrons required to kill the malignant tumor in vitro and the dose required to kill the mouse. For neutrons the ratio of tumor dose to mouse dose is less than the corresponding ratio for x-rays. If malignant tumors in general are correspondingly more sensitive to neutron radiation, neutrons will supersede x-rays in the treatment of cancer, and should make possible a much larger percentage of cures. You might tell Dr. Fleming that in view of this important possibility, we are definitely planning to go forward with the treatment of human cancer with our cyclotron.

Unusually, Chadwick[18] took about a month to reply to this letter, because he had been fully occupied in negotiating with Fleming and others; he had now rejoined forces with his old ally John Cockcroft, who had at last persuaded Rutherford that the Cavendish should have a cyclotron.

> George Holt Physics Laboratory,
> The University of Liverpool.
> 11th May 1936
>
> Dear Professor Lawrence,
> I was very interested to hear of your experiment with a malignant tumor. I think

> there are very great possibilities in the biological application of neutrons and of the artificially produced active elements. I am myself more particularly interested in the chemical applications than in the direct effects, although the latter may be of more immediate importance, and I do not intend to neglect them.
>
> We have not proceeded very far with the design of the magnet for the cyclotron. There are, at the moment, three magnets to be considered, one for the Cavendish, one for Fleming and one for me. The Cavendish will have a very large magnet indeed. They have plenty of money from the sale of Kapitza's apparatus. Their magnet will be built so as to be suitable for magnetic work as well as for the cyclotron. Fleming intends to build a small cyclotron for Metro-Vick.

Rutherford[3] had negotiated a deal with the Soviets in October 1935 to transfer equipment from the Mond Laboratory to Kapitza's new laboratory in Moscow in exchange for £30 000. The onerous task of arranging the actual transfer fell to Cockcroft*, who, according to Chadwick, had always been responsible for the day-to-day administration of the Mond Laboratory. Chadwick[5] felt that Kapitza took full advantage of Cockcroft's willingness to help, a view which was echoed by Rutherford in several letters to Kapitza in 1936. In January, after receiving a long shopping list from the Russian, Rutherford[19] told him that in addition to the apparatus agreed upon with the Soviets 'we have already either promised you or provided you with apparatus which is costing us a large sum. While I am anxious to help you as far as possible, I feel that these drains on the laboratory should come to an end.' After further criticism of the slow pace of the transfer, Rutherford assured Kapitza 'that but for Cockcroft it would have taken twice as long'. He[20] confided to Chadwick:

> Cockcroft has been extraordinarily busy this year transferring apparatus to Kapitza . . . I hear from Kapitza and he is quite well, but of course full of his own difficulties and inclined to criticize Cockcroft because he does not move faster, though as a matter of fact I do not think anyone else could have put the job through under a year or two! We have transferred an enormous amount of stuff to Kapitza, for he was continually increasing his demands beyond the original agreement, and at one stage I thought he would ask us to remove the paint from the laboratory walls—I cannot think of anything else that has not gone!

The Cavendish[3] were expecting to make a profit of about £17 000 from the transaction—a sum equivalent to several years' budget for the whole Laboratory. Their intention was to spend about £3500 on a large electromagnet which could be used both for a cyclotron and low-temperature research. In May 1936 came another windfall which transformed the finances of the Cavendish beyond recognition. Sir Herbert Austin, the motor car magnate, agreed to make the staggering gift of £250 000 to the Cavendish through the University Chancellor, Stanley Baldwin. Baldwin also happened to be the Prime Minister, and a few weeks later Sir Herbert became Lord Austin. Chadwick[21] sent a note of congratulation to Rutherford.

---

* John Cockcroft was also responsible, at that time, for designing the electromagnets for the various cyclotrons. He made freehand drawings of the pole pieces, which were elegantly shaped to match his mental image of the flow of magnetic lines of force.

> Everybody will be pleased that the Cavendish should receive such a magnificent gift. It is far more appropriate than having to collect a large number of small contributions, and it certainly must be a relief to you, as well as a pleasure to have this done properly.
>
> Begging, like swindling, is only respectable on a big scale.

In his reply, Rutherford[20] was less than jubilant and indicated that he missed the presence and counsel of his old deputy.

> You can imagine that the University has been worrying its head wondering what to do with the securities handed over by Austin, and to safeguard themselves against potential death duties. We will not know for a couple of years how the financial position stands, so I do not expect to do any additional building for some time to come. As I told you in my last letter, I would be very glad if you could give us any suggestion of what you think would be a good policy to pursue. You know the situation as well as, or better than, anyone, and may be able now from outside to view it with even better perspective.

Chadwick was to be the external examiner in the Cambridge Tripos exam at the end of the summer term, and it was agreed that these matters would be discussed then. He stayed with the Rutherfords at Newnham Cottage, and found his host in a brittle mood. It seems that the responsiblity of the huge Austin bequest was playing on his mind, especially as it brought the prospect of all sorts of large-scale changes at the Cavendish. Just at a time when he needed the advice of trusted colleagues, Rutherford seemed to be losing them at an increasing rate and found himself becoming isolated. The third loss in prospect was Mark Oliphant, who was being wooed by Birmingham University to take the chair in physics. Rutherford had initiated the move himself when the Dean of the Faculty of Science in Birmingham, Professor Moss, had visited the Cavendish earlier in the year. He had told Rutherford that they were looking for a new physics professor and Rutherford had sent him to talk to Oliphant.[7] At first Oliphant was not particularly interested, but he paid a visit to Birmingham and met Norman Haworth, the Professor of Chemistry, who began to persuade him that he could be happy there. Unsure of what to do, he had then written to Chadwick who had offered the following advice:

> To me all seems to depend on whether this is the kind of post you want—if you want a university chair and a laboratory of your own. If you feel lukewarm about this, then it would be a mistake to go to Birmingham or any other university. If you do want a chair, then I think arrangements can and must be made which should satisfy your obligations to Cambridge and to yourself.[7]

Oliphant then went to talk it over with Rutherford and mentioned that he had sought Chadwick's advice. He listened with growing surprise and asked: ' Do you mean to say that you think of accepting the post?' to which Oliphant replied 'Yes, I think I shall.' Rutherford grew red in the face, thumped the table, and literally shouted:'Then go! And be damned to you!' He then added, 'First Chadwick deserts me. Now you, who took his place, also desert me.'[22] Oliphant, who

had never before been on the receiving end of one of Rutherford's choleric

> outbursts. . . left his room greatly upset and worried. Shortly afterwards he came to my room where I was sitting in despair, and asked hesitantly, could I spare time to talk with him. His apology for his reception of my news was complete, reinforcing my distress at having upset him.[7]

Chadwick was to experience his own first verbal onslaught during his stay at Newnham Cottage that summer. He and Rutherford were sitting in the garden when Rutherford began attacking him for supporting Cockcroft's proposal to build a cyclotron at the Cavendish. By that time Rutherford knew that one was definitely going to be built in Liverpool and Chadwick[5] 'repeated the arguments for one in the Cavendish'. His central tenet was that 'further real progress in nuclear physics would demand considerable expenditure on equipment and space'. When Chadwick had presented this opinion the previous year before leaving for Liverpool, Rutherford 'had not objected', but 'perhaps on this occasion I expressed my views more definitely' and 'he was very angry with me; it was quite shattering.' Cockcroft had designed the electromagnet and Chadwick was only waiting for Rutherford's decision regarding the Cavendish 'before giving the order to Metrovick to proceed'. He told Rutherford that he would wait until the end of June, but no longer. Rutherford's last words on the matter were 'I won't have a cyclotron in my laboratory'.

The disagreement over the cyclotron was, according to Chadwick, 'mild compared with what followed' which was a tirade about Oliphant's proposed move. 'He then attacked me violently, saying I had encouraged or advised [Oliphant] to accept the chair in Birmingham—that is to leave him'. Chadwick tried to protest that the decision was Oliphant's alone, that he did not know what he would decide and that he 'had no intention whatever of trying to interfere.' Rutherford would not be placated and paid no heed to Chadwick's explanation. Chadwick returned to Liverpool feeling wretched, but it seems likely that Rutherford forgave him within hours and certainly bore no grudge.

By October, full cooperation between Liverpool and Cambridge seems to have been re-established and Rutherford[23] thanked Chadwick for obtaining copper for the electromagnet at cost price.

> It is very good of you to have taken this trouble on our behalf, and as you know, we on our side are only too glad to help you in any way we can in the design, etc. of the new magnet.
>
> Work is just starting up and I am as usual much oppressed with what I have to do in this term. I had a good holiday in the country and feel that I will need it before Christmas comes!

In December, Rutherford[24] invited him to stay at Newnham Cottage again for the annual Cavendish dinner. This was to be a special occasion to toast J.J. Thomson's eightieth birthday, but Chadwick, with the memories of his last unpleasant visit still fresh, was unable to attend.

Chadwick had lost his own senior lecturer, Norman Feather, back to Cambridge at the end of his first year in Liverpool. It is ironic in view of the quarrel at Newnham Cottage that Rutherford[20] had apologized to Chadwick for taking Feather away, and used the prospect of Oliphant's 'almost certain' departure from the Cavendish as the reason 'I have to look ahead a little'. As a replacement, Chadwick was fortunate to recruit E.J. Williams from Manchester. Evan James Williams[25] was the most Welsh of atomic physicists, and apart from being a first-rate experimenter he had an unusually strong grasp of theory. He had spent the year 1933–1934 as a Rockefeller Fellow working with Niels Bohr in Copenhagen, and Bohr greatly admired his abilities. He thought Williams 'to a quite unusual extent combined a thorough knowledge of modern atomic theories with a complete mastery of the experimental technique in this field'.[26] To Blackett, Williams strongly resembled Bohr 'in his rare gift of analysing in detail the mechanism of complicated physical processes, using a minimum of mathematical analysis and a maximum of physical understanding'.[25] Like Bohr, Williams made important observations about various collision phenomena involving subatomic particles, and during his year in Copenhagen had developed a model which allowed such problems to be understood without needing to resort to the complex mathematics of quantum mechanics. Williams had also published a number of seminal papers concerning the conservation of energy and momentum in high-energy atomic reactions, which was a subject of immense controversy in the mid-1930s. Referring to a 1934 letter to the *Physical Review*, Blackett said: 'Thus at this early date and more clearly than anyone else at that time, Williams had already come to the entirely correct conclusion that both the ionization and the radiation formulae must be considered valid up to extremely high energies. But it was not for several years that this became universally accepted.' A consequence of Williams' conclusion was that the penetrating particles observed in cosmic rays could not be high-energy electrons, as many had argued, but must have a greater mass. The prospect of the discovery of another type of fundamental subatomic particle was irresistible to Chadwick, and he set his accomplished new senior lecturer to work to start cosmic ray research in Liverpool.

Although Chadwick had never done any cosmic ray research himself, he had been interested in it since his days at the Reichsanstalt in 1914. At that time one of Geiger's young assistants, Werner Kolhörster, was making pioneering measurements at altitudes up to 10 000 m in a balloon. Kolhörster survived his balloon trips and the Great War, and worked closely for many years with Walther Bothe, who was such a prominent figure in the unfolding story of the neutron. At the international conference on β- and γ-ray problems that Chadwick had organized in Cambridge in 1928, it was announced that Bothe and Kolhörster[27] were working on a method to register cosmic rays by the coincidence of pulses in two Geiger counters and that they hoped to be able to study the penetrating power of the rays by this method. Chadwick[6] discussed this work with Bothe, at the conference. He would have also read the report of their completed experiment[28] the following year when they found that at sea level, 75% of the cosmic rays were ionizing particles which could pass through a block of gold 4 cm thick. On the basis of this experiment, Chadwick[6]

> thought (in fact [he] was convinced) that you could not explain all the properties of

> cosmic radiation on the basis of gamma rays of whatever energy and electrons; and there must be some particle with the same charge as the electron but with a much bigger mass—of the order of a hundred times the mass of the electron.

Chadwick intended to make this point at a meeting on the cosmic radiation at the Royal Society which was chaired by Rutherford. He had told Rutherford beforehand that he wished to make a contribution to the discussion: 'Rutherford called on me. By that time I had got very nervous about it indeed, and I just did not have the courage to say what I had intended to say, and I invented quite suddenly some remarks of a trivial nature.'[6]

E.J. Williams brought some of his own apparatus with him from Manchester and built a large, randomly operated cloud chamber for photographing cosmic ray tracks. Similar work was going on at the California Institute of Technology, and in the spring of 1937, Neddermeyer and Anderson[29] (who had won the 1936 Nobel Physics Prize for his earlier discovery of the positron) published evidence that 'there exist particles of unit charge, but with a mass . . . larger than that of a free electron and much smaller than that of a proton'. These cosmic particles would soon be christened *mesons*, and by the summer of 1937 Williams was finding evidence for them in Liverpool. He presented his results at two conferences in September, and Chadwick[30] sent news to Rutherford in October.

> E.J. Williams has just returned from conferences in Copenhagen and Moscow. He saw a fair amount of the Kapitzas, and both Peter and Anna seem happy and contented.
>
> Williams has been taking photographs of the cosmic particles, hoping by counting the ions to get information about the different kinds of particle which may be present. Some months ago he found a couple of tracks which may be due to the new kind of particle that Anderson wants. But they seem to be very few, and he has not yet been able to get enough evidence for any degree of certainty.

Although Chadwick believed in the meson's existence, he would not relax his exacting requirements for experimental proof. Williams himself thought there was some doubt about his early photographs (and even more uncertainty about some of the American ones). By February 1938, he and Chadwick were both satisfied, and E.J.[31] wrote to his mentor Niels Bohr.

> To be certain that the particle is not a proton, the conditions in the chamber must be such that there is no likelihood of the curvature being due to air currents. I am pleased to say that I have now got a photograph of a new particle track under such conditions that there can be little doubt on this point . . . The mass of the particle [from the measured ionization and momentum] is accordingly about 150 times electron mass ~1/10 proton mass. (My earlier photographs also indicate a mass in this neighbourhood.) Professor Chadwick, who has previously doubted this kind of evidence for a new particle (because of the effect of air currents), thinks that the present example is convincing. It will probably be published in due course.

The work was published, with Williams' assistant Pickup as co-author, in a note to *Nature*, 'Heavy electrons in cosmic rays'. Chadwick would not dream of including his own name on the paper since he had played no part in the actual experiment, and the success of his junior colleagues was ample enough reward for him. Williams went on to develop a theory of the scattering of fast electrons and cosmic ray particles, which allowed for the finite size of the nucleus and the shielding of the nucleus by the atomic electrons. This theory was published after he had moved on to the Chair at the University of Aberystwyth later in 1938.

E.J. Williams' career in Liverpool was almost as short-lived and spectacular as one of his own cloud chamber tracks. He was an irreplaceable genius, and after his inevitable departure, there was no one to continue the cosmic ray research. Chadwick was managing to retain a number of talented Liverpool graduates, who were mostly involved in work on the cyclotron; this was now being overseen by Bernard Kinsey, who had returned from Berkeley in the late summer of 1936. Kinsey found the rate of progress rather pedestrian compared with what he had been used to in Lawrence's laboratory. Chadwick[32] too was frustrated when he wrote to Lawrence on 7 August 1937.

> We have not got very far with our cyclotron. The steel blocks for the magnet have been finished for some time but are not yet here. We expect them in September with the coils, cooling boxes etc. from Metropolitan Vickers and we hope the complete magnet will be erected then. Kinsey is building the oscillator system here, and it is nearly ready. We could not afford to buy it from Metro-Vick. Although we are making quite a lot of the general apparatus ourselves, I am sure the cost of the complete outfit will be over £5,000. However, we can just manage it. I hope everything will be here shortly and I think we shall have the cyclotron in full blast before Christmas. I am very pleased with Kinsey. He has taken nearly all the cyclotron work off my hands. I think he has found the responsibility rather heavy at times but I am sure it has been good for him.

Chadwick was about to acquire another English cyclotron researcher from Lawrence, Harold Walke, who had been at Berkeley on a two year Commonwealth Fund fellowship. Lawrence recommended him warmly and Kinsey was used to working with him since their time in California had overlapped for a year. Chadwick thought he could 'fix him up for one year, but prospects after that are rather vague'.

Apart from trying to oversee the ambitious cyclotron project, revamp the department, and play a full role in general university matters, Chadwick found that he was acquiring an increasing number of outside commitments in the form of examining and invited lectures. The latter terrified him, but there were some that he just had to accept. In 1937 he gave the Kelvin Lecture on 'The elementary particles of matter' and in this reaffirmed his long held belief that 'the main problem in nuclear physics is the nature of the elementary particles and the nature of the interaction between these particles'.[6] His own discovery of the neutron had been a huge advance, but he felt that there were many quite fundamental things that were still not clear. For example, why did a neutron and a proton not combine or why did two neutrons not combine to form a neutral particle with double mass? It was to attack these basic problems that he wanted the cyclotron—he was not interested in bombarding elements and observing disintegrations and new isotopes. 'That was easy, inevitable. One could do that. It amused some people; it did not amuse me. It was unlikely, it seemed to me, that one could get any clue to the nuclear forces from experiments of that kind . . .[they were] very little more than descriptive botany.'[6]

There was one invitation to give a series of lectures that arrived in the autumn of 1937 that gave Chadwick unexpected pleasure. It came from Rutherford and involved the Scott Lectures to be given at the Cavendish the following summer. The invitation was particularly important to Chadwick because he took it as a sign that he had been truly forgiven after their temporary falling out, and he had now been restored to favour. There was a snag which he[30] wrote to Rutherford about on 11 October 1937.

> This is most unfortunate. I am very pleased to be asked to give the Scott Lectures, and I would be delighted to have the chance of spending a week at the Cavendish. But I have promised to go to Holland for the first few weeks in May to give a series of lectures at different places. They have been badgering me for a long time and last year I promised faithfully to go in the spring of 1938. I have only recently fixed the time for May. Perhaps you will give me another opportunity later on.
>
> I hope you will enjoy your visit to India. It will give our Indian colleagues great pleasure, and I am sure it is worth doing. I should like to go myself but I cannot ask for leave again so soon, for I was away most of last spring. Nor can I leave the cyclotron just when we hope to get it working. It is being erected now with a great deal of bustle and noise, and the first set of coils is just outside my window. The cost is proving rather more than I expected, but I think it will be worth it.
>
> Walke has just come here from California. I was sorry to find that in a mild way we had competed for him. I had actually written to him just before I got a letter from Lawrence recommending him rather strongly. He seems to have turned into quite a good experimenter. He has had a lot of experience in running the cyclotron, and I am sure he will be as anxious to place his experience at your disposal as at mine.

Chadwick closed by saying that he hoped to pay a brief visit to Cambridge to see Rutherford before he set sail for India. Lord Rutherford was due to preside over the meeting of the British Association for the Advancement of Science, which was going to be held there at the end of the year to mark the Jubilee of the Indian Science Congress. The reply was both relaxed and swift in coming.

> Cavendish Laboratory,
> Cambridge,
> 12th October 1937.
>
> My dear Chadwick,
>
> I am sorry to hear that you are unable to give the Scott Lectures this year, but I hope that you will keep yourself free for next time, and remind me of my offer well in advance.
>
> I was rather amused by Walke. Cockcroft and I received strong letters from Lawrence, more or less asking us to do whatever we could to find a place for Walke in the Laboratory, and apparently at the same time he had been writing to you and others. I am glad, however, that you were able to get Walke's services. He ought to be helpful to you in getting the cyclotron going. I presume that we are in

the same stage as yourself with regard to getting the magnet etc. going. The parts are all waiting in the Laboratory, and we are hoping to get them erected shortly. I have long since recognised that the cyclotron is a costly toy in this country, and I hope you will manage to get through without exhausting your funds.

We had Bohr along a few days ago, and he gave us a lecture on the development of his ideas on nuclear structure. To-day we have had Langmuir here, and he is talking tomorrow on his work on films.

We have got the high tension apparatus going O.K., and experiments are in progress. The modification of the Cockcroft set, up to about 600,000, is being run by another group in the high tension room.

I had a long letter from Kapitza some time ago, and have also been hearing about him from Bohr and Dirac.

I shall be interested to see the picture of William's high ionisation track.

I hope that we shall be able to see you during next month.

With best wishes,
Yours,
Rutherford [33]

Two days after this letter was written, Rutherford felt unwell and consulted his physician, Dr Nourse. The events of the next few days were relayed to Chadwick in this letter from Lady Rutherford,[34] which was written on Tuesday 19 October, from his beside in the Evelyn Nursing Home, Cambridge.

Dear Jimmy,

It's a long time since I called you that, but one clings so to old friends in times of trouble. My husband is only hanging by a thread. The operation was strangulated hernia with no gangrene which meant they didn't open the gut, but it was paralysed. They treated it and thought it would be all right. He only had a mouldy feeling on Thursday, doctor Friday morning and consultation with Ryle the Regius Prof. [of Medicine at Cambridge University]. He advised getting Sir Thos. Dunhill from London and they operated Friday night. Saturday went well, Sunday he started vomiting all day and they realised things were gone wrong or rather that the bowel had not recovered its elasticity. They have tried everything, the surgeon came up to operate again but wouldn't do it, he couldn't take the anaesthetic safely and the shock would have done more harm than good. He has intravenous saline going night and day and all sorts of injections. This morning I was told there was no chance. Today at 4, Dr. Nourse said he couldn't see that he was any worse than 8 am today and I have just been told that he is several ounces up on intake over output today so there is still a chance. He was so splendidly fit up to Wednesday. For some time he has had a slight bowel weakness and worn a small pad with plaster, but no internal trouble. He is a splendid patient but is terribly weary and tired of being worked at. The nursing could not be better.

Love to Aileen.
Yours very sincerely,
Mary Rutherford.

Rutherford died, presumably from septicaemia, later that evening. He was, in the words of Ralph Fowler,[35] his son in law, the outstanding experimental scientist of the age, and his unexpected death gave pause to every scientist in the world.

Many of his friends were gathered in Bologna, at the two hundredth anniversary celebration of the birth of Luigi Galvani, and learned of his death in a moving eulogy delivered by a tearful Niels Bohr. Chadwick was profoundly affected by the loss, and it remained an 'abiding grief'[5] to him that their last parting in Cambridge had been so bitter. Rutherford's ashes were buried with great pomp in Westminster Abbey, close to the tomb of Isaac Newton, in a ceremony which Chadwick found deeply moving. On the train back to Liverpool, he wrote an obituary for *Nature*.[36] After a masterly summary of Rutherford's career came the following two concluding paragraphs.

> Even the casual reader of Rutherford's papers must be deeply impressed by his power in experiment. One experiment after the other is so directly conceived, so clean and so convincing as to produce a feeling almost of awe, and they came in such profusion that one marvels that one man could do so much. He had, of course, a volcanic energy and an intense enthusiasm—his most obvious characteristic—and an immense capacity for work. A 'clever' man with these advantages can produce notable work, but he would not be a Rutherford. Rutherford had no cleverness—just greatness. He had the most astonishing insight into physical processes, and in a few remarks he would illuminate a whole subject. There is a stock phrase—'to throw light on a subject'. This is exactly what Rutherford did. To work with him was a continual joy and wonder. He seemed to know the answer before the experiment was made, and was ready to push on with irresistible urge to the next. He was indeed a pioneer—a word he often used—at his best in exploring an unknown country, pointing out the really important features and leaving the rest for others to survey at leisure. He was, in my opinion, the greatest experimental physicist since Faraday.
>
> I cannot end this tribute to Rutherford without some words about his personal qualities. He knew his worth but he was and he remained, amidst his many honours, innately modest. Pomposity and humbug he disliked, and he himself never presumed on his reputation or position. He treated his students, even the most junior, as brother workers in the same field—and when necessary spoke to them 'like a father'. These virtues, with his large, generous nature and his robust common sense, endeared him to all his students. All over the world workers in radioactivity, nuclear physics and allied subjects regarded Rutherford as the great authority and paid him tribute of high admiration; but we, his students, bore him also a very deep affection. The world mourns the death of a great scientist, but we have lost our friend, our counsellor, our staff and our leader.

Chadwick would often reflect on Rutherford's genius and as his closest collaborator would frequently be asked his opinion. He eventually created an image which satisfied him: Rutherford's mind was like the bow of a battleship, it had so much weight behind it that it had no need to be as sharp as a razor.[6]

References

1 Spruch, G.M. (1979). Pyotr Kapitza, octogenarian dissident. *Physics Today*, Sept., 214–20.

2 Boag, J.W., Rubinin, P.E. and Shoenberg D. (1990). *Kapitza in Cambridge and Moscow*, pp. 46–7. Elsevier Science Publishers B.V., Amsterdam.

3 Wilson, D. (1983). Kapitsa. In *Rutherford: simple genius*, pp. 496–537, Hodder and Stoughton, London.

4 Bullard, E. (1974). Patrick Blackett. *Nature*, **250**, 370.

5 Chadwick, J. (?1971). Undated notes on Oliphant's biography of Rutherford (ref. 14). CHAD II, 2/1, CAC.

6 Weiner, C. (1969). Sir James Chadwick, oral history. Niels Bohr Library, American Institute of Physics, College Park, Maryland.

7 Oliphant, M. (1972). *Rutherford—recollections of the Cambridge days*. Elsevier, Amsterdam.

8 Heilbron, J.L. (1974). *H.G.L.Moseley*, p. 170. University of California, Berkeley.

9 Lovell, B. (1975). Patrick Maynard Stuart Blackett, Baron Blackett, of Chelsea, 1897–1974. *Biographical memoirs of fellows of the Royal Society*, **21**, 1–115.

10 Pais, A. (1991). *Niels Bohr's times*, pp. 337–40. Oxford University Press.

11 Rutherford, E. (17/2/36). Letter to J. Chadwick. Rutherford's papers, University of Cambridge Library.

12 Chadwick, J. (25/2/36). Letter to E. Rutherford. Rutherford's papers, University of Cambridge Library.

13 Rutherford, E. (23/5/36). Letter to J. Chadwick. Rutherford's papers, University of Cambridge Library.

14 Lawrence, E.O. (31/1/36). Letter to J. Chadwick. Lawrence's papers, Bancroft Library, University of California, Berkeley.

15 Chadwick, J. (11/3/36). Letter to E.O. Lawrence. Lawrence's papers, Bancroft Library, University of California, Berkeley.

16 Chadwick, J. (1936). *Royal Society Grants*, p. 13. Royal Society, London.

17 Lawrence, E.O. (24/3/36). Letter to J. Chadwick. Lawrence's papers, Bancroft Library, University of California, Berkeley.

18 Chadwick, J. (11/5/36). Letter to E.O. Lawrence. Lawrence's papers, Bancroft Library, University of California, Berkeley.

19 Rutherford, E. (27/1/36). Letter to P. Kapitza. In *Kapitza in Cambridge and Moscow* (ed. J.W. Boag, P.E Rubinin and D. Shoenberg, 1990). Elsevier Science Publishers B.V., Amsterdam.

20 Rutherford, E. (14/5/36). Letter to J. Chadwick. Rutherford's papers, University of Cambridge Library.

21 Chadwick, J. (25/2/36). Letter to E. Rutherford. Rutherford's papers, University of Cambridge Library.

22 Oliphant, M.L. (1992). Letter to the author.

23 Rutherford, E. (6/10/36). Letter to J. Chadwick. Rutherford's papers, University of Cambridge Library.

24 Rutherford, E. (7/12/36). Letter to J. Chadwick. Rutherford's papers, University of Cambridge Library.

25 Blackett, P.M.S. (1947). Evan James Williams, 1903–45. *Obituary notices of fellows of the Royal Society*, **5**, 387–406.

26 Bohr, N. (9/12/45). Letter to P.M.S. Blackett. In *Niels Bohr, collected works*, **8**, (ed. E. Rüdinger), pp. 649–50. North-Holland, Amsterdam, 1987.

27 Skolbeltzyn, D. (1983). The early stage of cosmic-ray particle research. In *The birth of particle physics* (ed. L.M. Brown and L. Hoddesdon), pp. 111–19. Cambridge University Press.

28 Pais, A. (1986).The nucleus acquires a new constituent. In *Inward bound*, pp. 397–444. Oxford University Press.

29 Neddermeyer, S.H. and Anderson, C.D. (1937). Note on the nature of cosmic-ray particles. *Physical Review*, **51**, 884–6.

30 Chadwick, J. (11/10/37). Letter to E. Rutherford. Rutherford's papers, University of Cambridge Library.

31 Williams, E.J. (15/2/38). Letter to N. Bohr. Niels Bohr Archive, Copenhagen.

32 Chadwick, J. (7/8/37). Letter to E.O. Lawrence. Lawrence's papers, Bancroft Library, University of California, Berkeley.

33 Rutherford, E. (12/10/37). Letter to J. Chadwick. CHAD II, 1/17, CAC.

34 Rutherford, M. (19/10/37). Letter to J. Chadwick. CHAD II, 1/17, CAC.

35 Eve, A.S. (1939). *Rutherford*, p. 428. Macmillan, New York.

36 Chadwick, J. (1937). Lord Rutherford, obituary. *Nature*, **140**, 749–50.

10 The post-heroic age

Niels Bohr[1] said of Rutherford that, like Galileo, 'he left science in quite a different state from that in which he found it'. The statement is incontrovertibly true if one remembers that in 1895, when Rutherford first arrived at the Cavendish as a graduate student on an 1851 Exhibition Scholarship, the world knew nothing of X-rays, the electron, the proton, the neutron or radioactivity. He extended knowledge of all these phenomena and more by his own work, and in his lifetime also saw fundamental advances made by many of his students, for which some share of the credit is rightly given to him. To Rutherford himself, it seemed that he was living in 'the heroic age of physics'.[2] He had achieved a degree of public recognition and fame which went beyond his status as the Cavendish Professor and Nobel prizewinner. He had become the embodiment of science throughout the British Empire, and his elevation to the peerage served to confirm his rank as the equivalent of the Law Lords or Lords Spiritual*.

To Chadwick, it seemed almost an impertinence that anyone should replace Rutherford as Director of the Cavendish. Rutherford had confided in Chadwick[3] when he left for Liverpool in 1935 that he would be retiring in two or three years time, and would like Chadwick to succeed him. Chadwick was surprised and responded that he 'was not good enough to succeed him and at the same time I didn't think there was anyone good enough in this country' and he also said to Rutherford that 'it would not happen'. The university authorities obviously needed to take a more practical line and set out to find a new director. Chadwick[4] wrote to Feather on 23 January 1938: 'I saw in the Times that applications were asked for by February 1st. But I cannot imagine why it is thought necessary to ask for applications or how any man can think himself worthy.' The choice of Cavendish Professor is made by a group of electors, and there is no doubt that Chadwick's name was considered by them, although he never applied for the job. He[3] was asked informally to submit his ideas about the way the Cavendish should be developed, and replied that it should be expanded considerably and should diversify into molecular physics as well as continuing the tradition of nuclear physics.

---

* His biographer, Eve, tells a story of Rutherford dining at Pembroke College. He arrived early and struck up conversation with the only other person present, who was in ecclesiastical dress. He said 'I'm Lord Rutherford' and the man replied 'I'm the Archbishop of York'—at which point Rutherford did not think that either of them believed the other.

In the event, the man who had succeeded Rutherford to the Langworthy Chair in Manchester, Sir Lawrence Bragg, also followed him as the Cavendish Professor in Cambridge. Chadwick[5] again shared his feelings with Feather: 'I was by no means surprised, I almost expected it. I should like to have been asked, I still feel rather disappointed. But I was very relieved that I had not to leave here.' According to Peierls, most physicists had expected Chadwick to be appointed to the Cavendish Chair, and their rationalization was that the electors did not choose him because they were worried that he would not be able to cope with the social aspects of the position: playing host to visiting dignitaries and taking a full part, as Rutherford had done, in the affairs of the Royal Society and other learned bodies. Reflecting 30 years later, Chadwick[3] had this to say:

> I was not dismayed when I was not chosen. In fact, I was rather relieved, because I did realize my limitations . . . I was in a sense relieved and pleased, too, that Lawrence Bragg had been chosen, which I knew would happen. At the same time I must confess to a slight feeling of disappointment in spite of saying that I was relieved. The two things can exist together.

There was little time for Chadwick to dwell on his slight disappointment since there were growing demands for his services in Liverpool. In February 1938 the *Liverpool Daily Post* announced the formation of a commission of leading medical men and university academics under the leadership of Lord Derby to report on cancer research and treatment in the city. The remit of the commission[6] was broad and it was charged with paying particular attention to:

1. The best use, consistent with the welfare of patients, of the hospital accommodations and research facilities available in the voluntary and municipal hospitals in the Liverpool area and, in particular, in the Liverpool Radium Institute and Hospital for Cancer.
2. The most promising lines of investigation in regard to cancer (its causes and treatment) capable of being pursued in the Liverpool area.
3. The feasibility of co-ordinating, if that should seem desirable, the various activities under some unifying board or authority or by some other method.

The six man commission included three of the most formidable intellects to be found in Liverpool: the new Vice-Chancellor, Arnold McNair, a distinguished jurist who knew Chadwick well from Caius College, where he was also a Fellow; the Professor of Medicine, Henry Cohen, and the Professor of Physics, James Chadwick. He, of course, was excited about the possible medical applications of the cyclotron, especially in view of the encouraging snippets of information he was learning from Lawrence in Berkeley. There seemed to Chadwick several possible avenues to be explored using cyclotron products: there was the direct effect of neutrons on malignant tumours, but even more intriguing to him was the production of artificial radio-isotopes that could be used as tracer substances to study biochemical and physiological processes and which might themselves have some therapeutic use as anti-tumour agents.

One of his former students, Douglas Lea, with whom he had made an unsuccessful search for the neutrino a few years before, had left the Cavendish and was making a name as a pioneer in the new field of radiation biology. Rutherford 'showed no interest whatever in medical applications' and had regarded Lea as a loss to physics. Chadwick[3] now thought of attempting to recruit him to Liverpool because he intended 'to push the biological side strongly'. In the event, Lea did not come to Liverpool and Chadwick never appointed a scientist with a major interest in the biological effects of radiation. He was finding it difficult to hire and retain experienced staff. E.J. Williams held the Leverhulme Fellowship for about 18 months, which was longer than his predecessor Norman Feather, but left Liverpool early in 1938 to take up the chair at Aberystwyth. Bernard Kinsey was promoted in his place, although Chadwick would have preferred to expand the department. He[7] would have liked to have appointed Egon Bretscher, a Swiss experimental physicist then working at the Cavendish, but did not for fear that 'there would have been serious opposition—not I think from Faculty [of Science], but from higher up—on the grounds of nationality'. In Chadwick's opinion, the burghers of Liverpool were not as liberally minded as their pre-war counterparts in Manchester, where of course there had been a small but influential group of German descent. He also needed to replace James Rice, the reader in theoretical physics, who had died in harness in 1936. Rice had begun his career as a schoolmaster and always took a large teaching role in the department. Chadwick wanted to replace him with a more academically active theoretician. He first thought of Walter Heitler, then a lecturer in Bristol, but decided 'it was no use thinking about Heitler, because the appointment of a German would not have been agreeable to some of the people in the University'.[3] He then turned to Cambridge and approached Homi Bhabha, a young man from a wealthy Indian background, who was a member of Gonville and Caius College.

> Bhabha came up to see me. I would very much have liked Bhabha to come , but I thought he was too good. Well, you see, some teaching had to be done, and the quality of our students was not the quality of the students in Cambridge, and I thought that much of it would be drudgery to a man like Bhabha, who was a most exceptional man. He was a painter and a poet and had extremely wide interests—not merely interests but far more than that—and I didn't feel that however much I liked him it was fair.[3]

Ultimately, Chadwick settled on another talented Cambridge man, Maurice Pryce, who had studied under Fowler and Dirac. His name was put forward to Chadwick by Kinsey—they had been undergraduates together at Trinity College.[8] Although Chadwick had no illusions about the relative strengths of Liverpool students, the burgeoning research programme was beginning to attract some of the more able ones to stay on and do postgraduate research. John Holt[9] was one such and began his research career in 1938. In the best tradition of a Chadwick nursery course, he was set to work learning to make Geiger counters which had to be filled with argon gas and 'an essential whiff of alcohol vapour, without which they refused to function properly'. Geiger had sent Chadwick two original tubes a few years earlier, and Holt found that their performance could be improved by minor alterations, such as using end caps of brass rather than of ebonite. It was not long before Holt made his first significant scientific contribution, which resulted from the close cooperation that had built up between Liverpool and Lawrence's Berkeley laboratory. One of Lawrence's team, Luis Alvarez, had just described a novel form of radioactive decay called *K-electron capture.* Harold Walke brought back, by rail and sea, some new isotopes with long half-lives manufactured at Berkeley so that their properties could be investigated in Liverpool. Holt[9] found one isotope, vanadium 47, 'which emitted only soft X-radiation, the first example of a K-capture decay with no accompanying nuclear radiation. It was very convincing evidence for the new process.' He had the satisfaction of using his home-made equipment for the crucial observations. 'The matter was clinched by some absorption measurements which I was able to make with a counter having a window of Cellophane, which showed that the X-rays were characteristic of the daughter element titanium.' Cellophane, rather than the usual mica, was very transparent to the soft X-rays in question.

Chadwick continued to rely on Lawrence for information and guidance concerning the cyclotron; he peppered Lawrence with detailed questions[10] in preparation for an evening discourse he was scheduled to give at the Royal Institution at the end of May 1938. He wanted to know the maximum voltage and current that could be obtained, how rich it was as a source of neutrons compared with radium and beryllium, the weight of the magnet, the diameter of the pole pieces and the maximum strength of the magnetic field. After apologizing for asking so many questions (the answers to which he thought would be of interest to those more concerned with applications of the machine rather than the physics), Chadwick continued:

> I hope your new apparatus is really big. I feel that one ought to make a serious attempt to get up to 60 or 70 million volts (i.e. ~ 137 mc^2)—perhaps not just yet but one ought to prepare for it. I feel sure that with such particles we should begin to learn the true mechanics of the nucleus. Of course nature provides us with such particles in the cosmic rays but in most niggardly fashion. I think the phenomena in the cosmic rays point the way to us.
>
> I am afraid there is very little chance of being able to build a big cyclotron in this country for some time to come. We have had endless trouble in getting our small one. Even now we have not yet got the tank [vacuum chamber] from Metrovick. It is now a year since the drawings were finally completed and handed to them. For various reasons I cannot quarrel with them but I am sorely tempted to do so. Kinsey's patience was long ago exhausted.

He closed with some words of praise for the two members of his staff who had been trained by Lawrence.

> You will be pleased to hear that Kinsey is doing well. He has been extraordinarily useful and I have been able to leave most of the cyclotron work to him. He was very unsettled for some time—rather naturally, for the laboratory was in a very bad state and big changes have to be made cautiously here—but he seems

> much happier now. I am glad, for I like him very much and I think he has good prospects.
>
> Walke has joined in well. He is a nice lad and I am glad to have him and to learn my previous opinion of him was wrong. He is finding his past (i.e. his publications before he came to you) rather hard to live down. But he will do it.

The reason for the exasperating delays in the construction work was that Metropolitan-Vickers had become very busy with new defence contracts from the government. Chadwick did not know this until later, although he paid several visits to the factory to try to hurry them along. On these occasions, Chadwick[3] 'could see my bits of apparatus lying about the place waiting for their turn, but their turn never coming because they were big. I think there was only about one lathe in the place which would take them.' At the same time, he was upgrading the laboratory workshop in Liverpool, buying new lathes, but was short of skilled technicians. The workshop did manage to build the two 10 kW radiofrequency power oscillator valves that operated the machine, but like the vacuum chamber from Metrovick when it was finally installed, were a constant source of trouble because of vacuum leaks. The most reliable piece of equipment was the 12 kV DC power supply for the valves: this was a mercury pool rectifier set supplied by the German firm of Siemens. Chadwick[3] was able to obtain it 'at a very reasonable price' because 'at that time the Germans were very anxious to get hold of money from abroad'.*

Lawrence's programme was forging ahead and in his reply to Chadwick's enquiries, he mentioned[11] that his latest cyclotron was intended for 'the primary purposes of medical research, requiring openness and accessibility' in its design. He could not contain his excitement at the dramatic breakthroughs that might be achievable in this sphere.

> There can be no question at all as to the importance of the artificial radioactive substances and neutrons for medical research and therapy and I should think that your biophysical friends in London would undertake the construction of a cyclotron for this line of endeavor. As an illustration which should not be mentioned in public in as much as the experiments are still in progress and it will perhaps be another year before definite publications will be made, I would mention that at the present time my brother John Lawrence, is treating with radio-phosphorus a patient suffering from myelogenous leukemia , with remarkable results. Recently he had been studying leukemia in mice and found that radio-phosphorus is selectively taken up by not only the bones and lymphatic tissue but also, to an extraordinary degree by the diseased white blood cells. For example, he found that in the spleen of the diseased animal the uptake of radio-phosphorus is something like 5 times as great per gram of tissue as in the spleen of a normal animal. This suggested the clinical possibility of treating the human disease, and beginning early in January he gave, over a period of about two months, a patient suffering from myelogenous leukemia a total of 70 millicuries of radio-phosphorus.

* After the outbreak of the second world war, engineers from English Electric came to Liverpool to examine the rectifier because it was the only one of its kind in the country. It broke down only once during the war, when as Holt remarked, it had to be repaired 'with no help from the manufacturers'!

> At the beginning the patient's white blood count was 600,000 while the red cells were 2.5 million. Shortly after the beginning of the administration of the radio-phosphorus the white blood count steadily dropped, the myeloblasts falling much more rapidly than the other cells, while the red count steadily went up to normal. Several weeks ago the patient's blood picture had got to the point where it was not far from normal, the total white blood count being about 8,000, while the red count was 5 million, with less than half of one per cent of the white cells being diagnosable as diseased cells. The radio-phosphorus treatment has been stopped and now the patient is being watched to see what will happen next. Dr. John and all of the medical people feel that this patient's response to radio-phosphorus has been remarkable but they feel on the other hand that there is no evidence that the phosphorus has cured the disease, and I am afraid that if my brother knew that I had mentioned it to you he would scold me. However, I do mention it to you privately as an illustration of one of the first cases where one of the artificial radioactive substances, serving as a tracer element, has revealed fundamental information about a disease, that is to say—in this instance, a profound disturbance of the phosphorus metabolism of the diseased cells; and secondly, that radio-phosphorus has actually been used now in the treatment of disease with results at least as good, if not better than anything that has been achieved with x-rays or radium.

The letter reveals how closely Ernest Lawrence followed his brother's medical research and his remarkably detailed knowledge of the results; it would take little effort to turn the passage quoted above into a report that would be suitable for publication in a medical journal. Equally, one can sense the measured approach of John Lawrence, first making accurate observations in animals about the abnormal metabolism of phosphorus in malignant cells. Although the transition to a clinical application was extremely rapid, Dr Lawrence did not delude himself about the effectiveness of the new treatment and would make no major scientific report for two years.

In the period leading up to his Royal Institution discourse, Chadwick[7] showed himself to be on edge, complaining to Feather about 'too many university meetings and too many lectures, and all my weekends booked up so that we can't get away'. The address to the Royal Institution[12] was polished and well received. After reviewing the principles by which cyclotrons work, Chadwick told the audience that the machines in Liverpool and Cambridge were nearly ready for use. He showed pictures of the Liverpool machine under construction and rattled off some statistics:

> our Liverpool magnet contains 46 tons of iron and 8 tons of copper. (The copper was generously presented to me by the directors of British Insulated Cables Ltd.). The diameter of the pole faces is about 36 inches and the air gap between them is 8 inches. The power consumption in normal running conditions will be from 40 to 50 kW, under full load about 70 kW. The maximum field under the conditions of experiment, that is, with a working gap of 5 inches, is about 19,000 gauss.

He briefly recapitulated the advances in the field of artificial radioactivity since the Curie-Joliots' discovery of the phenomenon, and mentioned that about two hundred new isotopes had since been isolated, some of which had special interest. The applications of the cyclotron that Chadwick wished particularly to bring to their attention were those concerned with biology and medicine. He explained that neutrons from a cyclotron caused a more dense pattern of ionization in living tissue than did X-rays, and therefore might be expected to be biologically more active when given in equal doses (i.e. the same total energy absorbed). He also stated that some elements, boron for example, absorb slow neutrons very effectively. Although animal tissues did not ordinarily contain such elements in any significant amount, it might be possible to inject boron into tumours and then irradiate them with slow neutrons causing selective devastation of a small volume of disease.* Chadwick then moved on to the subject of *radioactive indicators,* which, despite Lawrence's enthusiastic reports about the possibilities of radioactive isotopes for cancer treatment, he correctly foresaw would have the greatest impact on physiology and medicine. The cyclotron could be used to produce isotopes of many of the elements commonly found in living tissue, which were chemically identical to their natural analogues and therefore handled by the body in an identical fashion. He cited experiments by his friend from Manchester days, Hevesy, who was now working in Bohr's Danish institute, and had shown by feeding radioactive phosphorus to rats that 'the mineral matter of bone is in a dynamic state, in which the bones are continually losing phosphorus atoms and taking up others which are later, in their turn, replaced.' To Chadwick 'it is clear that this method of radioactive indicators has many interesting possibilities, for its power and delicacy make it possible to attack problems which have so far been inaccessible to experiment'.**

Towards the end of the academic year, the Chadwicks had rented a small cottage in North Wales from a local farmer, which they hoped to use at weekends and for holidays with the girls. It seems that James, at least, got little enjoyment from it that first summer. In August 1938, he[13] still felt 'just completely worn out when even talking tires me'. By September he thought he was improving steadily, but in November told Feather that he was 'determined to be well for the beginning of next term. I diet, take medicines before and after meals, have injections once a week, go to bed at 10 o'clock and do as little work as possible. It sounds as if I had one foot in the grave but it is not quite as bad as that'.[14] He consulted Professor Cohen in Liverpool, but the exact diagnosis seems to have been elusive. He was never in robust health after his experiences in the camp at Ruhleben, and had some type of chronic digestive problem. There was also a tendency, from this time on, for Chadwick to become ill at times of personal stress and use it as a reason to withdraw from everyday life, sometimes for prolonged periods. He also suffered from chronic back pain, and 'My God, my back!' became 'almost a catch phrase

---

* This idea became the basis of *boron capture therapy*, which was tried in a limited way for patients with brain tumours about 20 years later, and is being looked at again in clinical research protocols in the mid-1990s.

** In 1945, Chadwick as chairman of the Nuclear Physics Sub-Committee was instrumental in persuading the government to centralize the production of radioisotopes for medical use, and the national radiochemical centre at Amersham resulted.

among the research students' in Liverpool.[9] Sometimes, he would forget himself and on one occasion was seen positively to skip down the corridor from his office, after using his terrible back pain as an excuse to terminate an interview with a research student, who had come to him to register a complaint. Such interviews, at the best of times, would be marked by long silences as Chadwick carefully considered his opinion and decided how to express it. This deliberate manner came to be valued by his junior colleagues as they realized that 'his judgement was invariably to be trusted'. Holt[9] summed up his style of leadership as follows:

> The role of leader in the conventional sense did not come naturally to him. He organised the Department as he wanted it to be, assigned people their roles, and then left them to play their part without interference. If there were serious problems or difficulties, of course he was there to advise and help, but one was made to feel responsible for one's own particular job, and such was the natural respect that one felt for Chadwick that one strove to do well and to earn his approval at the end.

The main forum for discussion in the department was at tea time, when all the research staff would sit round a long table in the small library outside Chadwick's office. This was the opportunity to dissect a scientific problem or any issue of general interest, but was not a time for ill-considered opinions when the Professor was present. Chadwick's views were always carefully thought out (indeed he would prefer to keep silent when he was unsure of his ground), and he expected others to adhere to the same exacting standards in their conversation. Any trespass against these would be reproved by a sardonic glance, rendering any verbal rebuke superfluous. As at the Cavendish, he would scrutinize all scientific papers before they were submitted for publication, and often helped his juniors rewrite large sections until the material was in a well-turned and unambiguous form. Chadwick's forbidding presence did not extinguish all the natural exuberance of his young staff. One afternoon he visited the basement to find a doctoral student, Gerry Pickavance, spreadeagled on the floor with his boilersuit firmly nailed to the boards so that all he could do was to lift his head. Assuming his most lugubrious manner, Chadwick approached the helpless figure and eyed him with curiosity. He then began asking Pickavance a series of questions about his research and progress with the cyclotron, before returning upstairs. On re-entering his office, he remarked to his secretary, Miss Lloyd-Jones, that 'the boys have been up to their tricks again'.[9]

The issue of *Nature* on 16 January 1939 carried a letter from Lise Meitner and her nephew Otto Frisch[15] headed, '*Disintegration of uranium by neutrons: A new type of nuclear reaction*'. Lise Meitner had ended her long and productive partnership with the chemist Otto Hahn in Berlin the previous summer. After the German annexation of Austria in March 1938, she was no longer protected by her Austrian nationality and became subject to the anti-semitic laws of Nazi Germany. Although she had grown up as a protestant, as Frisch[16] said 'her honesty did not allow her to conceal her Jewish descent' and with the help of several colleagues she was smuggled into Holland. From there she spent a short time as a guest of the Bohrs in Copenhagen before moving on to Sweden, where she had been given a position by Manne Siegbahn at the Nobel Institute in Stockholm. At Christmas time, she had received a letter from Hahn, telling her the results of some experiments that he had completed since she left. With another radiochemist, Strassmann, Hahn had found that when uranium is bombarded with neutrons one of the products was barium, a much lighter element far away from uranium in the periodic table. All previous scattering experiments involving α-particles or neutrons had given rise to, or been interpreted as giving rise to, products which were of similar atomic number to the bombarded element. Otto Frisch[17] had gone to visit her in Sweden to cheer her up and found her 'at breakfast brooding over a letter from Hahn.' He described the unfolding of their ideas.

> I was skeptical about the contents—that barium was formed from uranium by neutrons—but she kept on with it. We walked up and down in the snow, I on skis and she on foot (she said and proved that she could get along just as fast that way), and gradually the idea took shape that this was no chipping or cracking of the nucleus but rather a process to be explained by Bohr's idea that the nucleus was like a liquid drop; such a drop might elongate and divide itself . . . Lise Meitner worked out the energies that would be available from the mass defect in such a breakup. She had the mass defect curve pretty well in her head. It turned out that the electric repulsion of the fragments would give them about 200 MeV of energy and that the mass defect would indeed deliver that energy . . .

After a few days, Frisch had returned to Copenhagen where he was working and 'just managed to tell Bohr about the idea as he was catching his boat to the US.' Bohr's reaction was instantaneous—he actually struck his forehead and exclaimed: 'Oh, what fools we have been. We ought to have seen that before.' Frisch asked a visiting American biologist what term was used to describe the division of a cell into two daughter cells, and the word 'fission' in this physical context (not yet 'nuclear fission') first appeared in the Meitner–Frisch letter to *Nature*:[15]

> . . . an entirely different and essentially classical picture of these new disintegration processes suggests itself. On account of their close packing and strong energy exchange, the particles in a heavy nucleus would be expected to move in a collective way which has some resemblance to the movement of a liquid drop. If the movement is made sufficiently violent by adding energy, such a drop may divide itself into two smaller drops . . . It seems therefore possible that the uranium nucleus has only small stability of form, and may, after neutron capture, divide itself into two nuclei of roughly equal size . . . The whole 'fission' process can thus be described in an essentially classical way . . .

The heavy uranium nucleus plus the captured neutron have a combined mass slightly larger than the combined masses of the two fission fragments or nuclei. By Einstein's mass–energy equivalence, the fission process liberates a large amount of energy and the two fragments (both positively charged nuclei) fly apart with that energy. When Chadwick[3] read the letter, he immediately accepted the findings and the proposed new mechanism of fission because he knew Hahn and Meitner so well and trusted their work. He wondered whether he and Goldhaber might have made the observation themselves back in 1935, when they had bombarded uranium with slow neutrons. In fact, it seems extremely unlikely that they would have succeded in making the crucial chemical identification of barium, even if their experimental method had been modified to allow them to isolate it. Frisch had also been aided by Bohr's 1937 liquid drop model of the nucleus in making his own brilliant conceptual leap; so it was a regret which should not have troubled Chadwick for long. Nor did it, for Chadwick was busy with his own affairs, and at the time 'did not see any interesting consequences from it [fission] . . . It seemed to me that if something could be done with it, it would be a technical development rather than a search for new physical facts.'[3]

Although the cyclotron had played no part in the discovery of fission, there had been a remarkable convergence among the main nuclear physics laboratories across Europe during the late 1930s towards acquiring cyclotrons. As a result of Chadwick's endeavours, Liverpool was in the vanguard of this movement. In France, Joliot[18] had become convinced of the need for new particle accelerators in 1935; by 1937 he was installed as Director of the Laboratoire de Synthèse Atomique at Ivry and had commissioned the construction of a large magnet. With the help of one of Lawrence's acolytes, the French cyclotron started to produce a beam of deuterons within two years. In Copenhagen, Bohr had become very interested in the biological applications of radioactivity and had also secured a large grant to construct a cyclotron. He too borrowed a young physicist from Berkeley for a year, who spent 'most of his time with his feet upon a table, his chair tilted back, and on his lap a drawing board on which he drew unperturbably one component after another of [the] cyclotron'.[19] Similar activities were underway at the Cavendish, Birmingham, the Nobel Institute in Stockholm and in Leningrad. Strangely, there was none under construction in Germany; Chadwick's old friend Walther Bothe was given permission to commission a cyclotron at the Kaiser Wilhelm Institute in 1938, but the necessary resources were never provided. Other physicists in smaller laboratories wanted to join the cyclotron club, and one of these, Joseph Rotblat[20] from Warsaw, decided to come to Liverpool to see how to build one.

In many ways, the natural place for Rotblat to go would have been to Joliot-Curie's laboratory at Ivry. His professor in Warsaw was Ludwik Wertenstein, a pupil of Madame Curie: she had remained the honorary director of the Warsaw laboratory until her death in 1934 and Wertenstein knew her family well. Rotblat had learned French at school, but spoke no English. Paris seemed a more glamorous city than Liverpool, but Rotblat had the feeling that the Liverpool machine was at just the right stage of construction; if he went there, he could observe and participate in the crucial stages and then be able to take his expertise back to Poland. Physics in Poland was in a very poor state, but Rotblat was ambitious, and determined to establish a proper school of nuclear physics in Warsaw. Although they had access to only 30 mg of radium in solution, Rotblat[21] and Wertenstein pumped off the radon gas into a tube filled with beryllium powder, creating a minute neutron source. Using this simple device, they managed to compete with Fermi's prestigious group in the discovery of new radionuclides. They also made the first direct observation of the inelastic scattering of neutrons (where the neutron not only changes direction but loses energy in a collision with a target nucleus) and this was the subject of Rotblat's doctoral thesis. He had used gold as the original scatterer, but had also run experiments using a uranium target. As soon as he heard about fission in February 1939, Rotblat was in a position to replicate the Hahn–Meitner experiment, which he did within a week. He was particularly interested to see if more neutrons would be emitted than absorbed in the fission process and soon found that this was so. The same question had occurred to several other physicists[22] and three teams, headed by Joliot, Fermi, and Szilard, independently published affirmative answers in March 1939. Rotblat followed them in April with his own paper in which he suggested that about six neutrons are liberated per fission*. 'From this discovery', Rotblat[21] found, 'it was a fairly simple intellectual exercise to envisage a divergent chain reaction with a vast release of energy. The logical sequel was that if this energy were released in a very short time it would result in an explosion of unprecedented power.' Rotblat[21] found the thought so frightening that his 'first reflex was to put the whole thing out of my mind, like a person trying to ignore the first symptom of a fatal disease in the hope that it will go away. But the fear gnaws all the same, and my fear was that someone would put the idea into practice.'

Leaving Warsaw with a small scholarship which was meant to support him for a year, Rotblat was thirty years old and had never before travelled outside Poland. When he arrived by train at Liverpool's Lime Street station, he was shocked by the atrocious slums which provided the visitor with his first impressions of the city. His first impressions of the physics department were scarcely more favourable.[20]

> I expected that Chadwick would be the head of a modern department, and I would be impressed by the equipment and so on. But when I came, it did not take very long to realise that the Physics Department was two departments, cohabiting in the same building, but with very little contact between the two. One was the teaching side, the other was the research side. The teaching side was mostly lecturers who were left over from the days of Wilberforce . . . not concerned in research and ... quite happy to go on slowly, teaching the old fashioned type of teaching. Then quite separate, almost physically too because most of the other research was in the basement with the cyclotron—the most up to date equipment there. And the two parts hardly mixed. Chadwick had put in a few people, Kinsey, Kempton and Walke, and they were mostly involved with the cyclotron. But the only link, apart from Chadwick as a figurehead, was in the students who came back and began to work. Pickavance, for example, and afterwards Holt who came to work when the cyclotron started . . . I was shocked when I went to the teaching lab and discovered they had no a.c. [alternating current]. A teaching lab with no a.c., only d.c. Can you imagine in 1939 . . . how could you teach electricity? It was almost as though you ran a transportation firm and used a cart and horse.

Despite his lack of English, Rotblat was instantly accepted by the members of

---

* This estimate of six was too high, the yield was later generally accepted to be about 2.5 neutrons per fission.

the department and on his second weekend there was invited to Sunday tea by Chadwick. He thought that this was a normal courtesy and spent the afternoon in Aigburth with Professor and Mrs Chadwick and their two daughters. There were awkward silences and comical misunderstandings, as when Chadwick asked Rotblat[23] if his rooms were 'nice'; Rotblat thought he had said 'noisy', and replied: 'No, no, no—very quiet'. The tea passed pleasantly enough and Rotblat was astonished by the reaction when he went to work on Monday. He was besieged for details about the house and about Aileen because no other member of staff in Chadwick's four-year tenure had received such an invitation. Chadwick, unlike Rutherford, was never a gregarious man and kept his professional and personal lives completely separate. Aileen was a devoted wife and mother, but, unlike Lady Rutherford, did not see it as her role to entertain her husband's junior colleagues. She could not forget that she was from Liverpool's aristocracy and, according to Rotblat,[20] 'did not want to mix with the masses and the workers'. The essentially classless world of science was foreign to her. The family became very fond of Rotblat and would take him to the cottage in North Wales for weekends. There he became Chadwick's fishing companion and also found Aileen to be a warm person 'once you got over the class distinction and snobbery'.[20]

As a research student in his early twenties in Warsaw, Rotblat[23] had investigated some short-lived radioisotopes of silver. The lack of facilities in Wertenstein's department were such that he would perform the neutron bombardment in one place, then have to run down two flights of stairs and into another laboratory to test the sample in a shielded Geiger counter. He was nimble footed and used to take each flight of stairs in one leap so that the whole trip used to take seven seconds. After hundreds of jumps downstairs, he developed stress fractures in his tibia, and money was found from somewhere to build a chute to drop the silver foil from the neutron source to the Geiger counter. By the time he came to Liverpool he was a little older and wiser, and when Chadwick set him the task of investigating another very short-lived isotope, he devised a more sedate and elegant coincidence technique, which established that the isotope had a half-life of only a few microseconds. He completed the project within the first two months and Chadwick was duly impressed. He offered Rotblat the Oliver Lodge Fellowship which was the most prestigious in the department, having first been held by Charles Barkla. To Rotblat, at that stage, the promise of income meant more than the honour. His scholarship from Poland was worth £120 for the year, paid quarterly, and the Fellowship would exactly double his emolument. On hearing Chadwick's offer, Rotblat[20] said, 'Oh good, this means I shall be able to bring my wife.' Chadwick, who had not previously known of her existence, retorted: 'Good God, you won't be able to live on this.' But Rotblat was convinced they would manage and made plans to return to Warsaw later in the summer to fetch her.

In July 1939 the Liverpool cyclotron[24] produced its first beam of accelerated particles. The 50 ton electromagnet from Metropolitan-Vickers had cost £3123 and the total cost of the project (excluding salaries) amounted to £5184. This exceeded the funds obtained from the University and the Royal Society, and the balance was provided by Chadwick from his Nobel Prize money. Kinsey, with assistance from Walke, had provided the day to day supervision of the work; many of the engineering problems had been solved by Michael Moore, an Irishman in his early twenties, who had come to the department in 1937 after completing an apprenticeship with Metropolitan-Vickers. With promising research work being undertaken by Pickavance, Holt and Rotblat, Chadwick had every reason for quiet satisfaction and for the anticipation of an exciting era of advances in nuclear physics, radiobiology and medicine. The Liverpool cancer commission had become a major review and was taking up about half of his working time. As Rotblat had noticed, there were still great deficiencies on the teaching side, but before tackling these Chadwick was going to use the summer to recharge his personal batteries.

A holiday on a remote lake in northern Sweden was planned with another family from Liverpool, who owned a house there. There was a lake on the property stocked with the best trout in Sweden so that Chadwick would have the opportunity to indulge in his favourite pastime of fly-fishing. One or two friends asked him if he thought it was wise to travel to Europe with his family, in view of the increasingly unstable and threatening behaviour of Germany. Chadwick said that he did not believe that the British and Germans would go to war again. He may have been influenced by what he read in *The Times* or *Nature*, which had carried a confident editorial entitled 'The promotion of peace',[25] hailing the Munich agreement signed by Chamberlain and Hitler, as 'a step forward in the promotion of peaceful methods of settling disputes between nations'. It seems more likely that he was putting his trust in the decency of the German people, like those he had known in Berlin and Ruhleben, and could not believe the two nations would repeat the wanton destruction of two decades earlier.

Joseph Rotblat[20] returned to his native Poland in order to collect his wife and bring her to England with him. The idea of a new explosive weapon of unprecedented destructive power had been gnawing at his conscience ever since his arrival in Liverpool. He still regarded the prospect as abhorrent, but his concerns intensified with the growing signs that Germany was preparing for war. That summer he had read an article by S. Flügge in *Naturwissenschaften* which mentioned the possibility of nuclear weapons and this had plunged him into despair. He regarded the idea that he should utilize his scientific knowledge to produce such a device as morally repugnant, but gradually 'worked out a rationale for doing research on the feasibility of the bomb'.[21] He came to the view that 'the only way to stop the Germans from using it against us would be if we too had the bomb and threatened to retaliate'. He needed to talk about his ideas, but felt that his English was too halting to discuss such a sensitive issue with Chadwick or other colleagues in Liverpool. Once in Warsaw, he sought out his old professor, Wertenstein, and presented his dilemma. Wertenstein had not thought about such a weapon, but could find no scientific flaw in Rotblat's arguments. He would not advise Rotblat on the moral question.

The sense of impending catastrophe in Poland was unavoidable in August 1939; according to Rotblat[20] one could feel war was coming. His wife had been taken ill with appendicitis and was not fit to travel. He waited in Warsaw until the last minute hoping that she might be able to accompany him, but eventually decided that it would be better for her to travel later. He arrived back in Liverpool on the first of September, the day that Hitler's troops attacked Poland. Rotblat never saw or heard from his wife again. Warsaw itself held out for a month, and during those short weeks, as Rotblat wrote, 'the might of Germany stood revealed, and the whole of our civilization was in mortal peril'.[21] His scruples about weapons research were finally overcome.

References

This chapter contains material based on two interviews with Professor Joseph Rotblat, (refs 20 and 23). Transcripts for both interviews are available in the History of Science & Technology Dept of the University of Liverpool Physics Department.

1 Bohr, N. (1937). Lord Rutherford, obituary. *Nature*, **140**, 753.

2 Eve, A.S. (1939). *Rutherford*, p. 430. Macmillan, New York.

3 Weiner, C. (1969). Sir James Chadwick, oral history. American Institute of Physics, College Park, Maryland.

4 Chadwick, J. (23/1/38). Letter to N. Feather. Feather's papers, CAC.

5 Chadwick, J. (6/3/38). Letter to N. Feather. Feather's papers, CAC.

6 *Liverpool Daily Post* (7/2/38).

7 Chadwick, J. (21/5/38). Letter to N. Feather. Feather's papers, CAC.

8 Pryce, M. (2/7/94). Letter to the author.

9 Holt, J. (1988). *Seminar on James Chadwick.* University of Liverpool. See also Holt, J. (1994). Reminiscences and discoveries: James Chadwick at Liverpool. *Notes and records of the Royal Society*, **48**, (2), 299–308.

10 Chadwick, J. (16/4/38). Letter to E.O. Lawrence. Lawrence's papers, Bancroft Library, University of California, Berkeley.

11 Lawrence, E.O. (30/4/38). Letter to J. Chadwick. Lawrence's papers, Bancroft Library, University of California, Berkeley.

12 Chadwick, J. (1938). The cyclotron and its applications. *Nature*, **142**, 630–4.

13 Chadwick, J. (6/8/38). Letter to N. Feather. Feather's papers, CAC.

14 Chadwick, J. (24/11/38). Letter to N. Feather. Feather's papers, CAC.

15 Meitner, L. and Frisch, O.R. (1939). Disintegration of uranium by neutrons: a new type of nuclear reaction. *Nature*, **143**, 239.

16 Frisch, O.R. (1970). Lise Meitner (1878–1968), *Biographical memoirs of fellows of the Royal Society*, **16**, 405–15.

17 Frisch, O.R. (1967). Discovery of fission. *Physics Today*, Nov, 272–7.

18 Goldsmith, M. (1976). *Frédéric Joliot-Curie*, p. 61. Lawrence and Wishart, London.

19 Pais, A. (1991). *Niels Bohr's times*, p. 401. Oxford University Press.

20 Brown, A.P. (1994). Interview with Joseph Rotblat, London.

21 Rotblat, J. (1985). Leaving the bomb project. *Bulletin of the Atomic Scientists*, Aug., 16–19.
22 Graetzer, H.G. and Anderson, D.I. (1971). *The discovery of nuclear fission: a documentary history*. Van Nostrand Reinhold, New York.
23 Edwards, D. and King, C.D. (1992). Interview with Professor J. Rotblat, London.
24 Moore, M.J.(1962). 37" cyclotron—end of an era. *University of Liverpool Recorder*, **29**, 5–7. University Archives, University of Liverpool.
25 Editorial (8/10/38). The promotion of peace. *Nature*, **142**, 629.

11 The big question

Word reached the holiday party from Liverpool, a little late, through a local farmer who had heard the announcement on the wireless. To be trapped in Europe at the beginning of one world war may be regarded as a misfortune; to be caught there at the start of the second begins to look like carelessness. Disbelief, self-reproach, anger, and above all worry assailed Chadwick as soon as he heard the news. He and Aileen held many whispered conversations so as not to alarm the girls—producing just the opposite effect. One mistake he was not going to repeat was to sit tight, on the basis that, at worst, it would all be over by Christmas. They packed up immediately and travelled 500 miles south to Stockholm, only to find that all flights to London had been cancelled.

They booked into the Plaza Hotel and Chadwick contacted some of his friends in Stockholm. One was Lise Meitner, who was still unhappy and isolated in Sweden. She had been offered a position in Cambridge, and asked Chadwick how she should respond so as not to offend her Swedish hosts. Chadwick[1] advised her to tell Professor Siegbahn how matters stood, otherwise 'he might be justifiably annoyed at having been kept in the dark'. The last paragraph of Chadwick's letter, dated 11 September 1939 and written on hotel notepaper, hints at his own distraction:

> I hope I have not expressed my thoughts too clumsily but we are leaving in a few minutes and I have little time to choose my words. And I hope you will not think that I am interfering in your affairs. I am ready to do anything to help you.

He closed with best wishes and 'auf baldiges Wiedersehen'. The Chadwicks had managed to get a flight to Holland from where they hoped to take a boat to England. The next week was spent in a hotel in Amsterdam, where a fellow guest in the same predicament was the writer, H.G. Wells;[2] in 1913, he had published a prophetic novel, *The world set free,* in which the devastation from an atomic bomb had led mankind to a lasting rejection of warfare. Now he had abandoned a planned address to fellow writers at the PEN club congress in Stockholm on 'The honour and dignity of the human mind', because of the developing European crisis. Chadwick did not approach him, and it is unlikely that Wells knew he was sharing the hotel with the scientist who had discovered the neutron.

No one knew when there would be a sailing to England, and they dared not leave the hotel in case a message came through from the port. Eventually they boarded a stinking, rusty, tramp steamer and made an unmolested, but emetic, North Sea crossing. On his return to Liverpool, Chadwick found the department depleted of both students and some staff who had joined the military or had been seconded for government work. Kinsey, along with other scientists mainly from Cambridge and Birmingham, had been summoned to Kent even before war broke out in order to become familiar with the new radar chain that was installed along the south coast of England. Joe Rotblat was back in Liverpool, but destitute and heartbroken. His Oliver Lodge Fellowship was not due to start until October, and his quarterly grant from Poland had, of course, not arrived. As a result he had seven shillings and sixpence to his name, which would not even pay a week's rent for his digs, and he had been surviving on his landlord's charity. Chadwick's immediate solution was to appoint him as a lecturer, even though Rotblat's English was 'still very, very shaky'.[3] Rotblat demanded that he should be given time to prepare his lectures and Chadwick agreed that the course in nuclear physics to the honours school would not start until the first of November. Rotblat immersed himself in learning the language over the next six weeks and wrote out what he was going to say in his lectures. He decided not to approach Chadwick about the question that was preoccupying him until he had spent this period improving his English so that he would be able to convey clearly his thoughts about the potential application of nuclear fission.

Although Chadwick was mightily relieved to have brought his family home, he was not favourably impressed by the national mood. He wrote[4] to Feather in Cambridge on 15 October, telling him about their escape from Sweden, where he had been 'well away from rumours and excitements'; he expressed himself 'as very surprised at the state of mind of people here when I got back. They seemed to me to have lost hold—dithering about and feeding on their imaginations, instead of getting on with the job. So far I have resisted the fever and I am going on as usual.' Where he had been so reluctant to accept that another war with Germany was coming, now it had begun, Chadwick saw no hope of a diplomatic solution. Of course the public mood was fearful: the newspapers had been full of stories about the brutal invasions of Czechoslovakia and Poland; at home there were black-outs, restrictions on entertainment, the issue of gas masks and petrol rationing, all of which added to the general despond. The day before Chadwick wrote his observations to Feather, a U-boat had penetrated the defences of Scapa Flow and sunk the battleship *Royal Oak* as she lay at anchor. The tide of Nazi propaganda was in full flood, and Hitler attempted to strike new fear into the hearts of his enemies with the promise of a 'secret weapon' to which there was no counter.

There had been some earlier propaganda in the British press about a new and terrible weapon:[5] the *Sunday Express* in April had carried a sensational article on an unimaginably destructive bomb, containing only 1 lb of uranium, that would derive its power from the explosive release of atomic energy. Although there were a few exceptional scientists, notably the Hungarian Leo Szilard, who had imagined the possibility of a chain reaction as early as 1934 ('because I had read

H.G. Wells'),[2] the physics community at large did not believe that such a bomb was feasible. Following the *Sunday Express* article, it was, however, something which had registered with some members of the public and a handful of political figures. Chadwick received a letter from his old Cavendish colleague, Edward Appleton, who was the new Secretary of the Department of Scientific and Industrial Research (DSIR), asking him privately for his opinion on the matter. In his reply, dated 31 October 1939, Chadwick[6] had the following to say:

> I am not surprised that you are bombarded with the uranium bomb. I suspect that this is Hitler's 'neue Waffe gegen die es keine Verteidigung gibt'. It is not easy to say anything definite about it. It is certainly a possibility that under suitable conditions the uranium fission process might develop explosively. It is fairly well established that about 3 neutrons are emitted as a result of the fission process. If these liberated neutrons can be used to initiate new fissions then it is clear that the process will be cumulative and perpetuate itself. A very large quantity of uranium might be necessary—several tons—in order to catch the liberated neutrons.
>
> This is one limiting factor. Another is the fact that even in pure uranium a non-fission capture process occurs which will reduce the available neutrons. In impure uranium other processes will also be effective in reducing the tendency to a catastrophic extension of the fission process. Moreover, the neutrons which cause the fission process have thermal velocities (fast neutrons too but for these the cross section is much smaller); as the fission process extends, the consequent generation of energy in the material would increase the energy of the thermal neutrons and reduce their efficiency as agents of fission.
>
> The process is quite complicated in itself and I do not think it is possible at present to say more than that a catastrophic explosion is possible, for there is not sufficient information on which to base an opinion. I would not like to go further now. I will look more carefully into the whole question and see if I can get any indication from the experimental data which are now available. There may be new data which I can use.
>
> There is one point which must have occurred to you at once and that is, how to prepare a uranium bomb that will not blow up immediately. 'Hoist with his own petard' will almost certainly be the fate of the man first successful with this process. But I do not think that this difficulty is insuperable. The real trouble is the self-propagation of the fission process. I doubt very much its feasibility on general grounds, but I will look into it carefully and write again.

While the letter is sceptical, Chadwick is not dismissive and the seeds of doubt have been sown in his own mind. The leading German nuclear scientists—Hahn, Geiger, Bothe, Heisenberg—were men he knew well and for whom he had the greatest respect. If Hitler was indeed exploring the possibility of an atomic bomb, they could match any other group of scientists in the world in the theoretical analysis and experimental testing of the problem. Chadwick's first reservation concerned the amount of uranium that would need to be present in order to permit a sustained divergent reaction. If approximately three neutrons (a slight overestimate) were released during each fission process for each neutron consumed, those extra neutrons would have to cause the fission of other uranium nuclei and not just pass through the mass of uranium without further interaction. If the body of uranium were too small, the penetrating neutral particles would escape without resulting in any further fission and the reaction would not be sustained. There was another fundamental consideration which followed from a characteristically brilliant insight provided by Bohr[7] in February 1939. He had suggested, and it had become widely accepted despite the lack of direct experimental verification, that fission by slow neutrons was entirely due to the rare uranium isotope of weight 235, ^{235}U. This isotope comprises only 0.7% of natural uranium, the rest being ^{238}U. It therefore appeared that 99.3% of any mass of uranium would not take part in a chain reaction. The same obstacle was apparent to Einstein, who at Szilard's behest, had written to President Roosevelt at the beginning of August to warn him that a new type of extremely powerful uranium bomb was conceivable. Einstein thought that if such a bomb could be constructed, it might well be too heavy to be carried by air and would need to be transported by ship.

The next point covered by Chadwick in his letter is that not every neutron–uranium nucleus collision would result in fission—for example many neutrons might just lose energy by *inelastic scattering,* others would be taken up or *captured* by the uranium nucleus without fission. Impurities in the uranium would further reduce the efficiency of the chain reaction. There follows a cautionary remark alluding to the fact that fission depended predominently on slow (or thermal) neutrons; as fission energy was released some would be transferred to the neutrons which would speed up and become less effective agents of fission. The final caveat in the letter follows from the recognition that there are neutrons present in cosmic radiation, and a chain reaction in a critical mass of uranium could be triggered by a stray neutron from space.

Chadwick knew that there was a rapidly enlarging body of literature on nuclear fission, a topic which he had not initially considered to be of fundamental interest, and he needed to make a review of what had been published in the past six months. He found three recent papers which taken together gave fresh impetus to his thoughts on the bomb. The first of these was the definitive review of nuclear fission published by Bohr and John Wheeler,[8] an American co-worker, published in *Physical Review* on 1 September. Rudolf Peierls[9] had also become interested in the concept of critical mass and published a short theoretical treatment in the more obscure, but still influential *Procedings of the Cambridge Philosophical Society.* The third paper was from the group who had, at this stage, made the furthest inroads into the actual physical conditions necessary for a divergent chain reaction—Joliot and his colleagues in Paris.[10] They had addressed the obstacle identified by Chadwick in his letter to Appleton, namely that as the chain reaction built with the release of increasing amounts of energy, the neutrons produced would become too energetic themselves to propagate further fission effectively. The solution, it seemed to them, might be to mix the uranium with *heavy water* so that the neutrons would be slowed down again (or moderated) by collisions with deuterium atoms. Their paper was dated 19 September so that all three crucial pieces of work appeared within weeks of the outbreak of war, and two were published in the open literature after hostilities had commenced. Joliot's group were probably the most convinced that the uranium bomb was feasible, whereas Bohr[7] in December 1939 continued to think that chain reactions in uranium would be 'too short and rare for there to be any question of an explosion. The situation would be quite different if we had a sufficiently large quantity of ^{235}U available.'

While his reading heightened Chadwick's interest in the uranium fission bomb, the crucial stimulus came from Rotblat,[3] who approached him after work one evening in late November. By now he felt that his English was adequate to present his ideas to Chadwick without the risk of complete misunderstanding. The most original and crucial proposal that Rotblat made was to suggest that in order to produce an explosion, the chain reaction would have to be propagated by fast neutrons. Slow neutrons would lead to the release of heat, as in a chemical reaction, but not the instantaneous, catastrophic, outcome required. Chadwick sat and listened intently to the quietly spoken young Pole; he betrayed no flicker of interest and at the end of Rotblat's presentation just grunted. Rotblat was discouraged and thought the professor would take it no further. The lack of a discernible response was exactly the unsettling experience that Maurice Goldhaber had when he had suggested the photodisintegration of deuterium to Chadwick in 1935, and it again masked a deep level of interest in what was being proposed.

It is a moot point as to whether the same idea had already occured to Chadwick independently—his poker-faced reaction to Rotblat gives no clue. It seems probable that he had thought of it himself since he was always scrupulous about crediting the work of others at every opportunity, and did not do so for Rotblat in this case. Writing shortly after the war, Chadwick[11] had this to say:

> It was only the direct impact of war which made me put my mind to such questions. I then saw how simple the problem of producing a violent explosion really was, provided that a suitable material existed; that is, a material which would support a chain reaction with fast, not slow, neutrons so that a substantial part would react, and release large amounts of energy, before the system had time to fly apart.

Whatever the provenance of the hypothesis, the vital thing was to look for available experimental data which might throw light on the subject. There was almost nothing published, but the review by Bohr and Wheeler did mention some results obtained for the fission cross section of fast neutrons by the American physicist Merle Tuve at the Carnegie Institution in Washington. His measurements were very low suggesting a small probability that fission could be sustained in this manner. Chadwick[12] asked Pryce to make some calculations based on Tuve's numbers for the critical size of uranium that would be necessary, and the answer came out to many tons. Although Chadwick had great respect for Tuve who was a very good experimenter in his opinion, he began to think that Tuve's figure for the cross section was 'too damn low altogether'[12] because it implied that only a small fraction of the collisions between neutrons and uranium 235 nuclei would result in fission. Here is an example of intuition about the behaviour of nuclei of which Rutherford would certainly have been proud.

A few days later Chadwick sought out Rotblat[3] and they discussed what experiments needed to be done. Apart from the question about the fission cross-section for fast neutrons, there was little known about the energies of neutrons that would be produced by fission, or about the proportion of neutrons that would be captured without producing fission. One of Rotblat's main concerns was the inelastic scattering of neutrons which might cause them to slow down too much. Chadwick agreed that these were all critical questions and assigned Rotblat two research assistants. One was a Quaker and conscientious objector who had been ordered to work in Liverpool instead of going into the army. He was not supposed to undertake any military work, which arguably these neutron experiments were not at that stage; not for the last time Rotblat's conscience pricked him, but he had been instructed by Chadwick not to divulge the nature of the research to anyone else. Rotblat convinced himself that the experiments they were engaged in had some general scientific validity, and it was, therefore, permissible to involve the young Quaker. In the meantime, Chadwick[13] wrote a second letter to Appleton.

GEORGE HOLT PHYSICS LABORATORY,
The University of Liverpool.
5th December 1939

Dear Appleton,

The examination of the uranium fission experiments and the calculation of the possibilities of an explosive chain reaction has taken much longer than I expected. The calculation is in fact quite involved, and I had to ask Pryce to look into it. In the meantime two papers have appeared, one by von Halban, Joliot and Kowarski on experiments with a large quantity of uranium oxide, and one by Peierls on the calculation of conditions for chain reactions.

The conclusions I have reached are these. It seems likely that both types of fission—the one due to thermal neutrons, the second to fast neutrons—could be developed to an explosive process under appropriate conditions. The thermal neutron fission would require a mixture of uranium and hydrogen (i.e. water); otherwise the chain reaction would be prevented by a resonance capture process which takes place. Moreover, the materials would have to be free from rare earths even in very small quantities for many of these show marked resonance effects. Given these conditions, a chain reaction will probably occur, but, as far as I can estimate from the available data, a large quantity of uranium would be necessary—perhaps more than 1 ton.

The fast neutron fission is inhibited only, as far as is known at present, by the effect of inelastic collisions. Unfortunately, we have no data whatever about inelastic collisions in uranium; the data for other heavy elements are meagre, and it would be unsafe to speculate. I think one could say that this explosion is almost certain to occur if one had enough uranium. The estimates of the amount necessary vary from about 1 ton to 30 or 40, according to the data adopted in the calculations.

There is of course a general argument against the possibility of these explosive processes—that no such explosion has occurred in pitchblende mines or in laboratories where the ore is dealt with. This argument is open to objection. First, the concentration of uranium in the crude ore is not high, and the pitchblende is said to occur in strata of only about 10 cm thickness. Secondly, the presence of impurities of rare earths, etc.etc. It does seem quite possible that the necessary conditions for explosion may never be realised either for the fast or for the thermal neutrons without the aid of man.

I am sorry I can give no definite answer to this question, for it is a very interesting one indeed. The amount of energy which might be released is of the order of the energy of the well known Siberian meteor. The difficulty is really lack of data. Very few experiments have been made which throw any light on the actual mechanism of the fission processes; they have been mainly concerned with the radioactive products of the fission—interesting and important, but apt to degenerate into a kind of botany. I think it would be desirable to get some information on the mechanism, and if I can get enough uranium oxide I will do so. I have here a Polish research man who is very able and very quick.

Yours sincerely,
J. Chadwick

Copies of the Appleton–Chadwick letters found their way to the Committee of the Imperial Defence, where they were immediately seized on as a duplication of effort, 'an illuminating example of the scientists' right hand not knowing what the left hand was doing'.[14] Despite his position as a leading government scientist, Appleton had not realized that the uranium question had been under active consideration by various Whitehall committees since April 1939. An early note had been provided by Sir Henry Tizard, Rector of Imperial College and Chairman of the Air Defence Research Committee, as soon as evidence had appeared in the literature that a chain fission reaction might be possible. In his prescient opinion, Tizard[15] wrote that the chance against a huge release of atomic energy having military application was 'quite 100,000 to 1, but if the issue were succesful it might be of such great importance that the 100,000 to 1 chance should not be ignored'. G.P. Thomson (J.J.'s son) at Imperial College and Bragg, the new Cavendish Professor, had also been sounded out informally and they agreed that there was a much smaller than even chance of producing an atomic explosion. Finally, Professor Tyndall[16] of Bristol University had written a lucid memorandum for the Chemical Defence Committee in which he identified key obstacles such as the energy of fission neutrons, the separation of the isotopes of uranium and the detonation mechanism. Like Tizard, he warned that 'in view of the issues at stake one must not rely on any instinctive feeling one may have that the scheme will not work'.

The vital decision was made to unite all matters relating to uranium under the auspices of the Air Ministry, and specifically to refer them to the committee chaired by Tizard. Uranium ore was mainly extracted from mines in the Belgian Congo, and Thomson and Bragg were both anxious that the waste uranium oxide in Belgium, which remained after the extraction of radium, should be denied to the Germans and secured by Britain. In April, Tizard had suggested that the British Government should buy or place an option to buy 300 tons of the uranium oxide, but by May had decided that the level of scientific credibility did not justify an outlay of £70 000 and bought 1 ton instead for research purposes.[17] The experimental work was to be undertaken in Thomson's department at Imperial College, and at Birmingham University under Oliphant. Winston Churchill, who was still a backbench Conservative MP and not yet in the Government, was kept informed of all these developments by his personal scientific advisor, F.A. Lindemann*, Professor of Experimental Philosophy (Physics) at Oxford and a vexatious member of the Tizard Committee. Lindemann had obviously briefed his friend well, judging by the cogent letter Churchill[18] wrote to the Secretary of State for Air on 5 August 1939.

> Some weeks ago one of the Sunday papers splashed the story of the immense amount of energy which might be released from Uranium by the recently discovered chain of processes which take place when this particular type of atom is split by neutrons. At first sight this might seem to portend the appearance of new explosives of devastating power. In view of this it is essential to realise that there is no danger that this discovery, however great its scientific interest, and perhaps ultimately its practical importance, will lead to results capable of being put into operation on a large scale for several years.
>
> There are indications that tales will be deliberately circulated when international tension becomes acute about the adaption of this process to produce some terrible new explosive, capable of wiping out London. Attempts will no doubt be made by the Fifth Column to induce us by means of this threat to accept another surrender. For this reason it is imperative to state the true position.
>
> First, the best authorities hold that only a minor constituent of Uranium is effective in these processes, and that it will be necessary to extract this before large-scale results are possible. This will be a matter of many years. Secondly, the chain process can take place only if the Uranium is concentrated in a large mass. As soon as the energy develops it will explode with a mild detonation before any really violent effects can be produced. It might be as good as our present-day explosives, but it is unlikely to produce anything very much more dangerous. Thirdly, these experiments cannot be carried out on a small scale. If they had been successfully done on a big scale (i.e., with the results with which we shall be threatened unless we submit to blackmail) it would be impossible to keep them secret. Fourthly, only a comparatively small amount of Uranium in the territories of what used to be Czechoslovakia is under the control of Berlin.
>
> For all these reasons the fear that this new discovery has provided the Nazis with some sinister, new, secret explosive with which to destroy their enemies is clearly without foundation. Dark hints will no doubt be dropped and terrifying whispers will be assiduously circulated, but it is to be hoped that nobody will be taken in by them.

Appleton had passed Chadwick's first letter to Sir John Anderson,[19] the Lord Privy Seal, who had special responsibility for Civilian Defence. Unusually for a British politician, Anderson had a science background and many years earlier had carried out research work on the chemistry of uranium, while a graduate student

---

* Later Viscount Cherwell, but always known to Churchill as 'the Prof'. Chadwick had first met him at the Rubens Colloquium in Berlin before World War I.

in Germany. Chadwick's letter was also sent to Lord Hankey, Minister without Portfolio in Chamberlain's War Cabinet. Like Anderson, Hankey had been a career civil servant, retiring as Secretary to the Cabinet in 1938; among his new ministerial responsibilities was chairmanship of the Science Advisory Committee. Anderson[20] had commented to Hankey after reading Chadwick's October letter: 'I am quite sure that we need not regard the uranium bomb as a serious danger'. Hankey,[21] in turn, after studying all the Appleton–Chadwick correspondence wrote to the Minister for the Co-ordination of Defence, Lord Chatfield, with more complacency than was justified, 'I gather that we may sleep fairly comfortably in our beds, but that Professor Chadwick, with the assistance of a Polish Research Specialist, may do some further work on the subject.'

Although the Liverpool physics department had given up several members of staff, like Kinsey, it had not (unlike the Cavendish or Birmingham) been designated as a site for radar development or other war work. Several reasons can be advanced for this omission: it was still a small and generally ill-equipped department, and the research interests of the professor were relatively narrow and restricted to nuclear physics. But now there was interest in uranium fission, Liverpool did have one of the two functioning cyclotrons in the country, and Chadwick was the experimental scientist with the greatest experience and success in the field of neutron physics. He had assembled a small team of exceptionally able young research workers, who were keen to make their contribution and were not diverted by other work. The measuring of neutron cross-sections began straight away.

The cyclotron was proving to be a temperamental piece of equipment, which would often yield no beam at all. According to Holt,[22] the key to successful operation of the machine was a procedure known as 'shimming' the magnet. As protons spiralled out from the central source, they were contained in a low-pressure gas vessel or vacuum box, which was sandwiched between the two poles of a large magnet. Ideally the proton spiral should be kept in a horizontal plane midway between the two large magnets, but left to their own devices most of the protons tended to wander from the central plane to the top or bottom of the box, where they would be lost. The imperfect solution was to shape the magnetic field across the spiral so that its intensity decreased slightly with distance from the centre. This was done by inserting thin rings of iron, or shims, against the pole pieces of the magnets. It required considerable patience and skill to get the cyclotron running so that it would produce a beam current of a few microamps. Chadwick liked to try his hand on occasion to see if he could do better than the regular operators, and promised that when they managed to generate 50 μA he would bring some champagne to celebrate the event. The day eventually arrived and the champagne was consumed, suitably chilled, from laboratory beakers.

The maintenance of high vacua that had proved so challenging to the Cavendish workers a decade earlier had still not been solved completely, and leaks continually developed. Once located, they were sealed with an aromatic, red, resinous varnish. To protect the operators from neutron radiation, a shielding material rich in hydrogen was required rather than lead, for example, which would be used as a shield against X-rays. Chadwick persuaded a local margarine manufacturer to donate about ten tons of whale blubber in wooden boxes, which was ideal for the purpose of absorbing neutrons and could easily be stacked to form a wall. The only drawback was that occasionally one of the wooden boxes would split releasing its pungent greasy contents. On one celebrated occasion, Aileen was visiting the laboratory when this happened, and mindful of the wartime food shortages took some of the fat home to use in cooking. The fat was rancid and the recipe was not a success.

Every time an adjustment was made to the magnetic field or high-frequency oscillator, the power supply to the cyclotron had to be switched off. This was a tedious process and it was tempting to take short cuts. Harold Walke was an impetuous young man who wanted to do things as fast as possible.[3] Just before Christmas 1939 he entered the protective cage around the cyclotron without disconnecting the power supply, touched a live terminal and was electrocuted. His father, a sea captain, had landed in Liverpool the previous evening, and they were planning to travel home together to Devon the next day for the Christmas holidays. Walke had spent his last summer in Ernest Lawrence's department at Berkeley, and only returned to Liverpool in November. Lawrence had been singing his praises to Chadwick, saying how much he had accomplished during his stay and how eager he was to get back to move the work along in Liverpool. During the preceding weeks, Chadwick, Kinsey and Walke had sent two congratulatory telegrams to Lawrence, first when he was awarded the Hughes Medal of the Royal Society and then the 1939 Nobel Prize for Physics. Now Chadwick[23] sat down, with a heavy heart, on Christmas Eve to write a long letter to Lawrence relaying 'the sad and painful news'. In the letter, he revealed the affection he felt for his young assistants in a way which might have surprised them. Of Walke, he said: 'He was so simple and single minded, so enthusiastic and so happy. We were all very fond of him and to lose him in any way would have been a bitter blow, but to lose him like this and to feel to some degree responsible is extremely painful.'

Chadwick was by now in his late forties and by common consent was the leading experimental nuclear physicist in the country. In Liverpool he was surrounded by gifted but inexperienced men who were of a younger generation. There was no seasoned deputy who could oversee the research work in the way he had done at the Cavendish for Rutherford, and indeed there were now security implications which tended to preclude full and open disclosure. At this stage, security was purely an informal consideration to be decided upon by Chadwick as the head of department. It seemed to him that the most prudent approach was to avoid the creation of 'any undue curiosity'.[12] The worst thing, he thought, would have been to attempt a comprehensive muzzling of those working in the laboratory, but he discouraged loose talk even more than usual, and expected junior assistants to conduct experiments without being told why they were doing them. Rotblat quite openly included nuclear fission in his lecture course to the honours students, since he and Chadwick both reasoned that it was by then common knowledge, and to exclude it would be more suspicious than presenting it in a matter of fact way.

The day-to-day running of the cyclotron now became the responsibility of Gerry Pickavance, the postgraduate once nailed to the basement floor; he was assisted by Mike Moore, the Metropolitan-Vickers trained engineer. Pickavance knew the machine well since he had been responsible for the design and construction of its various control systems, and had set up the radiofrequency power supply. As a boy, Pickavance[24] had shown insatiable curiosity about things mechanical—he remained passionately interested in cars—and was a self-taught wireless expert; these practical accomplishments now proved invaluable. Only three weeks after Walke's death, Chadwick sounded pleased with the operation of the cyclotron, but worried about the broader picture of university departments in wartime. He wrote[25] to Cockcroft at the Cavendish:-

> I am now fairly satisfied that we have a real cyclotron. We can now take the [vacuum] tank down, put it together etc., and within an hour of getting the pressure down, we can get a beam—no cleaning or fiddling about with discharges...Our laboratories seem to be so disorganised that very little useful work of any kind—peace or war work—is going on. I am afraid if this lasts for any length of time, we shall be hopelessly behind the Americans.
>
> I am not really so pessimistic as this sounds, but I do think it should have been possible to keep some things going, and especially to retain at least a skeleton team at Cambridge.

He also expressed concern about Rotblat's future.

> I don't know if you have met Rotblat, a Pole who has been here about nine months. He is an extremely able man, one of the best I have come across for some years. There is a possibility that he may be called up for the Polish Army fairly soon and as he is Jewish, things may go hard with him. I should like to keep him here but it would not be safe. What I should like to do would be as it were, to book him a place on the research side, to which he could go at the end of this term. This would give him time to write up some finished work, to give his course of lectures, and to start some new work in which we are very interested.

Why he thought it would be unsafe for Rotblat to remain in Liverpool is unclear. The city did, of course, become the target of heavy German bombing because of its dual military importance as a port and the telecommunications centre for Atlantic shipping. The main way Rotblat's existence differed from those of other department members was that as a 'friendly alien', he was subject to various police restrictions such as not being allowed to ride a bicycle, to possess any maps, nor to be out of his residence between sunset and sunrise. If he wished to leave the city, he was supposed to obtain a permit. Chadwick happened to be friendly with the Chief Constable of Merseyside and would frequently invite him to lunch at the University Club. Sometimes Rotblat[3] would join them and they would all laugh at the latest breach of the rules that he might have committed. The prize occasion was probably when the cyclotron broke down and Rotblat, Pickavance and Moore decided to take a break for a few days since they were all exhausted. They set off for the Lake District in Pickavance's car, without obtaining any permit for Rotblat. After dusk they lost their way (there was a blackout in force and many roadsigns had been removed); Rotblat then compounded his sins by studying a map by torchlight—there may have even been a bicycle within his reach.

Although Chadwick's responsiblities were beginning to weigh heavily on him, there were still occasional reminders of happier days. In the New Year of 1940, he received unexpectedly a letter from Guiseppe Occhialini, Blackett's wonderful research assistant at the Cavendish in the early 1930s, who was now working in Sao Paulo. He had applied for a fellowship from the Guggenheim Foundation and given Chadwick's name as a referee. Employing an irresistible combination of idiosyncratic English and execrable typing, he wrote:[26]

> Please, reply them and lie like a gentleman. If I do not get the fellowship another fool will get it.
>
> It will maybe please you to know that the Graf Spee Business has done quite a lot here for british prestige. It was a nice bit of work.

The *Graf Spee* business was of course the scuttling of the German battleship off the coast of Argentina at the Battle of the River Plate—one of very few major losses sustained by the Germans in 1939. The testimonial for Occhialini[27] was succinct, positive and bore a characteristic flash of Chadwick humour.

> I have a very high opinion of Dr Occhialini's abilities. He is very fertile in ideas and he has an original cast of mind. He is also a very good experimenter, fully competent to test his ideas in practice.
>
> Dr Occhialini has a great vitality and enthusiasm for scientific research. In general matters, he is independent in thought and , though reasonable and considerate in the views of others, decided in his opinions. In fact he has a personality full of colour—and the colour is certainly towards the red end of the spectrum.

By the spring of 1940, G.P. Thomson's team at Imperial College had become doubtful about the prospect of atomic energy making any contribution to the war effort. Their attempts to produce a sustained chain reaction in uranium oxide, whether they tried with slow or fast neutrons, had been uniformly unsuccessful. They had slowed the neutrons using ordinary water and paraffin wax as moderators. In Thomson's mind there was at least some reassurance that the objective could not be achieved easily or quickly, and he wrote[28] to Chadwick on 10 March, seeking collaboration:

> If you are really thinking of coming in on the uranium bomb we ought to have a talk. We are coming to an end of one stage of the work and have pretty well concluded that it can't be done with uranium oxide and water or wax, but it is not far off. A detailed report for the Air Ministry is being written now which would tell you much more than I can in a letter.

Before the Imperial College report reached the Air Ministry, they received another short document from Oliphant in Birmingham which transformed the concept of a uranium bomb and gave the theoretical blueprint for its construction. Chadwick's first knowledge of this development came in another letter from

Thomson, dated 16 April 1940.[29] He explained that a sub-committee had been appointed 'to consider the possibility of constructing a uranium bomb by separating the isotopes [^{235}U and ^{238}U]. This is Oliphant's suggestion, and although at first sight it seems a bit wild, it is not so impossible when you come to look into it. I hope you will be able to join us.'

As Chadwick found out when he attended his first meeting of the sub-committee on 24 April, the latest suggestion was not Oliphant's, but originated from two men then working in his department, Otto Frisch and Rudolf Peierls. At this stage, Oliphant and most of his department were exclusively involved with radar research, and were developing new high-frequency generators and detectors. This work, spurred by Tizard, had the highest priority because of the threat of German invasion, but it was also top secret so that Frisch and Peierls, two scientists who had only left Germany comparatively recently, were not permitted to contribute to it. So Peierls[30] had been free to continue on his analysis of the critical mass of uranium, and the scales had fallen from his eyes when his newly arrived colleague Frisch had asked him, quite simply, what would happen if 'someone gave you a quantity of pure 235 isotope of uranium?' Within a few weeks they had written (and typed themselves for reasons of security) a short paper headed, '*On the construction of a "super-bomb"; based on a nuclear chain reaction in uranium*'. It would later be known as the Frisch–Peierls Memorandum,[31] and represents the first irrevocable step towards the nuclear bomb.

The force of the arguments presented by Frisch and Peierls was devastating. With no knowledge of the work that had been going on at Imperial College, they pointed out that it is impossible to make a chain reaction by mixing uranium with water. They also quantified the speed of the fission reaction, and reached the same conclusion that Rotblat had already impressed upon Chadwick, namely that 'no arrangement . . . based on the action of slow neutrons could act as an effective super-bomb, because the reaction would be too slow'. The second, revolutionary, suggestion was that since Bohr had put forward convincing arguments that the rare isotope ^{235}U has a generally higher probability for fission than the common isotope ^{238}U, the bomb should be fabricated using almost pure ^{235}U, 'a possibility which apparently has not so far been seriously considered'. They thought that a method of thermal diffusion, recently described by a German chemist, Clusius, could serve as the basis for separating out large amounts of ^{235}U, if the uranium could be first changed into the gaseous compound, *uranium hexafluoride.*

Frisch and Peierls conceded that 'the behaviour of ^{235}U under bombardment with fast neutrons is not known experimentally, but from rather simple theoretical arguments it can be concluded that almost every collision produces fission and that neutrons of any energy are effective'. Making a few reasonable assumptions in the absence of experimental data, they calculated the critical radius of a sphere of metallic ^{235}U, the densest form of the fissile material that could be obtained, and came to the chilling conclusion that 'one might think of about 1 kg as a suitable size for the bomb'.

The second part of the memorandum was non-technical, but equally thorough.

Having disclaimed competency to discuss the strategic value of such a bomb, Frisch and Peierls pointed out that as a weapon it would be practically irresistible. It would necessarily kill large numbers of civilians, both by its explosive might and by the spread of radioactivity, and 'this may make it unsuitable as a weapon for use by this country'. Although they had no direct knowledge of the scientific effort being applied to this end by the Nazis, 'since all the theoretical data bearing on this problem are published, it is quite conceivable that Germany is, in fact, developing this weapon.' Given that there was no effective shelter that could be offered to citizens against the bomb, 'the most effective reply would be a counter-threat with a similar bomb. Therefore, it seems to us important to start production as soon and as rapidly as possible, even if it is not intended to use the bomb as a means of attack.' Thus the concept of *mutual deterrence* was born.

Oliphant,[32] who had already had the advantage of discussing the memorandum with the two authors and had been convinced by their arguments, remembered that 'the Committee generally was electrified by the possibility, but Chadwick, who was also a member, was embarrassed, confessing that he had reached similar conclusions, but did not feel justified in reporting them until more was known about the neutron cross-sections from experiments'. The other members of the Committee, Thomson, Oliphant, Cockcroft and Moon, had all been at the Cavendish, and would have instantly recognized the reverberation of Rutherford in Chadwick's remarks—speculating without experimental evidence was like drawing a blank cheque on eternity. Equally, the fact that Chadwick was predisposed to agree with the technical points of the memorandum immediately increased its credibility in their eyes.

The point made by Frisch and Peierls that all the relevant theoretical data regarding fission were published in the open literature and could, therefore, be exploited by scientists of other countries, including Germany, was well taken. Cockcroft[33] had met Lieutenant Allier, an agent from the French Ministry of Armaments, in London earlier in April 1940, and learned from him that Joliot-Curie was working hard on the production of an atomic bomb. Allier had recently spirited 200 kg of heavy water from Norway to France in clandestine operation worthy of Hollywood. Having taken a taxi with his precious cargo onto the tarmac at Oslo airport, he feigned to board a flight to Amsterdam and at the last minute boarded a plane flying to Scotland. The Amsterdam flight was intercepted by German fighters and forced to land in Hamburg, where it was thoroughly searched.[34] Allier also learned from his Norwegian contacts that the German government had endeavoured to procure this stock, and had asked to be supplied with 2000 kg more. There was no doubt that the Germans were actively pursuing uranium research.

The committee met at the Royal Society, and it needed a name that could be used to refer to it without disclosing any inkling of its purpose. One item for discussion at an early meeting was a telegram which had been sent by Lise Meitner on behalf of Niels Bohr to say that he and his family were safe despite the German occupation of Denmark at the begining of April 1940. The telegram ended: 'Tell Cockcroft and Maud Ray Kent'.[5] Cockcroft had already written to Chadwick about it, suggesting that 'the last three words were an anagram for "uranium taken"'. Some time was spent trying to solve the enigma, but no plausible solution was found;* it was decided, however, that the M.A.U.D. Committee it should be. The other original, but 'junior', member of the Committee, Philip Moon, now worked in Oliphant's department, but had previously studied under Thomson at Imperial College after leaving Cambridge. The committee was expanded in July 1940[5] to include two more Cavendish notables, Blackett and Ellis, as well as Norman Haworth, Professor of Chemistry at Birmingham University and winner of the Nobel Prize in 1937 for his work on carbohydrates and vitamin C. It was Haworth who had convinced Oliphant to accept the Physics Chair in Birmingham. With a Secretary, Dr B.G. Dickins, provided by the Ministry of Aircraft Production, under whose auspices it met, the Maud Committee was a cohesive, experienced and intellectually heavyweight unit. At no stage did it ever threaten to degenerate into a senior common room clique; apart from their superlative scientific gifts, these were men whose feet were firmly on the ground, and whose inclination was towards action rather than talk.

Ironically, the committee were now barred from communicating with the two men who had provided the major catalyst for their activities. Although Peierls had recently become a naturalised British citizen, Frisch (an Austrian) was regarded as an 'enemy alien', and neither man was permitted to engage on work of a secret nature. Fewer than one in a hundred enemy aliens had been interned after the outbreak of war, but by the spring of 1940 the public clamour for much wider measures had become hysterical according to *The Times*. Oliphant was allowed to tell Frisch and Peierls that the committee were grateful for their memorandum, but they would not be consulted further. Peierls[30] was incensed, not through any personal vanity, but because he and Frisch 'had thought a great deal about the problems already and might well know the answers to important questions'. He also thought it ridiculous that officialdom imagined that they could be excluded from their own secret. In fact it was not just officialdom that was concerned with security—Lt. Allier had passed on a warning from Joliot-Curie in his April meeting with Cockcroft. Joliot-Curie knew that Frisch was now working in Birmingham and was very perturbed about the possible leakage of information to Germany, especially via Lise Meitner to Otto Hahn. It seems probable that Joliot-Curie was suggesting that such communication would be inadvertent and result from the active exchange of ideas between nephew and aunt, but the Secret Service were alerted nevertheless to stop 'any leakage to Germany which might be escaping from Dr Frisch'.[33] Peierls[30] did not know the identity of the chairman of the Maud Committee, but wrote him a letter, addressed 'Dear Sir', which again was hand-delivered by Oliphant. G.P. Thomson, who was the chairman, accepted the logic of Peierls' argument and convinced the Ministry that he and Frisch should be fully informed about progress.

* The explanation was prosaic. Maud Ray had been a former governess to the Bohr family and lived in Kent. The rest of her address had been lost in the transmission of the telegram.

Frisch and Peierls had concluded their memorandum by calling for immediate steps to be taken to separate a small amount of ^{235}U so that their ideas about the action of fast neutrons could be tested directly, and even before this could be accomplished they thought there were other experiments which could be done to support or refute their hypothesis. This work would need to be undertaken on a co-operative basis between several university departments, and it was plainly essential to have one man in overall control of the whole programme. To the other members of the Maud Committee, there was one obvious candidate for this position, fitted by his incomparable experience of marshalling the research programme at the Cavendish for over a decade. On 20 June 1940, Thomson formally asked Chadwick 'to take on the main part of the physical work'.[35] He accepted, mentioning that in addition to work that would be undertaken in Liverpool, he wished to make use of certain facilities at Bristol University and work to be done by Norman Feather and the Swiss physicist, Egon Bretscher, in Cambridge. The sharpness of Chadwick's critical faculties, his powers of communication and organization, and his instinctive feel for experimental method were now going to be tested as never before in the cauldron of the largest military–scientific endeavour in history.

During the two months since the Frisch–Peierls Memorandum had been presented to the Maud Committee, the passage of events in the wider context of the war had been tumultuous. Chamberlain's government had fallen on 10 May, and Churchill had succeeded him as Prime Minister. As May lengthened into June, the British Expeditionary Force and allied troops made their miraculous escape from the beaches of Dunkirk. These events were followed with fascination by the whole nation—the scientists as enthralled as everyone else. Oliphant[36] wrote to Chadwick on 11 June 1940:

> I am somewhat worried about the position with regard to the uranium supplies, which we understood were stored at Le Havre. The German mechanised divisions seem to be filtering round in that direction in an uncomfortable manner, and it seems desirable that the uranium should be removed to this country.

The fall of France was imminent and there was no time to rescue any material from Le Havre, but the heavy water had been taken from Joliot's laboratory to Bordeaux in readiness for transportation to England.[34] The Earl of Suffolk, an impossibly heroic and romantic figure, who happened to hold a first class pharmacology degree from Edinburgh, had managed to secure the apparently mundane appointment as the representative of the DSIR in Paris. In fact he held court, surrounded by beautiful women and flowing champagne at the Hotel Ritz, and decided that he would arrange to shepherd as many French scientists and engineers out of the country as he could manage. He decamped to Bordeaux, carrying a large cache of industrial diamonds from Antwerp that he had obtained at pistol point. He virtually commandeered a British vessel the *Broompark,* and ensured that she did not sail until he was ready, by keeping the crew plied with drink. The heavy water was brought on board in 26 cans by Halban and Kowarski, two researchers from Joliot's laboratory, who had been his co-authors on the important paper on nuclear fission published in 1939. Joliot also travelled to Bordeaux, but the Earl of Suffolk could not persuade him to leave French soil.

The *Broompark,* a drab ship normally used for carrying coal, eventually got up steam and left France with about 25 scientists (including Halban and Kowarski), and its small, valuable, collection of exotic substances. It slipped into Falmouth, the fishing harbour on the south west coast of England, on 21 June 1940. The 26 cans of heavy water were first transported to Wormwood Scrubs gaol, an eminently suitable site for safe keeping, but were then moved to the Library of Windsor Castle.[5]

By the end of the month, Great Britain stood alone and it seemed that the Germans would surely attempt a fullscale invasion before long. The French Minister of Armaments certainly believed this, and instructed Halban and Kowarski to go to North America to finish their work and to take the heavy water with them.

References

This chapter is the first to include any wartime papers. Margaret Gowing's official history, *Britain and atomic energy,* is indispensable; although her monograph did not list references, I was provided with a complete catalogue by Lorna Arnold, Historian for the UK Atomic Energy Authority. Where the documents are in the public domain, I have read them and cite them accordingly—**CAB** refers to Cabinet files at the Public Record Office in Kew (**PRO**). Some of the material used by Professor Gowing remains classified (as do Chadwick's papers at the UKAEA.), and so in places I have been forced to rely on her account (which I have always found to be impeccably accurate, when I have compared it to original documents).

1 Chadwick, J. (11/9/39). Letter to L. Meitner. Meitner's papers, 5/3, CAC.

2 MacKenzie, N. and MacKenzie, J. (1987). *The life of H.G. Wells: the time traveller*. Hogarth Press, London.

3 Brown, A.P. (1994). Interview with Joseph Rotblat, London.

4 Chadwick, J. (15/10/39). Letter to N. Feather. Feather's papers, CAC.

5 Gowing, M. (1964). *Britain and atomic energy, 1939–1945*. Macmillan, London.

6 Chadwick, J. (31/10/39). Letter to E. Appleton. CAB 104/186, PRO.

7 Pais, A. (1991). Fission. In *Niels Bohr's times*, pp. 452–72. Oxford University Press.

8 Bohr, N. and Wheeler, J.A. (1939). The mechanism of nuclear fission. *Physical Review*, **56**, 426–50.

9 Peierls, R.E. (1939). Critical conditions in neutron multiplication. *Proceedings of the Cambridge Philosophical Society*, **35**, 610–15.

10 von Halban, H., Joliot, F., Kowarski, L. and Perrin, F.J. (1939). Mise en évidence d'une réaction nucléaire en chaîne au sein d'une masse uranifère. *Journal of Physics*, **10**, 428–33.

11 Chadwick, J. (1946). The atom bomb. *Liverpool Daily Post*, 4 March.

12 Weiner, C. (1969). Sir James Chadwick, oral history. American Institute of Physics, College Park, Maryland.

13 Chadwick, J. (5/12/39). Letter to E. Appleton. CAB 21/1262, PRO.

14 Elliot, W. (1939). Committee of the Imperial Defence Memorandum. CAB 21/1262, PRO.

15 Tizard, H. (27/4/39). Note on uranium. CAB 21/1262, PRO.

16 Tyndall, A.M. (3/5/39). Review paper for Chemical Defence Committee. CAB 21/1262, PRO.

17 Tizard, H. (13/5/39). Committee minute. CAB 21/1262, PRO.

18 Churchill, W.S. (1939). Letter to Kingsley Wood. In *The second world war: the gathering storm*, p. 301. Cassell, London.

19 Wheeler-Bennett, J.W. (1962). *John Anderson: Viscount Waverley*. Macmillan, London.

20 Anderson, J. (28/11/39). Letter to Lord Hankey. CAB 104/227, PRO.

21 Lord Hankey (12/12/39). Letter to Lord Chatfield. CAB 21/1262, PRO.

22 Holt, J. (1988). *Seminar on James Chadwick*. University of Liverpool. See also Holt, J. (1994). Reminiscences and discoveries: James Chadwick at Liverpool. *Notes and Records of the Royal Society*, **48**, (2), 299–308.

23 Chadwick, J. (24/12/39). Letter to E.O. Lawrence. Lawrence's papers, Bancroft Library, University of California, Berkeley.

24 Holt, J. (1994). Thomas Gerald Pickavance, 1915–91. *Biographical memoirs of fellows of the Royal Society*, **39**, 305–23.

25 Chadwick, J. (8/1/40). Letter to J. Cockcroft. Cockcroft's papers, 20/5, CAC.

26 Occhialini, G. (28/12/39). Letter to J. Chadwick. CHAD I, 24/1, CAC.

27 Chadwick, J. (7/2/40). Letter to Guggenheim Foundation. CHAD I, 24/1, CAC.

28 Thomson, G.P. (10/3/40). Letter to J. Chadwick. CHAD IV, 2/10, CAC.

29 Thomson, G.P. (16/4/40). Letter to J. Chadwick. CHAD I, 19/7, CAC.

30 Peierls, R.E. (1985). War. In *Bird of passage*, pp. 145–81. Princeton University Press.

31 Frisch–Peierls Memorandum (1940). In *British scientists and the Manhattan Project*, (F.M. Szasz, 1992), pp. 141–7. St. Martin's Press, New York.

32 Oliphant, M.L. (1982). The beginning: Chadwick and the neutron. *Bulletin of the Atomic Scientists*, Dec, 14–18.

33 Cockcroft, J. (10/4/40). Report to Air Ministry. CAB 21/1262, PRO.

34 Goldsmith, M (1976). Fission – and heavy water. In *Frédéric Joliot-Curie*, pp. 77–96. Lawrence and Wishart, London.

35 Thomson, G.P. (20/6/40). Letter to J. Chadwick. CHAD IV, 2/10, CAC.

36 Oliphant, M.L. (11/6/40). Letter to J. Chadwick. CHAD I, 19/3, CAC.

12 The MAUD report

Well before he had been officially charged by G.P. Thomson with the task of co-ordinating laboratory research, Chadwick had begun the process on his own initiative. During his review of nuclear fission papers in the autumn of 1939, Chadwick remembered a letter that had been published in *Nature* back in July. The paper from Powell and Fertel[1] of the University of Bristol was headed: 'Energy of high-velocity neutrons by the photographic method'. Powell had devised a method of placing a photographic plate tangentially in a beam of protons, and realized that it could be used as an indirect way to measure neutron energies. The letter to *Nature* demonstrated that this was indeed the case, and it seemed to offer important advantages over the existing cloud chamber technique. The photographic method was as sensitive as the Wilson cloud chamber, but was much less cumbersome and offered the prospect of easier measurements of neutron energies.* This had obvious attractions to Chadwick given the urgent neutron work which needed to be undertaken as part of the experimental analysis of fission in uranium. Accordingly he wrote to Powell in November 1939 offering him the use of the Liverpool cyclotron in order to extend his measurements. Cecil Powell's reaction to the letter was eager excitement, and he told a friend that he was now 'at the centre of world physics'.[2]

An active collaboration was set up between Bristol and Liverpool, with Chadwick very much involved in the scientific work. Chadwick brought in Pickavance, who had been carrying out cosmic ray research for his Ph.D., using a high-pressure cloud chamber to study mesons. Pickavance went to Bristol in April to learn how to develop the photographic plates and measure particle tracks using a Leitz microscope. In Liverpool, a 'camera' with its photographic plate coated by silver halide emulsion was attached to the vacuum tank of the cyclotron. The camera was filled with various gases, such as neon, through which passed a beam of protons from the cyclotron. On 8 June 1940 a letter[3] appeared in *Nature* describing this work. The decision to publish was Chadwick's; he thought it would be a good cover to write up an innocuous piece of nuclear research rather than to have

* Powell's photographic techniques bore their main fruit after the war, when they were used to study cosmic rays and led to the discovery of the pion (or pi-meson) which was photographed decaying into a lighter particle, called the muon. For this discovery, Powell was awarded the 1950 Nobel Prize for Physics.

complete silence from Liverpool in the journals. The real work of the team was to study the inelastic scattering of neutrons, and it is remarkable that they should produce this other piece of quite respectable research in a few weeks.

Once the Frisch–Peierls Memorandum had been digested, there were still many unanswered questions, and a few crucial unresolved issues. The latter included the separation of uranium-235 from uranium-238, the preparation of pure uranium in the metallic form, and the measurement of neutron cross-sections and energies so that an accurate calculation of critical mass could be made. Chadwick and his physicists were centrally concerned with the last of these, but he also needed to be closely informed about progress on the non-nuclear aspects. From Liverpool, he actively supervised work on the nuclear constants between there, Cambridge and Bristol, and also entered into the network covering Oxford, Birmingham and Imperial Chemical Industries (ICI) as they grappled with the challenges of uranium chemistry and isotope separation. Nor should it be imagined that he devoted himself full time to Maud business—he was still expected to function as a university professor. In May 1940, for example, he was busy with external lectures and setting examination papers;[4] in June he spent several weeks examining undergraduates in Sheffield and Leeds as well as in Liverpool, complaining to Feather: 'I don't like examining, and it always reduces me to a rather feeble state'.[5]

The challenge of obtaining pure uranium in metal form had fallen to the Chemistry Department of Birmingham University. Oliphant, the Professor of Physics there, kept an eye on progress and told Chadwick on 1 June that 'the attempts to follow textbook methods and produce uranium metal by chemical processes have proved entirely abortive'.[6] Just three days later, Chadwick received a much more optimistic report from Haworth, the Chemistry Professor,[7] who wrote:

> Oliphant will have told you that we are on the verge of obtaining the required metal and that we are going all out to get it quickly. The older method has failed badly but I pin my faith securely on an electrolytic method published in 1930, the directions for which are so clear that I am sure little can go wrong.

He sent Chadwick the 1930 journal reference and urged him to acquire the necessary materials from ICI so that he could replicate the process in Liverpool. Chadwick[8] mentioned in his reply that he had just had a visit from Professor Otto Maass of McGill University who informed him that pure uranium metal had been prepared on a laboratory scale in Canada and he would arrange for three pounds to be sent to Chadwick immediately.* At the beginning of July, Haworth[9] triumphantly sent two bottles containing 110 g (approximately one-quarter pound) of 98% pure uranium metal made by the electrolytic process. He asked how much Chadwick required, and was told 'at least 30 pounds'.[10] By the end of the month, Hawarth[11] was becoming a little discouraged because he found that the electrolytic process could not be employed continuously to increase production; in addition the purest form of metallic powder was pyrophoric (caught fire when exposed to the air)

* Nine months later Chadwick had still not received any.

and needed to be handled carefully. He was also finding that contamination with calcium fluoride and other salts was difficult to remove. Chadwick[12] confirmed that he needed metal of high purity and in powder form so that it could be packed into tin containers and be used for neutron absorption and scattering experiments. Sensing the inherent difficulties in attempting to produce large quantities of a dangerous and elusive substance in a university chemistry laboratory, Chadwick[13] suggested in early September to the Ministry of Aircraft Production, who were loosely overseeing the Maud Committee, that the time had come to prepare the ground for production on a larger scale and they should officially approach ICI to manufacture uranium metal. As if to reinforce the wisdom of this decision, Haworth,[14] a fortnight later, sent Chadwick formal reports on the nature of 'metal X' (which was the familiar name given to uranium in Birmingham), in case 'our own records should be destroyed and some of the men with them' by German air raids. In the spring of 1941, Haworth[15] apologized to Chadwick and explained that he had not been able to provide the thirty pounds requested because ICI were taking most of his output. In fact ICI chemists had devised a method of producing pure uranium metal by this time so that Haworth's team were soon able to revert to basic research.

From the beginning of the Maud work, Chadwick was conscious that to be brought to fruition the combined achievements of the scientists involved would have to be successfully translated into an industrial setting. Lurking in the back of his mind was the disturbing thought that the Germans would probably be much more efficient at this transition from research to production than the British. In his quiet way, he began to put out feelers to Canadian friends and colleagues. Apart from the meeting with Professor Maass, Chadwick received a visit from an ex-Cavendish student George Henderson, now a professor in Halifax, Nova Scotia. Henderson was a trusted friend, and Chadwick confided in him completely about the prospects for a uranium bomb. When Chadwick was officially appointed to oversee the experimental research of the Maud Committee, G.P. Thomson asked him about those individuals to whom he had spoken about the project. Chadwick[16] supplied a short list of names and wrote about his recent contact with Henderson:

> I must confess that I have spoken fairly openly about the problem to Henderson because it seemed to me that if anything should prove possible on a practical scale it will have to be done in America. There are many ways in which the Canadians might be able to help us in addtion to supplying uranium metal.

Henderson had been in Liverpool at the time of Dunkirk, when the future of Britain was far from secure. He had two daughters slightly younger than the Chadwick twins, and suggested to James and Aileen that they might like to send Judy and Joanna to Canada for safe keeping. While Canada represented a safer haven than rural England with the threat of German invasion, there was the dangerous obstacle of the U-boats in the North Atlantic to be overcome first. The Chadwicks faced the wrenching decision that confronted thousands of English parents in

1940, and eventually decided that they would evacuate the girls to Nova Scotia. They sailed from Liverpool in July, at just about the time Churchill was letting the Home Secretary know that he 'entirely deprecated any stampede from this country'[17] in reference to a Government-sponsored scheme for the evacuation of children to North America. The twins zig-zagged across the Atlantic on a convoy ship, and soon adapted to the Canadian life quite happily. It was fortunate that Aileen still had some family in Liverpool because the twins had been the focus of her activities and she missed them terribly. James was working such long hours that the only way she could see him for even short periods was to visit the physics department. Aileen gradually became more outgoing and would also look in on the team labouring at the cyclotron in the basement; they, in turn, discovered she possessed a keen sense of humour which they came to appreciate 'when, as was often the case, things were not going well'.[18]

In his heart, Chadwick was more prepared to join forces with Canadian scientists at this stage than with the Americans. Canada was part of the Empire, their troops had been evacuated alongside the British at Dunkirk, and, more specifically, their nuclear physicists had been trained in the Rutherford schools, either at McGill or at the Cavendish. Although he recognized from 1935 onwards that the British supremacy in the field of nuclear research was being eroded, and while he enjoyed cordial relations with and greatly respected individual American physicists, most notably Ernest Lawrence, he did not like the change in order. For example, in a letter to Norman Feather in February 1940, Chadwick[19] wrote: 'I am much afraid that the C. [centre] of G. [gravity] of nuclear physics will move to U.S.A.'. Oliphant, an unswerving supporter of Chadwick's and enthusiastically pro-American from the first, urged him to accept the Maud appointment and also to visit the USA to get work started there. With the invasion of England apparently imminent, Oliphant[20] thought 'if things go really badly with this country there is a great deal to be said for investigating any possibility which offers a chance of hitting back from the New World'. Chadwick[21] replied:

> I have heard nothing about your suggestion concerning America, but I have been convinced for some time that it will be necessary to work in collaboration with the Americans, especially if we reach any technical development.

Since the outbreak of war, there had been almost no active exchange of scientific information between Great Britain and the United States. Transatlantic crossings were infrequent and hazardous. America was not yet a combatant, which raised issues of security regarding research with military applications. In the late summer of 1940, Churchill, anxious to draw the Americans into the war, decided to send a scientific mission led by Tizard to make the Americans aware of the remarkable inventions and weapons developed in Great Britain.[22] Tizard's deputy on this trip was Cockcroft, and while radar-related topics such as the cavity magnetron for generating microwaves were top of the agenda, he was also alert to any developments in the atomic field. He visited various universities and attended a meeting of the Briggs Uranium Committee, which Roosevelt had set up in October 1939 after his letter of warning about the fission bomb from Einstein. Whereas the Maud Committee comprised active scientists of the first rank, its American counterpart was rooted in officialdom, with a chairman, Lyman J. Briggs, who had started as a government soil scientist in 1896 and had risen to become Director of the National Bureau of Standards, and two other members who were military ordnance experts with no special knowledge of nuclear physics.[23] Cockcroft thought the uranium work in North America deserved a more vigorous attack than it was receiving, and reported back to the Maud Committee that all the research seemed to be several months behind comparable investigations in Britain and was proceding with less urgency.

While Chadwick was concerned with the whole range of experimental work for the Maud Committee, he was especially charged with establishing the cross-section of ^{235}U for fast neutrons so that an accurate estimate of the critical mass for a bomb could be made. He wrote[24] to Peierls, after studying the Frisch–Peierls Memorandum, pointing out that they had assumed a fission cross-section for fast neutrons eight times higher than Tuve's experimental results; Peierls[25] replied that Frisch was attempting to measure this in Birmingham. Chadwick[26] had also written to Feather at the Cavendish on 7 June 1940, asking him to look into the same question:

> I want to know the cross-section for fission of uranium 235 for fastish neutrons. I cannot provide any uranium 235 at the moment, so the measurement has to be made on ordinary uranium. It will probably be advisable to use neutrons of energy about .5 to .6 MEV, and they must be fairly homogeneous. Slower neutrons would give a large cross-section for uranium 235 and faster ones a large cross-section for uranium 238. The indications are that the cross-section for 238 is very small for .5 to .6 MEV, and so we should get a reasonable approximation to the value for U 235 . . . The indications to which I referred above are some measurements of Tuve reported in Bohr and Wheeler's paper . . . These were reported last June but they have not been published. There is no other reason to suspect them, and Tuve is a reliable worker. I will write to find out whether he still holds to these measurements and how he made them.

Feather replied a week later to say that he was setting up a fission cross-section experiment with his colleague Bretscher, and he expected it to be difficult 'if the effective cross-section is as low as the value you quote'.[27] The behaviour of the rare isotope, 235, in the fission process could only be inferred indirectly by experiments with ordinary uranium which was 99.3% ^{238}U: no ^{235}U had yet been separated.

Tuve evidently reassured Chadwick about the accuracy of his cross-section estimates, but at the beginning of 1941 an American report on fission cross-section was passed from G.P. Thomson to Chadwick which contained revised figures and caused consternation. Although it contained many of the data that he was interested in, Chadwick[28] told Thomson that he would continue his own efforts for two reasons:

> First, that the cross-section which Tuve now gives for 0.5 MV neutrons is 10 times

> his former value, which in a private letter he stated to be reliable, and secondly because it will be necessary to repeat the work for 235.

In fact, just as Tuve revised his previously low estimate upwards, Frisch was producing experimental evidence to show that the guess made by him and Peierls in their memorandum was a little high, so the numbers were converging. Frisch had begun his experiments in Birmingham, expecting at any moment to be dragged off to an internment camp.[29] In July 1940, it was agreed that he should transfer from Birmingham and join the team in Liverpool. He had also been exploring the possibility of separating the uranium isotopes by thermal diffusion using uranium hexafluoride, a stable but extremely corrosive gaseous compound, and made a preliminary visit to Liverpool with Peierls to discuss this with Chadwick. On being shown into the Professor's study:

> [Chadwick] came in after a short while, sat down at his desk and started to scrutinize us, turning his head from side to side like a bird. It was a bit disconcerting, but we waited patiently. After half a minute he suddenly said 'how much hex do you want?' That was his way; no formalities, straight down to brass tacks.[29]

A few weeks after Frisch's arrival, the bombing of Liverpool started in earnest. Initially he lived in a boarding house and would spend the nights with his fellow lodgers huddled under the staircase or in the basement as they listened to the cacophany of sirens, anti-aircraft fire and the whistling of falling bombs. In the morning, they would emerge to find 'well-known buildings were now empty shells, a few walls still precariously standing' until with the swing of the demolition crew's steel ball 'they collapsed into a cloud of dust and rubble'. Chadwick went out secretly himself to test some of the early bomb craters with a Geiger counter, fearful that the Germans might attempt to disseminate large quantities of radioactive matter in the explosions. He carried a permit from the Chief Constable of Liverpool requesting that police on duty should give him every possible facility in making his tests.[30] Fortunately, he found nothing.

Whereas Rotblat had been accepted into the department immediately, Frisch experienced some of the Liverpool xenophobia that Chadwick had worried about when making earlier appointments. The presence of an 'enemy alien' seems to have especially provoked Moore, the cyclotron engineer, to spiteful behaviour. Frisch ignored this for a while, but eventually he and Moore 'had quite a row, after which we became the best of friends'.[29] Chadwick used his friendship with the Chief Constable of Liverpool to exempt Frisch from the curfew in force for enemy aliens and to allow him to use a bicycle; his influence did not extend to one zealous police constable who stopped Frisch riding home from the laboratory on a summer evening with no lights on his bicycle—he was subsequently fined ten shillings, five for each lamp! In the department Chadwick assigned John Holt to be Frisch's research assistant, and they soon became known as 'Frisch and Chips'. The duo began by constructing a small Clusius separator to explore the thermal diffusion approach that Frisch and Peierls had cited for uranium isotopes in their memorandum. They were given some of the first samples of uranium hexafluoride gas produced by ICI and were frustrated to find they could achieve no measurable separation of ^{235}U from ^{238}U by this technique. Their lack of success worried Chadwick and he revealed his pessimism to his old friend from Manchester days, Charles Darwin, who was now the Director of the National Physical Laboratory. He wrote[31] on 1 November 1940:

> I am a little surprised to hear that G.P. Thomson takes a rather sanguine view about the uranium problem. My view is that the information we have now certainly leads one to believe that it is possible, but the technical difficulties may be very serious indeed. Whether possible or not, the problem is so important that we must know the answer. I am inclined to hope that the difficulties will prove unsurmountable for I am afraid we may be a long way behind the Germans on the technical side, and that is the side which will count.

Rudolf Peierls[32] had thought of another possible method of isotope separation, again starting with the dangerous and corrosive hexafluoride which is the only gaseous compound of uranium. His idea was to apply gaseous diffusion, which had been devised by Herz in Germany about ten years earlier, but which had never been used to separate gases differing in their respective masses by only 1%. The principle involves forcing molecules of gas through microscopic holes in a permeable barrier: the uranium-235 atoms would pass through the holes marginally faster than the heavier uranium-238 atoms, and if the process were to be repeated many thousands of times a gas enriched in ^{235} U would eventually be obtained. Peierls first mentioned the idea to G.P. Thomson, as Chairman of the Maud Committee, and he suggested the name of a scientist who might carry out the experimental work. Peierls thought the man in question was 'an exceedingly nice person but did not have the required energy or drive'. He wanted the job to go to Franz Simon,[33] another German refugee who had been at the Clarendon Laboratory, Oxford, since 1933 and was by now a naturalized British citizen. Simon was Reader in Thermodynamics and, together with several other German physicists given refuge by Professor Lindemann, he had formed an outstanding school of low-temperature physics. Peierls decided to go to Liverpool to enlist Chadwick's help.[32]

> He received me in his usual, somewhat distracted manner. When he wanted to consult some correspondence relating to our conversation, he would open a drawer in his desk containing a jumble of letters, each neatly in its envelope, and after a while he would find the right one.

Chadwick accepted Peierls' point of view at once, and wrote to Thomson to suggest that experiments on gaseous diffusion should start immediately. Since Chadwick had been put in charge of all experimental work the choice of laboratory was really his, and he asked Thomson whether he could give permission to Peierls to discuss this with Simon in Oxford, 'who will do the experiments much better than I can'.[34] Permission was granted and Simon was officially sanctioned by the Maud Committee to undertake the work.

Peierls[32] agreed to prepare a detailed summary document for the September 1940 meeting of the Maud Committee. He wrote a ten-page overview of the 'Uranium problem', and with some help from Frisch supplemented it with a series of nine papers giving a rigorous theoretical analysis of the different subjects, including the critical mass for a bomb, how such a mass might be safely assembled, the efficiency of such a weapon, and the separation of ^{235}U. As an intellectual exercise this was formidable enough, but when one considers the wartime conditions under which it was produced, it becomes scarcely credible. The nine papers were replete with mathematical formulae which Peierls had to write into the typed script by hand. The typing was carried out by Oliphant's elderly secretary, Miss Hytch, at Birmingham University, but because the building in which she worked housed the radar research, Peierls was not allowed to enter it. He dictated the text into an Edison-type machine with wax cylinders from which she typed. The secret papers were sent to London by train for the meeting, but the postal services were in such disarray that the batch was inextricably buried in a stack of mailbags at Euston Station. Miss Hytch had kept the stencils and was able to reproduce a second collection of rather smudged papers which Peierls brought to the meeting. Despite their lack of preparation time, the Committee members cross-examined Peierls in detail and his conclusions stood. According to Peierls, Chadwick's influence on the committee was very important: 'he would always stress the basic issues, and was never afraid of supporting unusual solutions for unusual problems'.

In December 1940, only four months after he was initially approached, Simon fully vindicated Peierls' faith in him by submitting his own report, *Estimate of the size of an actual separation plant.* This was a remarkably complete treatment of the scientific, engineering and economic aspects of the isotope separation problem and predicted that 1 kg of 99% pure uranium-235 could be produced per day from an industrial plant covering about 40 acres. Simon thought if every effort were made production could start within 18 months, and believed that 'the scheme is, in view of its object, not unduly expensive of time, money and effort'.[33] He had calculated a capital cost of about £4 million and annual running costs of £1.5 million which turned out to be correct within a factor of two or three.

When Frisch was making his own estimates of cross-sections for neutron-induced fission shortly before transferring from Birmingham to Liverpool, he had been assisted by Ernest Titterton,[35] a Birmingham graduate recalled from his school teaching job by Oliphant to join the team developing radar. Titterton was exceptionally talented, especially in the construction of electronic apparatus, and made an ionization chamber for Frisch that contained about 1 g of uranium.[29]

> [Frisch] kept complaining that the chamber from time to time produced a pulse which looked just like a fission pulse but surely couldn't be; there was no source of neutrons. Or was there? We went so far as to search the laboratory to see if by chance a small source had been left in a drawer. We tested the electronics and tried all kinds of improvements. Nothing made any difference. . .

In the end, having excluded all the other alternatives that he could think of, Frisch advanced the novel conclusion that the uranium would occasionally undergo fission *spontaneously,* that is without the need for a neutron to trigger the process. He and Titterton were unable to publish their important discovery because of the need for secrecy, but it immediately became another issue of concern for the Maud Committee. Was spontaneous fission the latest example of natural radioactivity, akin say to the emission of an α-particle, or was it a process induced by a stray, cosmic neutron? Once in Liverpool, Frisch put this question to his new assistant, John Holt, who, in order to answer it:[18]

> spent the night below ground level in a cubicle off the platform in James Street underground station, with a large multilayer fission chamber and amplifier. It was necessary to work at night because of interference from the electric trains . . . the platforms were covered with people sleeping out of reach of the air raids.

Within a few weeks a Russian journal (written in German) arrived in England with a paper on spontaneous fission from two Leningrad physicists, Flerov and Petrzhak. Chadwick heard about this and asked Simon if he could borrow the journal from the Bodleian Library in Oxford to read the paper himself. The request was denied, as Simon wrote:[36] 'The Library could not give permission to send the Russian journal to you—such a thing was not done during the last five hundred years'. To photograph the article would have taken too long, so Simon had a student copy it out by hand to send to Chadwick.

Frisch and Chips set about confirming the former's Birmingham results for fission cross-sections with neutrons of different energies. They again used the uranium coated ionization chamber which was placed in the beam of neutrons from the cyclotron. Work would occasionally be interrupted by bombing raids, but the cyclotron was relatively well protected by its basement site and never took a direct hit. In March 1941, a parachute landmine descended into the courtyard of the physics department watched by Maurice Pryce, who was a theoretical physicist by day and fire watcher atop the university clocktower at night. Frisch[29] came in for work the next morning, saw the damage and decided to go back to his lodgings until the building was repaired.

> Then I saw that there were some people around, doing things such as nailing large sheets of cardboard over the broken windows; had I better stay and see if I could help? Who, I asked, was in charge; this was spontaneous cooperation, people just doing what was needed. I asked one of them whether there was anything I could do, and he said 'Yes, go over to the engineering department and see if you can scrounge some more hammers!' This I did and then spent the rest of the day helping to nail cardboard over the windows and over the holes that had been torn in the roof. Fortunately it was a fine day; at the end of it the building was usable, and the next morning we were back at work. I remember thinking at the time that if that sort of thing had happened in Germany everybody would have gone home and waited until the appropriate authority had detailed a brigade of repairmen to come and repair the building properly.

In the first week of May, Liverpool was bombed seven nights in a row, gutting the centre of the city. Three thousand civilians were killed or injured, and over seventy thousand were made homeless. A munitions ship bound for the Middle East exploded in the docks, adding further damage and disrupting shipping; for a while the tonnage landed was cut to a quarter. Along with many other householders, James and Aileen converted a ground floor room in their house into a bedroom, rather than risk sleeping upstairs. The air raid sirens were ceaseless, and the Chadwicks soon stopped bothering to go to their appointed bomb shelter—it was just too disruptive. At the university, it seemed the windows were blown in every night. Chadwick recalled,[37] 'we put up cardboard shutters. But during every night they would be blown in again. We'd just put them up in the morning and go on.' During the May blitz of Liverpool, the city was hit by 859 high explosive bombs, 69 parachute mines, oil bombs and numerous incendiary bombs.[38] Nor were all the obstacles to progress dropped by the Luftwaffe. Chadwick was responsible for a huge load of quite tedious administrative matters of the 'pay and rations' variety. He also had to ensure that his young staff like Holt and Pickavance were not called up for military service, and wrote numerous letters to the Polish authorities in London so that Rotblat was not pressed into their military forces in exile. In the summer of 1941, he finally succeeded in having the Movements Restriction Order lifted from Rotblat and Frisch, and wrote a charming letter thanking the Ministry for 'making honest men of them again'.[39]

The Whitehall bureaucracy could be exasperating, and did not seem to appreciate the scale of the project that they were nominally managing. The Ministry of Aircraft Production (MAP) had imposed a £10 limit on the direct purchasing of equipment; items more than £10 had to be pre-approved by the ministry. Chadwick wrote to Dickins, the MAP-appointed Secretary to the Maud Committee, objecting to this arrangement: 'If you really must insist upon a limiting sum, I would suggest £20 (The principle is still wrong but I shall seldom have to observe it)'.[40] He wrote again to the Ministry in January 1941[41] pointing out that the petrol allowances were hampering their work with ICI. He and others from the department needed to visit ICI at Widnes, a town close to Liverpool, in order to discuss uranium metal and hexafluoride production, but could not get sufficient fuel for even this short journey. There was no swift dispensation granted, but a series of letters requesting details about the horsepower and registration numbers of the cars—answers which Chadwick patiently provided. A month after his urgent request, extra petrol allowances were issued with numerous provisos. That the department was able to function as well as it did at the scientific level owed much to Chadwick's quiet, thoughtful leadership. Frisch[29] appreciated the absence of overt security restrictions:

> One surprising feature of the way our work was conducted was the lack of any outward sign of secrecy; no guards, no safes. The institute was apparently devoted to various kinds of innocent research and of course very largely in the teaching of physics to students. Rotblat even included fission in his lectures and mentioned the possibility of a chain reaction, but in such a casual way that nobody would have thought that it might lead to an important development in weaponry, let

> alone that we were actually working on that. I think it caused some worry among the more security minded people; but it certainly worked. Any spy could have walked into the laboratory and asked questions, but of course we wouldn't have talked about our work to strangers, and moreover it didn't occur to any spy that we might deserve his attention . . . Chadwick often gave me a chance to discuss things with colleagues elsewhere, putting no great trust in the bogus security which relies on compartmentalizing knowledge, on letting every scientist know only what he needs to know. That kind of security leads to inefficiency.

Rotblat[42] was more aware of Chadwick's self-imposed isolation, which at times made him feel like the 'lab boy'. 'He used to speak very little. He would only talk to me about experiments; he would not take me into his confidence about what was going on outside.' Chadwick was treading a very demanding path. He found some of the official restrictions, particularly as they applied to aliens like Peierls, Frisch and Rotblat to be absurd and distasteful in view of the personal losses they had suffered; as Frisch noted, he rejected the idea of strict compartmentalization as anathema to science, but there needed to be some limitations. This was wartime, he was the head of a secret, national, experimental weapon programme, the outcome of which could have a critical influence on the course of the war. He had to be mindful of security considerations, and was adding to an already heavy responsibility by, in effect, following his own code of practice. As the Liverpool work reached preliminary, but fateful conclusions, he became desolate. He told Charles Weiner in an interview over a quarter of a century later:[37]

> I remember the spring of 1941 to this day. I realised then that a nuclear bomb was not only possible—it was inevitable. Sooner or later these ideas could not be peculiar to us. Everybody would think about them before long, and some country would put them into action. And I had nobody to talk to. You see, the chief people in the laboratory were Frisch and Rotblat. However high my opinion of them was, they were not citizens of this country, and the others were quite young boys. And there was nobody to talk to about it. I had many sleepless nights. But I did realise how very, very serious it could be. And I had then to start taking sleeping pills. It was the only remedy, I've never stopped since then. It's 28 years, and I don't think I've missed a single night in all those 28 years.

One memorable conversation between Chadwick and Rotblat was occasioned by a letter that appeared in *Physical Review* in June 1940. It was written by McMillan and Abelson,[43] describing their recent work at the Berkeley cyclotron which had resulted in the discovery of a new radioactive element 93. Prior to this, uranium, with an atomic number 92, was the heaviest element known, but by bombarding uranium under controlled conditions in the cyclotron, they had isolated a new, unstable element, which would later be called neptunium. At the end of their letter, they mentioned that element 93 decayed by β-particle emission into an even heavier transuranic element, 94, with a half-life of 2.3 days. Element 94 (later called plutonium) was an α-particle emitter, but appeared much more stable with a half-life 'of the order of a million years or more; the same experiment showed that if spontaneous fission occurs its half-life must be even greater'. Rotblat,[44] with characteristic perspicacity, saw that element 94 would probably be a fissile material under neutron bombardment, and might be a realistic alternative to ^{235}U as a nuclear fuel. Rotblat wanted to produce element 94 with the Liverpool cyclotron and then investigate its fission properties.[22] Chadwick was in a quandary because the whole thrust of the British project at that time was to separate ^{235}U as the atomic fuel, and the measurement of its properties was the first priority. Calculations were made by Pryce and others that showed the yield of element 94 from the Liverpool cyclotron would be so low as to make measurements of its fission properties impractical so the idea was dropped. There was another aspect of the whole affair that bothered Chadwick. He wrote[45] to G.P. Thomson in January 1941 complaining about the lack of restraint being shown in the US:

> Far too much is being said and published, even now, about the uranium problem, and I think some effort should be made to stop it. Even the negative statement about 94 in the letter of McMillan and Abelson should never have been published.

At Chadwick's urging,[22] an official protest was sent to the United States about publication of the McMillan–Abelson letter, which might have had some effect: plutonium was chemically isolated at Berkeley early in 1941, but the work was voluntarily withheld from publication until 1946.

At the meeting of the Maud Technical Committee on 9 April 1941, Chadwick reported that results so far obtained in Liverpool 'indicate that the quantities determining the size of the bomb are favourable, and it seems probable that we can trust to theory in places where we cannot test at present'.[46] It now seemed that the critical mass of ^{235}U might be 8 kg or less. The Committee were told that an American scientist, Alfred Nier, was able to separate a few micrograms of pure ^{235}U in his laboratory so that some limited determination about its neutron cross-section could now be made directly. Chadwick accepted that the Americans should make these measurements first in case the material was lost during a transatlantic crossing, but urged that the need for haste should be impressed on them. At this stage, the Maud Committee with its supporting network of British universities were the only runners (apart from the dark, Nazi, horses) in the race for the atomic bomb; work in the USA was proceding at a leisurely pace and was taken up with the possibility of a fission power plant based on natural uranium and a moderator such as heavy water to slow the neutrons. Joliot-Curie's group had been the pioneers in this field, and they had taken out patents on the design and use of nuclear reactors in 1939.[22] This was not for personal gain, but as a way to guarantee the priority of France in this technology of the future and also as a method to generate money for research. It was now apparent that apart from producing energy on a large scale, such a uranium 'boiler' would also yield small amounts of the new transuranic element, number 94, which was itself likely to be highly fissile.

Halban and Kowarski, who had carried the cans of heavy water over from France on the *S.S. Broompark* were a contrasting couple, both of whom had recently become naturalized French citizens. Hans von Halban was the senior by age and temperament: born in Austria, he had studied in Zurich, married a Dutch woman and settled in Paris. He was by nature charming, assertive and impatient. Lew Kowarski was a mild, shambling, Russian bear, who had started his career as an industrial chemist.[47] The two had continued their research into the uranium–heavy water system at the Cavendish, and Chadwick was well aware of their progress as a result of frequent letters from Norman Feather. In December 1940, Feather[48] had written to say that he and Bretscher were hoping to join forces with the two Frenchmen to produce large amounts of element 94 and to measure its fission cross-section. 'Incidentally', he asked, 'I have been wondering what the mechanism is of putting results like the above in cold storage for publication some time. I think they are sufficiently interesting for us to fix a lien on them if possible. Perhaps you have made parallel progress—I should like to hear if you have.'

Chadwick was in no mood to encourage academic discourse—the need for results was pressing and he was still troubled by the publication of McMillan and Abelson's original letter about element 94. He replied:

> I am glad to hear you are pushing on with the investigation of element 94. I have decided to do nothing on this for the time being as there are more urgent things we might do. Whereas 94 may prove important for the 'boiler' . . . I do not think you can establish any claim on ideas about 94, however. They are probably widely held. Certainly they have been in mind here for several months. It is the results which count, and if you get them, even if they cannot be published now, that is all that matters. A brief summary can be communicated to the MAUD Committee and things can be straightened out after the war.[49]

Clearly irritated, Chadwick also asked Feather what he meant exactly by 'fixing a lien'. Feather,[50] undaunted, pressed his case. He told Chadwick that 'Halban has shown that the boiler *will* go with heavy water—so that, on that score it is no longer a matter of "ifs".' He continued:

> Now about my suspected opportunism. No, I never had any intention of fixing on to any ideas. Your remonstrance might almost have been quoted from the Rutherford–Harkins* correspondence which I know too well not to have learnt my lesson from it.

Feather feared that he had been writing very quickly and might have left himself open to misinterpretation, but he was proud of the Cambridge work and although he had no intention of publishing anything now, he was just 'enquiring about cold storage possibilities'. Chadwick was not going to be mollified:[51]

---

* Feather is referring to a complaining letter sent by an American chemist, W.D. Harkins, to Rutherford in February 1921 about the latter's Bakerian Lecture of the previous year, in which he had proposed the neutron as a nuclear particle. Harkins had been speculating about 'atoms of zero atomic number' (i.e. particles carrying no charge) for some time and was annoyed that Rutherford had not acknowledged his contribution. Rutherford was not impressed by Harkins' claim of priority because he had made no active attempt to test the concept experimentally. In reference to Harkins, he commented to his friend Boltwood at Yale that: 'Harkins is a man of intelligence, but I wish he did more experimenting and spent less time in theorising and in endeavouring to cover every possible idea.' In his reply (below), Chadwick gives vent to the same frustration about Feather.

> Perhaps I worded my remarks badly but I assure you I do not harbour 'suspicions' of any kind. You wrote that you wanted 'to fix a lien' and I thought and still think that you cannot do that. What you can do is the experiment.
>
> With regard to 94, I think I am aware of most of the facts and the possibilities . . . We [in Liverpool] expect that the fission cross section of 94 ought to be very high and we thought that even spontaneous fission might occur. I also expected the boiler to work with heavy water and we calculated that with 10,000 kW dissipation one might create about 10 gms [of 94] per day. But in spite of this and in spite of Rotblat's keen interest I felt that the first question [for] *our* programme was the properties of 235 . . . 94 is not likely to be more potent than 235 as a military weapon. The problem we really have to answer is whether the U phenomenon can furnish a military weapon. I do not think the boiler helps this problem directly but the reason I supported von Halban's expts.—against a good deal of opposition in the committee—was because I felt we must have all the information we can get, whether referring to fast or to thermal [neutron] processes, and because I had in mind even at that time the possibilities of 94 . . . There is the possibility that the cross-section of 235 may not be as high as we think. If this turns out to be the case and if the separation [of the uranium isotopes] proves very difficult, it may not be impractical to try to make 94, if its cross-section is greater.

He did encourage Feather to publish some unrelated work on β-ray investigations in the belief 'that a few publications in nuclear physics would act as a better cover than a conspiratorial silence'. Feather, his tail firmly between his legs, sued for peace:[52]

> I am only sorry that my use, in successive letters, of two rather inexact phrases—'fix a lien' and 'suspicions of opportunism'—should have made you think that my attitude or feelings are in fact other than what they are or should be.

At the April 1941 meeting of the Maud Committee,[46] a telegram was read from Ralph Fowler, Rutherford's son in law, who was then in charge of the British Central Scientific Office in Washington. He reported that the Americans were on the verge of constructing a large scale heavy water production plant, expected to cost a million dollars. A full report had been requested from Halban on the work in Cambridge, but not received. Fowler thought the report 'accompanied by a critical study by Chadwick might carry [the] necessary weight for [the] decision'. This led to a general discussion regarding Anglo-American cooperation. Oliphant thought the time had come to make a decision about the siting for large scale work; he supported the idea of sending Halban and his team to Canada and thought the British should be pressing for full co-operation with both the Americans and Canadians. Chadwick disagreed, saying the time had not yet arrived to take a strategic decision about moving work to the USA. He thought the British should push on with the isotope separation plant, which was the only large scale construction plant necessary for the bomb; if a heavy water plant would interfere with this main priority, it should be shelved since it was not of military importance. Simon pointed out that the efficiency of the isotope separation plant should be known within the next two months and was, like Chadwick, inclined to defer any decision. Cockcroft, who like Oliphant had visited the United States and knew the enormous capacity waiting to be exploited there, wanted complete Anglo-American co-operation. In the end the Committee resolved that experimental work should continue in the UK; the Americans should be asked to press on with heavy water production, and when there was sufficient available, Halban's team should be transferred there; that full exchange of information, both by visits and on paper, between the UK and the USA was essential.

Chadwick arranged to visit the Cavendish early in May 1941 to inspect the work of Halban and Kowarski. He wrote a preliminary note to G.P. Thomson on 5 May[53] to tell him that he thought their measurements were reliable and their claims justified; a two page report followed the next day, with apologies for lack of polish. It was a concise account of the experiments, examining potential sources of error and suggesting some minor improvements that could be made in methodology. He concluded, however, that Halban and Kowarski's results could be accepted with confidence, and their claim that the fission chain is potentially divergent in their setup of ordinary uranium and slow neutrons moderated by heavy water was justified.[54] Thomson and Cockcroft thought that Chadwick should take his report to the United States in person as the most certain way to ensure that large-scale production of heavy water would be undertaken there. Chadwick said it would be very awkward, but he would go if it were essential. In the event, the Americans were happy to accept his word on the subject and proceeded with the heavy water plants.[22]

Halban too was grateful for Chadwick's endorsement and wrote to him on 24 May:[55]

> I would like to thank you once more for the kind way in which you have taken care of our cause; this gives us complete confidence that we shall in the end get the possibility to go on with our work.

As the uranium 'boiler' seemed a practicable proposition, the significance of the French patents increased. The British Government recognized the contribution of Joliot's team, but at the same time were concerned that the French were stealing a march; inaction served to mask their dilemma. Halban wrote to Chadwick again on 5 June 1941:[56]

> I think the patent situation ought to be cleared up before we go to the States if the British interests are to be safeguarded. So far the legal department of the Ministry have been rather bureaucratic and slow-moving in this matter.

Chadwick[57] had expressed a view on these matters to Feather earlier in February:

> The difficulty about patents stared us in the face six months ago and I recommended that a definite agreement should be reached before any work was begun. Apparently no one believed that it would be necessary but it seemed to me almost certain that the chain process would go under laboratory conditions.

Now Halban was coming to see Chadwick to seek a resolution of what was becoming an increasingly tangled web, and Feather,[58] who was directly involved by this time, wrote to Chadwick on 23 June:

> The patents question arose for the rest of us when it became clear that bodies like ^{239}X and ^{233}U, which we were studying were likely to be of importance for this machine [the boiler]. Then Halban with a degree of spontaneity which I have come to regard as genuine and arising from a real sense of fairness, rather than from mere opportunism (as suspicion first suggests), said that obviously we must share in all the patents taken by Kowarski and him from now on. In France they had always been partnered by Joliot or Perrin or both, and they wanted to enter into the same kind of departmental partnership here. So Bretscher and I drew up an agreement with Halban and Kowarski to take all patents on our present work or its issue in common, assigning a mutually agreed share of benefit to each side on each patent separately . . . I know you had the feeling from the first that the Ministry should enter into a definite agreement about patents with the workers concerned—and now they have at least made a move to do so . . . however big the thing might become (and there is a remote possibility that it might, I imagine), we do not intend to amass large fortunes personally.

Feather explained that in France, Halban and the other joint patent holders had made provision that any revenue should be largely used to endow scientific research, and he would want to do the same. Halban,[59] frustrated by the lack of an official response explained to Chadwick that he and Kowarski had drawn up their own plan for the dispersion of any royalties that accrued. There would be a three way split, and if future royalties exceeded £100 000 in 1939 currency, 10% would go to the original group of French scientists, 89% would be earmarked for future scientific research and the remaining 1% or so would be shared by any other scientific collaborators. He asked Chadwick and Cockcroft to act as trustees for this scheme. Chadwick thought the principle 'most reasonable'[60] and agreed to be a trustee, subject to legal advice. Wrangling over the French patents continued for most of the war and in the later stages became a bone of contention with the Americans, as we shall see. Halban and Kowarski's scheme took no regard of the possibility that nuclear power might actually lose money, and in the words of Margaret Gowing 'the excitement over the French patents could be counted as one of the first unfulfilled rosy prophecies about the simplicity and profitability of nuclear power.'[22]

Chadwick's responsibilities seemed to increase with no let up. His primary scientific focus was on the work being carried out in Liverpool, and he regularly informed the Maud Committee of work in progress there. At the end of June 1941, he listed a five-point programme of activity:[61]

1. Experiments on fission cross-sections, not quite finished.
2. Investigation of critical conditions for the bomb assembly by means of model assembly, probably by an optical analogy suggested by Frisch.
3. Investigation of the energy spectrum of fission neutrons. Some work on neutrons over 3 MV; may be important investigations under 3 MV.
4. Measure cross-section of separated ^{235}U (awaited from the US).
5 Scattering of neutrons in uranium—more experiments may be necessary.

In addition, he continued to be in close touch with Powell in Bristol, constantly urging him to send results and warning that other things that Powell thought of pursuing 'can and must wait'.[62] He would write detailed letters to Powell, about once a fortnight, analysing data and making new suggestions. On 17 June 1941, he asked: 'Can you send me any results about the scattering in uranium. We are waiting anxiously, and Pryce can get no further with some important calculations.'[63] When Powell[64] sent data on neutron cross-sections and the ratio of elastic to inelastic scattering (with the hopeful enquiry 'is this the information you want?'), the results were eagerly devoured by Chadwick. He replied by return, thanking Powell for 'something to be going on with'.[65] He had expected the results for elastic scattering, but was surprised at the figure for total cross-section of uranium oxide, which was 'considerably higher than our previous result'. Here was Chadwick practising his accustomed role as scientist and mentor, but he now had to deal as well with the French patents issue, and the stream of letters he received from Feather and others. There were also the site visits to the Cavendish and Birmingham, which were major undertakings, since he had insufficient fuel to drive there, and the railways were in chaos due to the mass transportation of troops and the disruption caused by bombing. He described one Saturday trip to Birmingham from Liverpool to Feather: 'What a day. Eight hours railway with all the world on the move—platforms three or four deep.'[66] In addition there were monthly progress reports from all the Cambridge scientists to vet, which he did with great diligence, unerringly spotting errors as well as opportunities for new advances. He was about to surpass all these efforts with a climactic burst of work which he would repeat periodically over the next few years, taking himself to the limits of his physical endurance.

At the meeting of the Maud Committee on 2 July 1941,[67] G.P. Thomson reminded those present that a report on their activities was needed, and he presented a preliminary draft. Chadwick's first reaction was that it was essential to keep the civil and military applications—the boiler and the bomb—entirely separate, which Thomson had not done. In this he was supported by Lindemann, who went further and would restrict the report to the bomb only; the 'Prof.' also warned against intruding into the political and economic aspects. Many committee members thought the report should clearly reflect their view that this was an essential weapon because of its unprecedented destructive power, and Chadwick stated as a corollary that steps should be taken to control all stocks of uranium forthwith. After debating these various amendments and points of emphasis, it was decided that Chadwick would be the best person to redraft the report.

He returned to Liverpool and immersed himself in the task of writing. He received some help from Frisch and would also show some draft passages to Rotblat for comment, when they were germane to his work. Rotblat[42] realized that he was only seeing fragments of the whole, and that Chadwick was concealing as much as he could. As Chadwick[30] wrote many years later in response to an enquiry: 'If you will read Part 1 of the report, you will see that no-one else could have written it.' Part 1 was a general statement on the use of uranium for a bomb, and in its economy of style and clarity, it is indeed typical of Chadwick's writing. The dispassionate language does not disguise the chilling implications of the report's contents, and reflects the absolute determination of the Maud Committee that this weapon should be available to the British in their struggle for survival.[22]

> We should like to emphasise at the beginning of this report that we entered the project with more scepticism than belief, though we felt it was a matter which had to be investigated. As we proceeded we became more and more convinced that the release of atomic energy on a large scale is possible and that conditions can be chosen which would make it a very powerful weapon of war. We have now reached the conclusion that it will be possible to make an effective uranium bomb which, containing some 25 lb of active material, would be equivalent as regards destructive effect to 1,800 tons of T.N.T. and would also release large quantities of radioactive substances, which would make places near to where the bomb exploded dangerous to human life for a long period. The bomb would be composed of an active constituent (referred to in what follows as ^{235}U) present to the extent of about 1 part in 140 in ordinary Uranium. Owing to the very small difference in properties (other than explosive) between this substance and the rest of the Uranium, its extraction is a matter of great difficulty and a plant to produce 2¼ lb (1 kg) per day (or 3 bombs per month) is estimated to cost approximately £5,000,000, of which sum a considerable proportion would be spent on engineering, requiring labour of the same highly skilled character as is needed for making turbines.
>
> In spite of this very large expenditure we consider that the destructive effect, both material and moral, is so great that every effort should be made to produce bombs of this kind. As regards the time required, Imperial Chemical Industries after consultation with Dr Guy of Metropolitan-Vickers, estimate that the material for the first bomb could be ready by the end of 1943. This of course assumes that no major difficulty of an entirely unforeseen character arises. Dr Ferguson of Woolwich estimates that the time required to work out the method of producing high velocities required for fusing is 1–2 months. As this could be done concurrently with the production of the material no further delay is to be anticipated on this score. Even if the war should end before the bombs are ready the effort would not be wasted, except in the unlikely event of complete disarmament, since no nation would care to risk being caught without a weapon of such decisive possibilities.
>
> We know that Germany has taken a great deal of trouble to secure supplies of the substance known as heavy water. In the earlier stages we thought that this substance might be of great importance for our work. It appears in fact that its usefulness in the release of atomic energy is limited to processes which are not likely to be of immediate war value, but the Germans may by now have realised this, and it may be mentioned that the lines on which we are now working are such as would likely to suggest themselves to any capable physicist.
>
> By far the largest supplies of Uranium are in Canada and the Belgian Congo, and since it has been actively looked for because of the radium which accompanies it, it is unlikely that any considerable quantities exist which are unknown except possibly in unexplored regions.

The general overview went on to consider the method of fusing the weapon, its probable explosive effect, and the costs of the whole project. There was also a paragraph referring to the American efforts, the bulk of which, it was suggested, had been concerned with the production of energy and only to a much lesser extent with the bomb. There followed a long review of the technical evidence used to support the generalized statements of Part 1. As he had indicated he would at the last Maud meeting, Chadwick completely separated the bomb from the boiler, and the latter was presented as the second, shorter, part of the report. He completed the complex redrafting of the whole report in less than two weeks, despite being 'busy on the State Bursary scheme'[68] for the university, and when he finished said his mind was completely addled. He wrote[69] to Feather on 16 July 1941:

> I have been working about 20 hours per day, but now I have got off the reports both on bomb and boiler and today I have had a long discussion with G.P. [Thomson] and Peierls. With ten minutes to spare , I will attempt to answer your point.

At the end of August, he was exhausted and looking forward to a short holiday in the cottage in Wales. 'I am hoping to go away tomorrow or Monday for a few days—a week, perhaps a day or so longer if I am not called to London. I shall be glad to put off the harness for I am feeling very down. I hope to get a little gentle fishing.'[70]

References

1 Powell, C.F. and Fertel G.E.F. (1939). Energy of high-velocity neutrons by the photographic method. *Nature*, **144**, 115.

2 Frank, F.C. and Perkins, D.H. (1971). Cecil Frank Powell, 1903–1969. *Biographical memoirs of fellows of the Royal Society*, **17**, 541–55.

3 Powell, C.F., May, A.N., Chadwick, J. and Pickavance, T.G. (1940). Excited states of stable nuclei. *Nature*, **145**, 893–4.

4 Chadwick, J. (16/5/40). Letter to N. Feather. Feather's papers, CAC.

5 Chadwick, J. (17/6/40). Letter to N. Feather. Feather's papers, CAC.

6 Oliphant, M.L. (1/6/40). Letter to J. Chadwick. CHAD I, 19/3, CAC.

7 Haworth, W.N. (4/6/40). Letter to J. Chadwick. CHAD IV, 2/10, CAC.

8 Chadwick, J. (11/6/40). Letter to W.N. Haworth. CHAD IV, 2/10, CAC.

9 Haworth, W.N. (1/7/40). Letter to J. Chadwick. CHAD IV, 2/10, CAC.

10 Chadwick, J. (2/7/40). Letter to W.N. Haworth. CHAD IV, 2/10, CAC.

11 Haworth, W.N. (30/7/40). Letter to J. Chadwick. CHAD IV, 2/10, CAC

12 Chadwick, J. (1/8/40). Letter to W.N. Haworth. CHAD IV, 2/10, CAC.

13 Chadwick, J. (5/9/40). Letter to G.P Thomson. CHAD I, 19/7, CAC.

14 Haworth, W.N. (9/9/40). Letter to J. Chadwick. CHAD IV, 2/10, CAC

15 Haworth, W.N. (7/3/41). Letter to J. Chadwick. CHAD IV, 2/10, CAC

16 Chadwick, J. (24/6/40). Letter to G.P. Thomson. CHAD I, 19/7, CAC.

17 Churchill, W.S. (1949). *The second world war: their finest hour*, p. 570. Cassell, London.

18 Holt, J. (1988). *Seminar on James Chadwick*. University of Liverpool. See also Holt, J. (1994). Reminiscences and discoveries: James Chadwick at Liverpool. *Notes and Records of the Royal Society*, **48**, (2), 299–308.

19 Chadwick, J. (24/2/40). Letter to N. Feather. Feather's papers, CAC.

20 Oliphant, M.L. (22/6/40). Letter to J. Chadwick. CHAD I, 19/3, CAC.

21 Chadwick, J. (27/6/40/). Letter to M.L. Oliphant. CHAD I, 19/3, CAC.

22 Gowing, M. (1964). *Britain and atomic energy, 1939–1945*, p. 64. Macmillan, London.

23 Hewlett, R.G. and Anderson, O.E. (1990). *A history of the United States Atomic Energy Commission: the new world*, p. 19. University of California Press.

24 Chadwick, J. (6/6/40). Letter to R.E. Peierls. CHAD I, 19/6, CAC.

25 Peierls, R.E. (12/6/40). Letter to J. Chadwick. CHAD I, 19/6, CAC.

26 Chadwick, J. (7/6/40). Letter to N. Feather. Feather's papers, CAC.

27 Feather, N. (13/6/40). Letter to J. Chadwick. CHAD IV, 1/8, CAC.

28 Chadwick, J. (21/1/41). Letter to G.P. Thomson. CHAD I, 19/7, CAC.

29 Frisch, O.R. (1979). *What little I remember*. Cambridge University Press.

30 Chadwick, J. Undated papers CHAD II, 1/19 and CHAD III, 4/5, CAC.

31 Chadwick, J. (1/11/40). Letter to C. Darwin. CHAD I, 12/2, CAC.

32 Peierls, R.E. (1985). War. In *Bird of passage*, pp. 145–81. Princeton University Press.

33 Kurti, N. (1958). Franz Eugen Simon, 1893–1956. *Biographical memoirs of fellows of the Royal Society*, 4, 225–56.

34 Chadwick, J. (2/8/40). Letter to G.P. Thomson. CHAD I, 19/7, CAC.

35 Newton, J.O. (1992). Ernest William Titterton, 1916–90. *Historical Records of Australian Science*, **9**, 167–85.

36 Simon, F.E. (19/2/41). Letter to J. Chadwick. CHAD I, 19/8, CAC.

37 Weiner, C. (1969). Sir James Chadwick, oral history. American Institute of Physics, College Park, Maryland.

38 Chief Constable of Liverpool (1947). Letter to J. Chadwick. CHAD II, 1/17, CAC.

39 Chadwick, J. (16/7/41). Letter to B.G. Dickins. CHAD I, 12/2, CAC.

40 Chadwick, J. (25/9/40). Letter to B.G. Dickins. AB, 1/21, PRO.

41 Chadwick, J. (21/1/41). Letter to B.G. Dickins. CHAD I, 12/2, CAC.

42 Brown, A.P. (1994). Interview with Joseph Rotblat, London.

43 McMillan, E and Abelson, P.H. (1940). Radioactive element 93. *Physical Review*, 57, 1185–6.

44 Rotblat, J. (1985). Leaving the bomb project. *Bulletin of the Atomic Scientists*, Aug., 16–19.

45 Chadwick, J. (6/1/41). Letter to G.P. Thomson. CHAD I, 19/7, CAC.

46 Minutes of Maud Technical Sub-Committee (9/4/41). AB 1/347, PRO.

47 Goldschmidt, B. (1990). *Atomic rivals*. Rutgers University Press.

48 Feather, N. (15/12/40). Letter to J. Chadwick. CHAD IV, 1/8, CAC.

49 Chadwick, J. (20/12/40). Letter to N. Feather. Feather's papers, CAC.

50 Feather, N. (26/12/40). Letter to J. Chadwick. CHAD IV, 1/8, CAC.

51 Chadwick, J. (31/12/40). Letter to N. Feather. Feather's papers, CAC.

52 Feather, N. (6/1/41). Letter to J. Chadwick. CHAD IV, 1/8, CAC.

53 Chadwick, J. (5/5/41). Letter to G.P. Thomson. CHAD I, 19/7, CAC.

54 Chadwick, J. (6/5/41). Report on the experiments of Halban and Kowarski. CHAD I, 28/6, CAC.

55 Halban, H. (24/5/41). Letter to J. Chadwick. CHAD IV, 1/9, CAC.
56 Halban, H. (15/6/41). Letter to J. Chadwick. CHAD IV, 1/9, CAC.
57 Chadwick, J. (3/2/41). Letter to N. Feather. Feather's papers, CAC.
58 Feather, N. (23/6/41). Letter to J. Chadwick. CHAD IV, 1/8, CAC.
59 Halban, H. (11/8/41). Letter to J. Chadwick. CHAD IV, 1/9, CAC.
60 Chadwick, J. (9/10/41). Letter to H. Halban. CHAD IV, 1/9, CAC.
61 Chadwick, J. (30/6/41). Letter to B.G. Dickins. CHAD I, 12/2, CAC.
62 Chadwick, J. (25/3/41). Letter to C. Powell. CHAD I, 19/2, CAC.
63 Chadwick, J. (17/6/41). Letter to C. Powell. CHAD I, 19/2, CAC.
64 Powell, C. (28/6/41). Letter to J. Chadwick. CHAD I, 19/2, CAC.
65 Chadwick, J. (1/7/41). Letter to C. Powell. CHAD I, 19/2, CAC.
66 Chadwick, J. (14/6/43). Letter to N. Feather. Feather's papers, CAC.
67 Minutes of Maud Technical Sub-Committee (2/7/41). AB 1/347, PRO.
68 Chadwick, J. (14/7/41). Letter to N. Feather. Feather's papers, CAC.
69 Chadwick, J. (16/7/41). Letter to N. Feather. Feather's papers, CAC.
70 Chadwick, J. (30/8/41). Letter to N. Feather. Feather's papers, CAC.

13 Transatlantic travails

Several copies of the Maud Report were sent to the United States with the aim of fostering collaboration. One copy went to the inert Lyman Briggs, who as Oliphant[1] subsequently discovered, locked the report in his safe without showing it to other members of his Uranium Committee. Fortunately, copies were also sent to an activist, Vannevar Bush—an inveterate inventor and academic engineer, who had persuaded President Roosevelt to place him at the head of the National Defense Research Committee (NDRC) in June 1940.[2] By now he had become the director of a new, overarching, agency, the Office of Scientific Research and Development (OSRD). Bush had been succeeded at the NDRC, by James B. Conant, an organic chemist, who had visited both Oxford and Cambridge in the early 1930s, when he found himself 'on the point of becoming an Anglophile'.[3] Bush and Conant were two middle-aged Yankees who had both spent a period working as government researchers in World War I before making successful careers as academic scientists in Boston: Bush had risen to become Dean of Engineering at the Massachusetts Institute of Technology before moving to Washington in 1939, and Conant was President of Harvard University. They were alarmed by events in Europe and were opposed to the United States' isolationist stand; they seized on the Maud Report as a realistic opportunity for a scientific project which could be developed during the present war and which the United States could not afford to ignore. Although Bush and Conant recognized the quality of the British Maud Committee and the thoroughness of their report, they decided to subject the proposals to an independent scientific review by leading American experts under the auspices of the National Academy of Sciences to establish an incontrovertible basis for future progress.[2]

The first response to the Maud Report from America to reach London was from Charles Darwin, now Director of the British Central Scientific Office in Washington. He had previously written in early July 1941 to report his impressions of a meeting of the Briggs Uranium Committee, which he had attended. It had lasted over five hours with hardly a mention of the salient problem of isotope separation: 'The plain fact is that they are very nearly stuck on that side', was his verdict.[4] In a handwritten letter dated 2 August 1941 to Lord Hankey (Chairman of the Cabinet Scientific Advisory Committee), Darwin[5] allowed himself to bring up the morality of the enterprise, something which the Maud Committee had studiously avoided. He also said that he had recently been approached by Bush and Conant on the subject of atomic bombs, and there did seem to be the promise of movement. He thought that decisions needed to be made in government circles about the large expenditures involved, and raised the question whether such a weapon would ever be used:

> Are our Prime Minister and the American President and the respective General Staffs willing to sanction the destruction of Berlin and the country round, when, if ever, they are told it could be accomplished at a single blow?
>
> It appears to Bush and Conant, and I concur, that the time is ripe for a full examination of whether the whole business should be continued at all.

According to Darwin, Bush and Conant favoured a joint project between the two governments; he thought it 'fairly clear' that the plant would have to be built in the US or Canada. Darwin recommended that a small secret conference should be set up 'with men of balanced judgement'; his suggestion for the British representatives were Chadwick, 'who is an authority on the general subject of nuclear physics', Professor Simon from Oxford, because of his expertise on isotope separation, and G.P. Thomson.

Within a few days of receiving Darwin's letter, Hankey also heard from Tizard and Sir John Anderson, who was now Lord President of Council in the War Cabinet. Tizard[6] admitted to still being a sceptic about the whole concept and could not believe that 'the physics was settled'. He shared Darwin's concern about a very big and speculative industrial undertaking in wartime Britain, and agreed that Chadwick and Thomson should go to the United States. Anderson[7] was also of the opinion that not all the practical obstacles to building a uranium bomb could necessarily be overcome, and was 'quite certain that the production of the first bomb would not, under the most favourable conditions, be achieved until long after the dates indicated in the reports'. He thought it was essential to proceed, but that it would be foolish to risk war damage so that the plant must be based in North America; he was also concerned that speculators in uranium might drive up its price. Anderson was additionally exercised by the likely impact of the bomb on world affairs after the war.

The Maud Committee was officially under the aegis of the Ministry of Aircraft Production (MAP), and the Ministry's Director of Scientific Research, Dr David Pye, was also a member of the committee.[4] He had made his own assessment of the report before passing it to his minister, Colonel Moore-Brabazon, who forwarded the report to Lord Hankey with his own personal comments. He urged that apart from the immediate military importance, Hankey's committee should consider the possible long term benefits from nuclear power and the unprecedented political clout that would devolve to America and Britain if they were sole possessors of the atomic bomb. Hankey, in his reply to Moore-Brabazon on 27 August,[8] divulged a certain mistrust of the Americans and displayed a firmly imperial attitude:

> In the present war we have had valuable support from the Americans but have ourselves had to do all the fighting. Very close association with the Americans over about 30 years leads me to doubt whether they will ever join us actively in policing the world. I would not put it beyond possibility that the Isolationists

> might get on top as they did after the last war. If that were so, and this scheme were a joint project carried out in America, it might place us in a very embarrassing position.

Similar sentiments were conveyed the same day to Churchill by Lord Cherwell[9] (the ennobled Prof. Lindemann). He had attended many meetings of the Maud Technical Committee as an observer and had studied the report. After telling Churchill that 'people who are working on these problems consider the odds are 10 to 1 on success within two years', but that he 'would not bet more than 2 to 1 against or even money', Cherwell strongly recommended that the project must go forward. When it came to location, Cherwell was 'strongly of the opinion that we should erect the plant in England or at worst in Canada'. He went on:

> The reasons against this course are obvious, i.e. shortage of manpower, danger of being bombed etc. The reasons in favour are the better chance of maintaining secrecy (a vital point) but above all the fact that whoever possesses such a plant should be able to dictate terms to the rest of the world. However much I may trust my neighbour, and depend on him, I am very much averse to putting myself completely at his mercy and I would, therefore, not press the Americans to undertake this work; I would just continue exchanging information and get into production over here without raising the question of whether they should do it or not.

Cherwell also suggested that a Cabinet Minister should be appointed to coordinate and take general responsibility for the programme. Churchill[10] responded decisively with the following minute to General 'Pug' Ismay on 30 August.

> Although personally I am quite content with the existing explosives, I feel we must not stand in the path of improvement, and I therefore think that action should be taken in the sense proposed by Lord Cherwell, and that the Cabinet Minister responsible should be Sir John Anderson.
>
> I shall be glad to know what the Chiefs of Staff Committee think.

At the beginning of the year, Churchill had given Anderson responsibility for harnessing the economic engine of the country to the war-making machine. Citing his 'energy, mature judgement, and skill in administration . . . his long experience as a civil servant at home, and as Governor of Bengal',[10] Churchill found himself relying increasingly on Anderson to control economic and domestic matters so that he could be free, for better or worse, to involve himself in military affairs. Anderson[11] was an undemonstrative Scot, whose natural grandeur and ineffable wisdom had earned him the sobriquet of 'Jehovah' throughout Whitehall. He welcomed his new task because he had a natural affinity and respect for scientists, and regarded it as a privilege to attend meetings at the Royal Society and to rub shoulders with Nobel Prizemen.

When Chadwick came to Whitehall in mid-September, he was unaware of the letters, memos and minutes on the Maud Reports that were swirling around the place like the early leaves of autumn. G.P. Thomson, the Maud Committee chairman,[4] had suddenly departed for the United States in August to be with his wife, who was seriously ill. As chief draughtsman of the Reports, Chadwick was therefore the first scientist sent for to give evidence to Lord Hankey's Scientific Advisory Committee (SAC). Hankey's committee comprised five distinguished scientists, three of whom were from biological fields and only one, Sir Edward Appleton, was a physicist who knew Chadwick well. Before the committee[12] called Chadwick in, they first interviewed David Pye from MAP and he was adamant that the technical basis of the project was so speculative that the UK could not afford to devote the necessary resources and it should be undertaken in the US. In his own presentation, Chadwick unwittingly supported the minority view of Lord Cherwell quoted above. He considered that time and secrecy were the key factors, and favoured a British project because he believed the Americans were lagging in the scientific aspects. He was also doubtful about Canada providing a suitable site for the plant because of the lack of large industrial centres with the necessary power supplies. Chadwick thought the engineering requirements of an isotope separation plant were similar to those for turbine production at which the British were technically superior. He had no doubt that a satisfactory bomb could be produced in Britain within two years, especially if ICI were pressed to make a maximal effort. The fact that no significant quantities of ^{235}U had yet been isolated did not deter Chadwick in his estimate of the prospects for a bomb, and he told the committee that the evidence was overwhelming that, where fast neutrons were concerned, the rare isotope would behave just like natural uranium. He urged that a prototype, twenty-stage, isotope separation plant should be constructed as soon as possible; the Maud Report had barely alluded to the alternative fissionable material, plutonium, and Chadwick was not asked about this at all.

By the next day, Chadwick[13] was having second thoughts about his evidence and wrote the following letter from Liverpool:

> Dear Lord Hankey,
>
> On thinking over my remarks at your meeting yesterday, I feel that I may have stressed too much my own opinion that the large scale plant for the M.A.U.D. project should be erected in this country. The view of the majority of the M.A.U.D. Committee, perhaps a large majority, was in favour of America and as the representative of the M.A.U.D. Committee I ought to have put forward their view more definitely.
>
> I should admit too that this was also my own view until a short time ago, when considerations of secrecy and the fear that sabotage might prove a greater damage than air raids changed my opinion.
>
> Yours sincerely,
> J. Chadwick

With hindsight, it is hard to understand why Chadwick thought any site in the United Kingdom would be more immune to sabotage than one in the United States. It must be remembered that the Americans, at this stage, were not belligerents and did not seem disposed to enter another European war; Chadwick may have been concerned about the possibility of a strong pro-German faction in the US, whereas the population of Great Britain were absolutely united in their struggle against the Nazis. He really did not believe that the project was a secret in the international community of scientists (during his presentation to the SAC, there was discussion about the danger that Germany might be proceeding very quickly with a bomb of their own); but it had proved possible to keep the activity of the Maud Committee remarkably secret so that even Appleton and Tizard had been completely in the dark until the report surfaced. His evidence to the committee does reveal him to be positive in outlook, and with an enduring belief in the ability of British scientists and engineers to tackle such an immense challenge successfully. He had tended to gloss over some of the unsolved technical problems such as the fusing of the bomb. This question had been considered in the Maud Report:[4]

> The problem of fusing is not so much how to start the explosion at the required instant, but how to prevent it from occuring prematurely, for trigger neutrons are always present. It is therefore necessary to manufacture the bomb in two or more parts, each of less than critical size, which are kept at a safe distance and which are only brought together by an auxillary mechanism when the explosion is wanted.
>
> The assembly of the parts should be done as rapidly as possible for the following reasons. As the parts approach one another, critical conditions for a chain reaction will be reached at a certain point and any trigger neutron coming after that can cause an explosion. The violence of the explosion depends however on the rate at which the neutrons multiply and will be diminished if it occurs before the parts of the bomb are completely assembled. A premature explosion though less efficient is still of very great violence.

It seems Chadwick repeated the point to the committee that even a partial release of the energy from a uranium bomb would have devastating results, and by implication this would be a sufficient demonstration of its awesome power to an enemy. Pye did not like Chadwick's evidence at all and told him so, in no uncertain terms, by letter on 19 September.

> I feel, myself, convinced that the [Maud] committee is over optimistic in its view about the time scale . . . You may feel that a Heath Robinson kind of bomb is all that is required but, in fact, even this would mean a difficult piece of engineering and necessary safety considerations during the steps prior to final assembly would be certain, in my view, to mean much slower progress than the committee's report suggests.[14]

Chadwick's ambiguous feelings about how efficient a nuclear bomb he would like to see are also evident in a letter to Sir Edward Mellanby, one of the SAC members, who had asked why the Maud Report did not consider the effects of neutron radiation and radioactive contamination that would result from a uranium bomb. Chadwick[15] responded that: 'We think that the most important effect of the X bomb is its explosive action. In most circumstances the chief object of a bomb is the destruction of buildings, machines, etc. rather than human life, although this had its point too.' This seems an extraordinarily detached view for a survivor of the Liverpool blitz. It may represent an attempt to appease his conscience, but it was true in the case of Liverpool that the major effect had been to threaten the North-Western Approaches to Atlantic shipping; it was an opinion destined to become less tenable as all sides rained down terror from the skies onto civilian populations.

Chadwick was correct that the material and human destruction caused by the atomic bomb would be mainly due to its explosive action and the associated firestorm, but it seems indefensible now that radiation effects received so little consideration. He did point out to Mellanby that the dosimetry of neutron radiation was very difficult because of scattering effects and it seemed that 'the fatal area due to neutron radiation would probably not extend beyond the area of complete destruction'.

The SAC held seven meetings within as many days, and most of their witnesses contradicted Chadwick's initial presentation. Sir Henry Tizard agreed that the engineering skills required for isotope separation were similar to those for turbine construction, but was absolutely opposed to diverting turbine designers away from aero-engines in order to pursue 'a project the prospects of which were doubtful'.[16] As far as the 'boiler' was concerned, Tizard saw no useful application and thought it would be more expensive than current methods of energy production. He was in favour of the bomb work being transferred to North America and recommended that the governments of the UK, Canada and the USA should act in concert to prevent the boiler from falling into the hands of private, commercial interests, just in case it did have a future. On the question of the timescale, both Tizard and Blackett supported Lord Cherwell's contention that two years was a gross under-estimate of the time required; they pointed out that the work would be dangerous and complicated, and unforeseen difficulties were bound to arise.

By 22 September the SAC[17] produced its report and attached 'the very highest importance' to the bomb. They thought the estimate of two years to produce the weapon was optimistic and projected that it could take five. They called for continued research on the neutron cross-section of ^{235}U, design of the fusing mechanism, and suggested that the Medical Research Council should consider the radiobiological consequences. Once the design of a pilot isotope separation plant was completed, construction of a twenty-stage unit should begin straight away, and some consideration should be given to siting this in Canada. Preparatory steps for construction and installation of a full-scale production plant should be started as soon as possible so that once the final requirements were known, there would be no delays. In view of the bomb risk to what would be a ten-day, continuous, separation process, they argued strongly that the full-scale plant should be assembled in Canada, 'the necessary components being manufactured in the United States'. The finished report was sent to Sir John Anderson on 25 September, together with a letter from Hankey recommending that both the power project and bomb be put under the Department of Scientific and Industrial Research (DSIR). This arrangement was adopted, and Anderson set about finding a man with sufficient technical knowledge and the personality to direct the group of scientists who had made the Maud Committee so effective. He eventually chose Wallace Akers, the Research Director of ICI, and together they christened his organization Tube Alloys, a meaningless name with 'a specious air of probability about it'.[4]

Even before the SAC report was ready, the Prime Minister had discussed the concept of the bomb with his Chiefs of Staff, who, not surprisingly, were enthusiastic about the scheme. They urged great secrecy and also wanted the project to be developed in Britain—prejudices which coincided exactly with Churchill's. The SAC envisaged making geographical use of North America, and collaboration with the Americans and Canadians on the power project in particular, but implied that the British would retain independent control of the development of the bomb. When President Roosevelt wrote to Churchill on 11 October 1941: 'It appears desirable that we should soon correspond or converse concerning the subject which is under study by your M.A.U.D. Committee and by Dr Bush's organisation in this country in order that any extended efforts may be coordinated or even jointly conducted', he had to wait two months for a perfunctory reply.[4]

At the SAC meeting on 16 September, Chadwick had been questioned about enemy activity in this sphere, and agreed to provide Lord Hankey with a list of those German scientists whom he would expect to be playing leading roles in the work. As he explained to Franz Simon at Oxford, 'I suggested that perhaps the easiest and least obtrusive way of getting a line on this was to find the whereabouts of those men who would be employed on such work'.[18] Simon replied[19] that the same idea had occurred to Peierls and his new assistant in theoretical physics, Klaus Fuchs. The two had made a special visit to London the previous week to go through the German periodicals to find where people were working and what they were publishing. Their findings were inconclusive, but they did find that both nuclear fission and isotope separation were being mentioned freely as late as the summer of 1941[20] (which Peierls thought might be a double bluff), and they were able to place some personnel from published university calenders. Chadwick[21] sent his list of German scientists to Hankey; it was headed by the name of Heisenberg and included some of his old friends from Berlin days—Hahn, Geiger, Bothe—amongst those who followed. He told Hankey: 'If enquiry shows that some of these men are now working in collaboration in one or two places, we can, I think, infer that the enemy is seriously pursuing the Maud project'.

Lord Hankey passed on Chadwick's list to Brigadier S.G. Menzies, head of the Secret Intelligence Service (SIS), cryptically known as 'C'. After three months of investigation, the Secret Service had only managed to track three scientists down, and C's initial reply[22] to Hankey was comical:

> Curiously enough, the first named on the list was Professor W. Heisenberg of Leipzig. I have had a reply through Switzerland that this man was, during the summer of 1939, working in the Cavendish Laboratory at Cambridge . . . I am asking MI5 if they can confirm the statements, and if the man ever left England.

After this deeply unimpressive start, C soon wrote again[23] to say they had established that most of the scientists on the list were still at their usual places of work, including Heisenberg, who had visited Bohr in Copenhagen in 1941. More worrying was the discovery that Bothe and his team had taken over Joliot's cyclotron in Paris. By the end of January 1942, C amplified[24] this message to say that in addition to five German physicists, Joliot himself was also working under Bothe; Professor Hahn had been sent to Paris as well, 'but refuses to do war work'. Geiger was still working in Berlin, but visited Paris. 'Professor W. Heisenberg still lectures in Leipzig, but goes three days a week to the Kaiser Wilhelm Institute, Berlin . . . Professor K. Clusius, who is still at Munich, has been entrusted with the separation of uranium 235 from the mother substance, of which large quantities are received from Joachimstel in CzechoSlovakia.' The circumstantial evidence that the Germans were striving to produce a bomb could hardly have been more convincing.

Despite the remarkable advances that the Maud Committee made in its short span, many uncompleted tasks remained. Frisch confessed to Chadwick[25] in September 1941: 'I am a bit worried about the amount of work ahead.' He thought more staff, 'even raw recruits', should be found for Liverpool and expressed the view that 'with the official support we are now hoping to get, it must be possible to get another half dozen people'. In fact, Chadwick found himself as the Physics Professor continually responding to claims on his staff from the outside; despite his key contributions on theoretical aspects, Maurice Pryce was seconded for other work. David Pye from MAP wrote to Chadwick requesting that Liverpool University should release engineers, mathematicians and physicists. Chadwick replied laconically:

> The Honours School in Mathematics is rather exceptional this year, for it consists only of three women . . ..As regards physicists, the position is even worse for the men who would normally have been available this June [1942], all chose to join the Services last June, and to delay the completion of the Honours course until after the war.[26]

Even when he did manage to provide occasional graduates to industry, things could rebound unexpectedly. One former student wrote in December 1941 to complain to Chadwick about the way he had been treated at Ferranti's, the defence contractor. Chadwick was livid: 'I was most disagreeably surprised at the tone of your letter, which is rude and offensive . . . You joined the radio training scheme, you accepted a grant to enable you to attend the summer school, and you are under an obligation to undertake the work assigned to you.' The miscreant then sent Chadwick a long tale of woe, explaining how things had gone awry, and was rewarded with a note of paternal sympathy: 'I am very glad you explained matters at length. From what you say I can quite understand how unsatisfactory you felt your position to be and how you came to revolt against it . . . ADVICE . . . Keep in touch and I will do anything I can to help.'[27] On another occasion, he received a letter from one of the students who had left to join the army; he had been discharged after losing a foot and was going to seek a job as a teacher. He asked Chadwick for a letter of reference. Instead of simply acceding, Chadwick showed a deep concern in his reply,:

> I should be very glad if you could come to see me, for then I should know better what kind of employment you are looking for. Some day towards the end of the week would suit me very well.[28]

The young man might well have thought this was just a kind, slightly vague, response from his old professor, who was not very busy now the war had depleted his department, but even so the effect on his morale is easy to imagine.

In a similar vein, Chadwick received a letter from an RAF doctor concerned about barotrauma, which was causing ear problems in fighter pilots, asking about the 'flutter valve' action of the Eustachian tube. He had been given Chadwick's name by Professor Cohen, the Liverpool physician, whom he had 'bumped into on a train'. In his initial reply, Chadwick made some preliminary observations, but said he would have to consult medical colleagues on 'some points about the anatomy of the ear on which I am still not clear'. Within two weeks he had sent the RAF squadron leader a detailed opinion with diagrams and reprints of relevant articles.[29] If you want a favour, ask a busy man.

When Roosevelt wrote to Churchill in October 1941 proposing a joint Anglo-American effort to develop the bomb, he did so after being briefed at the White House by his scientific chief Vannevar Bush.[2] Earlier in the summer, the British scientists were puzzled by the lack of any communication from Bush or other top-ranking American scientists about the Maud Report. In late August, Mark Oliphant flew to the US in an unheated bomber to discuss the latest developments in radar, and was charged by the other Maud Committee members to find out what fate had befallen their report. It was he who discovered that Lyman Briggs, 'this inarticulate and unimpressive man',[1] had locked the papers away for security. Oliphant was 'amazed and distressed' and made it his business to inform the other Uranium Committee members that the bomb should be their absolute priority. He even flew from Washington to Berkeley in order to convince Lawrence of the crucial nature of the Maud Committee's findings. Once persuaded, Lawrence became an irrepressible agent himself, goading the other members of the American scientific elite into action.[2] As a result of the combined lobbying, Bush decided to send two members of the Uranium Committee on a visit to Britain so that they could gather firsthand information about work in progress. The two selected were George B. Pegram, whose Columbia University department was now home to Fermi and Szilard, and Harold Urey who had won the Nobel Prize for his discovery of heavy water.

When he heard that Pegram and Urey were coming to visit England, Chadwick[30] wrote at once to MAP expressing his wish to meet them. He wanted to get clear 'what is being done in the U.S.A.—from reports [they] seem only interested in the boiler, not seriously considering the bomb project'. He also wanted to establish what the Americans knew of the British work: 'Oliphant thinks they are not aware'. He closed his letter with a reminder about secrecy:

> There has been so much loose talk both here and in America that the enemy must be well aware that we are engaged on the uranium problem. We must take every step to prevent them from learning that we hope to proceed to a manufacturing stage.

This homily was probably prompted by a report only five days earlier, on 13 October 1941, in the *Daily Telegraph* about a speech Kapitza had made in Russia. Kapitza had stated that the 'probability of an atomic bomb being used in this war, unless it lasts a very long time, is small. It is worthwhile mentioning,' Kapitza continued, 'that theoretical calculations prove that . . . it could easily destroy a large capital with a population of several millions.'

By the time Pegram and Urey arrived in England in October 1941, the illustrious Maud Committee had been supplanted by Tube Alloys, controlled by DSIR through the person of Wallace Akers, who had been seconded from ICI. The transition was not handled diplomatically—some of the Maud team were not informed of the change until December and grew restive at the apparent lack of activity. Chadwick was officially informed by Appleton, the head of DSIR, on 25 October, but did not like the tone or the content of his letter. In his reply, Chadwick[31] issued a warning against usurping the scientists' leading role in the new organization. A few days after sending this letter, Chadwick attended the first meeting of the Technical Committee of Tube Alloys under Akers' chairmanship; the other members of the committee were Halban, Simon, Peierls, and Dr Slade—an ICI scientist.

30th October, 1941

Dear Appleton,

I received your letter of the 25th of October a few days ago but I wished to consider it at leisure before I replied, and the visit of Pegram and Urey has naturally taken up much of my time.

I am more than surprised that all decisions about the new arrangements for the M.A.U.D. work should be taken without some reference in some form to the M.A.U.D. Committee. I was fortunately able to have a short talk with Akers last week and the rough outline of the new scheme which he gave me in the short time available seemed not only workable but a great improvement on the old scheme. If I understand the proposal rightly, I think it would probably be welcomed by those working on the problem, provided the membership of the new committee is suitably chosen. Your letter, however, does not put the matter in the form which Akers gave it and I fear your remarks may give rise to some misunderstanding. For example, you say 'Mr. Akers, with the Lord President's approval, has been asked to take charge of all the scientific and technical work, and he will be assisted in this by Mr. Perrin. I shall keep in close touch with every aspect of the programme and I have appointed Mr. Jackson of my own staff to deal more particularly with administrative questions'. I must say at once that I should not be prepared to have the nuclear research which is necessary for the M.A.U.D. project directed by anyone but myself; and I must add that the ideas which led to the project, originating from myself on the one hand and Peierls and Frisch on the other, came from outside the original committee and the work in my department was for some considerable time independent of the Ministry of Aircraft Production. Mr. Akers' version of the new scheme was that he, acting not as a director of the I.C.I. but as an official of your Department, would be chairman of a new committee, of which Perrin and Jackson would be secretaries. This puts quite a different aspect on the matter, and I should be ready to continue to carry out the nuclear research and be responsible to this committee for it, provided of course that a suitable committee is elected.

As I said at the meeting of the Hankey Committee, I myself and at least the majority of the M.A.U.D. Committee were of the opinion that the work should be under government control, and that the scientific workers should work in intimate collaboration with I.C.I. representatives but not under them. The Akers version would, I think, meet these points.

There is little to be gained by further remarks now, and I ask for fuller information about the new arrangements. As I have already said, I am much afraid that the very curt statement in your letter will have produced undesirable reactions and I hope that a fuller account will do much to allay them.

Yours sincerely,
J. Chadwick

Chadwick's misgivings were loudly echoed by Oliphant, who 'registered a full-throated protest'.[4] He told Appleton he could 'see no reason whatever why the people put in charge of this work should be commercial representatives completely ignorant of the essential nuclear physics upon which the whole thing is based'. Oliphant[32] communicated to Chadwick his frustration with the Tube Alloys reorganization and the reluctance of the British to commit themselves to whole-hearted partnership with the Americans.

The Americans will undoubtedly go right ahead with both projects [bomb and boiler], and there is little doubt that they with their tremendous resources will achieve both before we have fairly begun. It seems to me far wiser to work in completely with them . . .

I only hope that you the only man in the country who really understands the problem will be allowed to have a real and deep influence on the scientific development.

Chadwick tried to convince Oliphant that the new Tube Alloy arrangement would be satisfactory, although he agreed that Appleton's handling of the matter had been 'autocratic and discourteous'.[33] He thought there was a determination to advance the bomb project with all speed, and while he had no reservations about full collaboration with the US, he did not share Oliphant's view about them finishing before we have begun: 'We are some way ahead and we shall remain ahead'. This was too much for Oliphant,[34] who replied by return:

I still feel that you in common with many other people in this country, underestimate seriously the extent of the American effort. I am extremely sorry that you have not gone to the States yourself for I am sure the picture which one gets in that way is rather different from that obtained from visiting Americans . . .

The visiting Americans, Pegram and Urey, had of course come to Liverpool to meet Chadwick, and were impressed by his near certainty that if pure ^{235}U were made available, a bomb of devastating power could be produced within a relatively short time. By the time they returned to the USA at the end of November 1941, the Committee of the National Academy of Sciences had submitted their report to Bush, essentially confirming the positive tone of the Maud Report about a weapon based on a critical mass of ^{235}U, predicting a 'fission bomb of superlatively destructive power'.[2] Bush sent the report to the President on 27 November, informing Roosevelt that he was forming an engineering group and organizing scientific research towards the mastery of the physical and chemical processes necessary to produce the components of the bomb, and to assemble them successfully. At this stage, as we have seen, Roosevelt had already written to Churchill with the intention of fostering an Anglo-American partnership. There was such high esteem for their British counterparts amongst the American scientists (particularly Conant, Pegram and Urey who had visited wartime Britain and witnessed its siege), that Oliphant's notion of complete co-operation was fully viable. If Chadwick had heeded his pleas and the two men together had pressed for such an arrangement, they might, by converting Lord Cherwell, have been able to persuade Churchill to overcome his own and the Chiefs of Staff's reticence; it seems likely that Sir John Anderson would have been a supportive and influential ally. As it was, Chadwick was not yet ready to abandon his conviction that British science was as good as any in the world and could pull off this project without outside help, and he did not rally to Oliphant's call. Indeed so deaf was he to Oliphant's entreaties that in early 1942, when Akers led a party of scientists from the Technical Committee of Tube Alloys (Simon, Halban and Peierls) to the US,[4] Chadwick chose to remain in England. The British visitors were given free range over the burgeoning American programme, no site was off limits to them and they were able to talk to all the American scientists they wished. The American effort under Bush was already more co-ordinated, and now the US had entered the war, it was fired with new vigour and determination. The opportunity for full partnership was already fading, and would disappear as the US Army took over control from the scientists and as millions of taxpayers' dollars were invested in a secret, all out, effort.

Whereas the National Academy of Sciences Committee, like the Maud Committee before them, had concentrated almost exclusively on the prospect of separated ^{235}U as material for the bomb, work had continued at Berkeley on element 94 (plutonium).[2] Lawrence, together with Arthur Compton from Chicago, convinced Bush and Conant that this newly created element might offer the shortest route to a bomb and should not just be considered as potential fuel for energy production. Chadwick had supported the Cavendish work on plutonium, and a year earlier had raised the relative merits of the two materials with G.P. Thomson. In January 1941, he told Thomson[35] that on his calculations of neutron cross-sections for 235, 94 was not likely to be more potent than 235, but might be just as good, and the choice between them would depend on the relative ease of preparation. He had then been forced by limited resources to concentrate on ^{235}U; now the Americans decided to broaden the experiment to include element 94 so that from the beginning of 1942, there were two types of bomb in prospect. Lawrence and Compton immediately organized teams of scientists to pursue both alternatives with early promising results*. Seaborg continued his work on element 94 and in a secret report[36] with Perlman in November, described how they had managed to produce

* Lawrence had also thought of an alternative method of uranium isotope separation in early 1941. Known as the *electromagnetic* method, Oliphant had been enthusiastic when told about it during his summer visit, and Lawrence was now planning to convert his 37-inch cyclotron at Berkeley into a giant mass spectrograph in order to test the idea.

microgram quantities of the substance by bombarding uranium with high-energy deuterons. They also devised several chemical processes by which element 94 could be extracted from the uranium and fission products of a reacting nuclear pile. Their studies on microscopic quantities of plutonium showed that its fission properties 'are somewhat superior to those of ^{235}U for the object in mind'.

In the spring of 1942, Bush approached the US Army to provide the gigantic engineering support that was going to be necessary to translate the scientists' predictions into a usuable weapon. By June, the US Army[37] was responsible for 'all large-scale aspects' of the atomic energy programme, and the Manhattan Project was born. The OSRD would continue to direct the scientific research, now with a budget of US$31 million for the fiscal year 1943 and Bush's forceful recommendation to the Secretary of War, Henry Stimson, that 'nothing should stand in the way of putting this whole affair through to conclusion, on a reasonable scale, but at the maximum speed possible, even if it does cause moderate interference with other war efforts'.

The gap between the American and British programmes was increasing week by week. In July, Anderson sent a 'very urgent' memo to Churchill[38] in which he stated that it had become clear that the production plant necessary for a bomb would have to be on such a huge scale that its erection in Britain during the war was 'out of the question'. 'Even the erection and operation of a pilot plant,' he continued, 'would cause a major dislocation in war production. Meanwhile, the Americans have been applying themselves with enthusiasm and a lavish expenditure, which we cannot rival'. 'With some reluctance', Anderson and the Consultative Council of Tube Alloys recommended moving the design work and personnel concerned to the United States, acknowledging that a combined Anglo-American effort was the most fruitful alternative. Indeed Anderson recognized the danger that the 'pioneer work done in this country is a dwindling asset and that, unless we capitalize it quickly, we shall be rapidly outstripped. We now have a real contribution to make to a merger. Soon we shall have little or none.' With Churchill's approval, Anderson[39] wrote to Bush on 5 August 1942 giving up all ideas of constructing plants in the UK and suggesting that the pilot gaseous diffusion plant should now be built in the United States. If this were done, he assumed that Bush 'would be willing to add British members to Dr Conant's Executive Committee'. Anderson also informed Bush of the intention to send Halban's team to Canada to continue their research there. In a separate letter written on the same day, Anderson explained the history of the French patents to Bush and the accommodation that had been reached with the British Government. He also raised for the first time the question of international control of atomic power and bombs, saying:

> it is not safe to allow the problem of the international control of nuclear energy to be left to be dealt with by any general solution covering other fields in which American and British inventions have been temporarily pooled for wartime use.

Bush replied separately to each letter on 1 September,[40] with equivocation that bordered on evasion. He supported the transfer of Halban's team to Canada, no doubt calculating that this would make a valuable asset more accessible to his own scientists. He thought patent control should be firmly in government hands, but thought it was too early to be setting up international controls. On the major question of moving British research and building the gaseous diffusion plant in the US, he did not conceal disappointment at the lack of recent progress in England:

> I had much hoped, from discussions that I have had previously, that this plant would be by now in operation, so that we would have a definite indication of the relative promise of various methods.

He was content to give only a general assurance on full scientific cooperation, including the future transfer of men to the US, once the course of the overall project had become clearer. Conant saw the draft replies to Anderson and sent Bush a congratulatory note exclaiming: 'Both letters are masterpieces! You should have been a lawyer or a diplomat!'.[41] A month letter, Bush wrote another holding letter to Anderson[42] suggesting that any decision about integrating the two programmes should be deferred until the Spring of 1943, when the British and American ideas about isotope separation could be compared. He hoped that Akers might soon visit the US again so that they could explore ways of co-ordinating research. Anderson attempted to inject a sense of urgency, at least on the question of patent controls, when he cabled a memorandum to Bush on 15 October:[43]

> I am glad to see that we agree on the extreme importance of proper control in this field but note that we differ on the question of speed of action. I am still convinced that all possible steps should be taken now by the respective Governments to secure control of patents in our two countires and Canada whether the inventions arise from work under contract or otherwise. It is for this reason that we have acquired all rights in the patent applications filed by Halban and his present colleagues and have done all possible at this stage to secure control outside France of the two original French patents.

Bush had become more interested in the tactics of negotiation than in the content of any British communication. He disregarded Anderson's plea that the area of international patents be specifically dealt with in any agreement made between the Prime Minister and the President. The failure to do so was to cause another serious rift between the Allies in 1944, and Chadwick would play a central role in restoring goodwill.

In the last week of September 1942, Bush had attended a fateful meeting in Secretary Stimson's office at the War Department, where a soldier from the Army Corps of Engineers, who had been promoted brigadier general that day, first cast his substantial shadow over Anglo-American relations.[37] Leslie R. Groves, the son of an army chaplain, had spent his youth living on different military posts in the American West where the frontier spirit was still alive and his boyhood imagination was stirred by tales of the Indian Wars. He was now 46 years old and although a graduate of West Point and MIT, his career had only started to prosper as the budget of the Army Corps of Engineers had recently swollen. In 1940 he had been elevated to the rank of colonel in order to supervise construction of the huge Pentagon complex in Washington. Now appointed to take charge of the Manhattan Project, he had again persuaded his superiors that in order to be taken seriously by 'the many academic scientists involved in the project', he would need a general's star.[44] A few hours after receiving his momentous appointment, Groves went to the meeting convened by Secretary Stimson at the War Department; also present were General George Marshall, the Army Chief of Staff, two other senior generals, Bush, Conant, and Harvey Bundy—Stimson's special assistant. While Conant may have been on the point of becoming an Anglophile during his sojourns in Oxford and Cambridge, Groves became a confirmed Anglophobe within hours of arriving in London on his only visit there as a young officer twenty years earlier. Showing no bashfulness in new, exalted, company, Groves immediately transcended his status as a military engineer, and asked about the relationship between American and British scientists. With unerring shrewdness, Groves had already arrived at the conclusion that 'in the future the British effort would probably be limited to the work of a very small number of scientists without any significant support from either the British Government or industry'.[44] With unrestricted exchange, the Americans would inevitably pass a great deal more information to the British than they could expect to receive in return. In the face of Groves' blunt logic, there was no support forthcoming for full collaboration with Britain, and it was agreed to make no decision until Stimson had consulted the President.

In contrast to the development of a military–industrial–scientific complex in the US, Tube Alloys' work in England remained in the realm of the universities, and even worse was becoming invested with petty, academic, politics. The contagion was particularly bad in Cambridge, and the hapless Feather proved ineffective in containing it. A persistent voice of discontent was Chadwick's ex-lecturer from Liverpool, Bernard Kinsey, who thought the leisurely, amateur style of the Cavendish was ridiculous in the face of the Nazi threat. He had found it difficult to readjust to the British way of muddling through after his years spent in California with Lawrence. After working on radar during the early years of the war, he transferred to Cambridge in January 1942 to take charge of the cyclotron. As he confessed to Chadwick[45], his experience with the radar project had been frustrating and in particular he was disillusioned with the hierachical, civil service approach to wartime research. Chadwick tried to encourage him:

> I do not think you have any need to feel uneasy about the change from Radio to this work. I agree, of course, that the radio work is important, but I think it is being rather overdone and there are quite enough men employed on this side. Further, I think one is justified in saying that radio will not win the war, for it is almost entirely defensive in its aspect. I must admit that our work may not lead to very much at any rate in this war, but it is obvious that it has great possibilities.[46]

Kinsey was soon complaining about the slow rate of progress and insisting that more assistance should be provided to do the donkey work since there was 'no room for half measures in any kind of war work'.[47] In August, Kinsey submitted a memo [48] requesting six more scientific workers, a secretary, new phones to connect the cyclotron with the research group of Halban and Kowarski (also working at the Cavendish) plus new equipment and workshop facilities that would cost six thousand pounds. Kinsey did not think that at the present rate, any results of value would be produced within two years, and the decision should be made to either 'drop this work altogether' or enlarge the staff. As if to emphasize his point, the Cambridge cyclotron broke down and took three weeks to repair. Feather, who had inundated Chadwick with letters asking for advice about how to organize the Cavendish, now forwarded Kinsey's memo with some additional complaints of his own. It was all too much for Chadwick, who told Feather that his actions were ill-considered, and reminded him of the subsidiary role of Cambridge:

> . . . the work was begun in Cambridge with two objects. 1/ to repeat some observations we had already made, notably the measurements on the fission cross-section, and later also to make experiments on the delayed neutron effect. 2/ to act as a reserve to us in case we were damaged by enemy action. In Bretscher's case, some simple measurements were required for the fast neutron work and help was to be given to Halban in obtaining some data necessary to him.
>
> Your work was clear and specific. In my opinion, you had sufficient men to do the work; the getting of material was entirely your own affair. To complain of lack of drive from above or a failure to provide the wherewithall to carry out the work seems to me fantastic.
>
> Consider the cyclotron experiments. The primary object of getting the Cambridge cyclotron on to this work was to check our measurements on the fission cross-section. I told Kinsey that the work was not easy and that I did not expect results in less than 3 months; but the last time I saw him I pointed out to him that it was time something materialised. If there is any lack of push, the failure is in the Cavendish. If you and Kinsey cannot do the experiments, it is unfair to put the blame elsewhere.[49]

He concluded sharply: 'We all have to work under rather difficult conditions at present, but I do not think that your difficulties are to be compared with those with which we have had to contend in the past.' Chadwick cooled off and attempted to take a closer interest in Kinsey's work, even making a visit to the Cavendish to see the situation for himself. Following his inspection he reminded Kinsey that he was only seeing part of the overall scheme and urged moderation in his demands, warning 'I should certainly oppose the accumulation of a lot of material which is never going to be used, especially when there is such an urgent demand from the Services.'[50]

The next flare up at the Cavendish came only one month after Chadwick's visit, and was rooted in personal antipathy between Feather and his Swiss colleague, Egon Bretscher. Bretscher had written to MAP, behind Feather's back, about the running of the Cavendish in general and Feather's role as team leader in particular. Feather complained to Chadwick about the disagreeable situation:[51]

> Poor man, he has a tortured soul! But what are we to do about it? It does seem that you are the only honest man left in Bretscher's eyes—so do come at the earliest opportunity and knock our heads together, if you think it necessary. But why

> can't he see the difference between ambition and the excitement over scientific discovery—or between neglect born of malice and neglect occasioned merely through the overwhelming pressure of work?

In another corner of the Cavendish, there was a bitter rift developing between the two Frenchmen, Halban and Kowarski. Halban was a member of the Technical Committee of Tube Alloys, had visited the US with Akers and had secured for himself a senior position in a new Anglo-Canadian programme based in Montreal; he seemed intent on using these advancements to humiliate Kowarski. Kowarski had stayed at the Cavendish, where he was generally liked and respected, while Halban had toured North America, and was offered a transfer to Montreal on terms which he found insulting. Chadwick's response[52] brings to mind the image of a grandfather despairing of younger generations:

> I am overwhelmed with troubles of one kind or another... I cannot understand why Bretscher should resent any discussion of the work, and when I come to Cambridge I intend to make it quite clear that collaboration is necessary and that you [Feather] have a right to know what is going on. We want as much exchange of ideas as we can get, and we cannot have any approach to rigid compartmentalisation.
>
> The Kowarski affair is only one of various troubles which have arisen recently. I think that Kowarski has been treated harshly and somewhat tactlessly and I sympathise strongly with him. At the same time, however, I am convinced that nothing but harm can come from drastic or violent action.

There were inherent frustrations, too, from the secret nature of the project. Chadwick wrote to Rutherford's old collaborator, the chemist Frederick Soddy, requesting the loan of an isotope that he had prepared in a strong solution. He could tell Soddy[53] that he required the 'preparation for work which has a high military importance', but he was 'therefore not at liberty' to reveal exactly the objectives he had in view. Soddy, a man of tetchy nature, replied: 'I am a little loathe to part with these [quantities] I prepared myself on some fool wild goose chase'.[54] Chadwick wrote again[55] to assure him the experiment he had in mind 'is quite a simple one and is not a fool wild goose chase'. He was forced to disclose to Soddy that he wanted the material as a source of energetic α-particles to bombard 'certain light elements' to see 'how many neutrons are produced' before a small amount was grudgingly sent.

By the autumn of 1942, Chadwick knew from Anderson that the Americans were turning sour on the idea of full collaboration. Apart from the diffident replies Anderson had received from Bush, there had been a perspicuous letter home from Professor Simon who had gone to the US in September 1942 to discuss the building of a pilot gaseous diffusion plant and found no enthusiasm. 'I would guess', he wrote,[4]'that the real reason is that the Army people want to run the thing 100 per cent American or something similar.' The invitation from Bush to send Akers over for talks was the only positive sign on the British horizon and was accepted with alacrity.

Akers arrived to a scene transformed from the one he had left earlier in the year. Groves and the military were now driving the project and setting priorities. He had been persuaded by J. Robert Oppenheimer, a theoretical physicist in his late thirties who had been working at Berkeley, to set up one central laboratory devoted exclusively to the theoretical and experimental aspects of building atomic bombs. Although Oppenheimer was dazzling, he was junior to many of the other scientists working in the field and was known to be left wing in his leanings. There were many reasons not to appoint him to run the laboratory: Groves ignored them all because he had made up his mind that Oppenheimer had the drive and ability to do the job. Groves had only been in control for a few weeks, and told Akers that he had inherited 'practically no organisation at all'.[56] He was 'especially horrified at the number of people who knew, or thought they ought to know, about large fields in the project outside their own particular work'. As a result Groves was determined to go much further in the direction of compartmentalization—although the British thought the existing degree of subdivision in American efforts already hampered efficiency. His intention, according to Akers, was 'to divide the work into as many separate compartments as can be devised and . . . only one person in each cell will be able to see over the top' and even that one person would see into a minimum number of other cells. Although Groves was prepared to allow visits by some senior British scientists, there would be heavy restrictions on their movement. At this stage Groves was proposing to keep the theoreticians under Oppenheimer in a state of quarantine from the outside world; if Peierls wished to visit the group he would have to remain shut up with them for the rest of the war. Chadwick could meet the group still in Chicago working on neutron physics, but would not be permitted to see Oppenheimer and the other theoreticians; 'in the circumstances', Akers thought, 'it's a good thing he is not coming immediately'.

At the end of November, Akers went to Chicago where he was joined by Halban down from Montreal. Halban presented a review of progress with heavy water, and Akers suggested there could be some collaboration on this between Chicago and Montreal. A large-scale production facility could then be erected in the US; from the scientists' view, Compton approved of the plan, but Groves ominously 'reserved judgement'. The most interesting news that Akers[57] picked up from Compton was that it would be only a matter of days before Fermi would achieve the first ever divergent chain reaction with his uranium-graphite pile. This did come to pass on Sunday 2 December 1942, the day Akers left Chicago. It seems that he was not invited to stay to witness the experiment. After Fermi's historic demonstration, Compton phoned Conant at Harvard to announce 'the Italian navigator has just landed in the new world'.[2] The production of plutonium in quantities sufficient for military purposes now seemed a reality. Groves had already approached the Du Pont chemical company to design and procure a plutonium separation plant to exploit Seaborg's laboratory discoveries. But there was still a major question over the suitability of plutonium in Chadwick's mind, and his concern was enough to shake the Americans when conveyed to Conant by Akers. Chadwick's thought was that the α-particles emitted by the plutonium would in turn produce neutrons in light-element impurities that might be present in the material extracted from the pile. These neutrons could then fission the plutonium and spoil the weapon before it could be assembled. An incredulous Conant checked this theory with Lawrence who agreed it was possible. Writing to Conant a week later Lawrence had become a little more optimistic even though 'no one like the du Ponts has produced in quantity such pure materials'.[58] However, he wondered 'if the production of let us say such pure metals would be a commonplace in industry if the product had a market at several million dollars a pound!' It was Chadwick's deep insight into nuclear reactions that made him an exception to Groves' theory about the one-sided nature of scientific interchange; the Americans still wanted to be able to talk to him and Peierls face to face.

Chadwick wrote to his old colleague Ralph Fowler on 3 December to say that he was 'feeling extremely anxious about the future of the Tube Alloys project'.[59] From his perspective, the Americans were 'making a tremendous effort, and we must keep our end up or we may find ourselves excluded from the final achievement.' The exclusion was already happening, and in another delaying manoeuvre, Akers was asked to spell out the British requests with supporting reasons. He wrote a six-page letter to Conant[60] urging that there should again be full disclosure of information to the British. He cited Chadwick's position as an example of the unsatisfactory nature of the prevailing arrangements.

> It seems impossible for us to send Professor Chadwick to America unless there is some revision in the policy here . . . he would not be allowed to have discussions with the American theoreticians, although the experimental work he directs in England [has the] sole object of providing data for theoreticians.

Chadwick thought that his time would 'largely be wasted' in the current circumstances. Apart from the wastage of scientific talent, Akers advanced political reasons why the Americans would want Britain to be an atomic power in post-war Europe, unless of course they intended to secure a world monopoly. Groves tried to convince him that the only American concern was security, but said if Chadwick and Peierls came quickly, they might have a last chance of meeting Oppenheimer's group before they were closeted in their secret camp. Akers was frustrated both with the Americans, who were paranoid about the 'wily British' obtaining and exploiting 'the secrets and know-how of the innocent American inventors', and with the inadequate resources in London where 'we have got to make up our minds whether this project is to go on as a minor and spare time occupation of DSIR or if we are to try to make some sort of show in comparison of the Americans'.[61]

Groves had convinced the two chief scientists, Conant and Bush, of the need for extraordinary security and although they favoured some limited interchange, they too were now ready to contemplate a complete cessation. Roosevelt was briefed in detail about the problem over Christmas and decided on a policy of limited exchange of information that could be used to win the war.[2] The schism that the British had felt coming for months was officially marked in a letter from Conant on 2 January 1943 to C.J. Mackenzie, Dean of the Canadian National Research Council. The letter[62] explained how the new policy 'from the top' would apply to

Halban's heavy water team now in Montreal. Conant stated that 'we are to have complete interchange on design and construction of new weapons and equipment *only* if the recipient of the information is in a position to take advantage of it in this war.' This was followed by a terse memorandum[63] from Conant on 7 January listing those areas on which there would be no further communication: electromagnetic separation, production of heavy water, fast neutron reactions and the production of uranium hexafluoride. There would be some exchange on gaseous diffusion, the use of heavy water and basic science.

The British and Canadians were appalled by the new policy and Conant's interpretation of it. Sir John Anderson expressed the mood of outrage succinctly in his minute to Churchill on 11 January 1943:[64]

> It appears that this principle is being interpreted to mean that information must be withheld from us over the greater part of the field of Tube Alloys. At the same time, the United States Authorities apparently expect us to continue to exchange information with them in regard to those parts of the project in which our work is further forward than theirs.
>
> This development has come as a bombshell, and is quite intolerable. I think that you may wish to ask President Roosevelt to go into action without delay. I feel sure that it cannot be his intention that we should be treated in this way, and I suggest you might urge him to give instructions that collaboration between us should go ahead on a fully reciprocal basis.

Still a cable arrived from the US on 16 January asking if Chadwick and Peierls would be prepared to visit under the conditions described. Chadwick flatly refused and when told, Groves thought his attitude 'most unfortunate'.[65] Simon wrote to Chadwick from Oxford suggesting there were 'many different lines of action by which we can bring this American mess into some kind of order'.[66] His leading proposal was to build a diffusion plant in England so that 'the Americans have, according to their own rules, to give us again all information on all types of plants . . . including all the military aspects'. The idea was taken up by the Technical Committee of Tube Alloys, and for the first time a serious study of costing and manpower requirements was undertaken.[4] Nevertheless, it was apparent to most that the impasse could only be overcome by the direct intervention of Churchill.

The tide of collaboration had turned against the British at some indefinable point between Roosevelt's letter to Churchill of October 1941, and Anderson's to Bush in August 1942. In the autumn of 1941, Churchill was preoccupied with the North Africa campaign which thus far had been dominated by Rommel. There was still the threat that the Germans would invade England, and as Churchill wrote he 'was in charge of a struggling country beset by deadly foes'.[67] Clearly there was a limited amount of attention that he could pay to the Maud Report, and his comparison with 'existing explosives' implies that he may not at once have grasped the magnitude of the project. Until the US entered the war after Pearl Harbour in early December, it seems improbable that he or the Chiefs of Staff would have been prepared to relinquish any control of a potentially war-winning weapon to a neutral power.

The impact that any single scientist could have on these events was plainly minor. The only real opportunity for intervention that Chadwick missed was to go with the Akers team in early 1942. Curiously, he would have been the only scientist in the party who did not speak English with a Germanic accent, and as we have seen, he had a formidable reputation with his American colleagues so that his opinions would have carried weight. It surely would have been a more productive use of his time than remaining in Liverpool, and as Oliphant pointed out would have opened his eyes to the scale of the American effort. Even if the British and Americans had agreed on full partnership in the Spring of 1942, it seems unlikely that the course of the bomb development would have been materially altered since the rate limiting steps from that time forward were related to technology and production rather than to science. The postwar picture, though, might have looked quite different.

References

In addition to cabinet papers (CAB), I have made use of the atomic bomb (AB) and premier's (PREM) collections at the Public Record Office. American papers relevant to the Manhattan Project are to be found under R[ecord] G[roup] 77, at the National Archives 2, in College Park, Maryland.

1 Oliphant, M.L. (1982). The beginning: Chadwick and the neutron. *Bulletin of the Atomic Scientists*, Dec, 14–18.

2 Hewlett, R.G. and Anderson, O.E. (1990). *A history of the United States Atomic Energy Commission: the new world*. University of California Press.

3 Conant, J.B. (1970). *My several lives*, p. 105. Harper and Row, New York.

4 Gowing, M. (1964). *Britain and atomic energy, 1939–1945*. Macmillan, London.

5 Darwin, C. (2/8/41). Letter to Lord Hankey. CAB 104/227, PRO.

6 Tizard, H. (5/8/41). Letter to Lord Hankey. CAB 104/227, PRO.

7 Anderson, J. (6/8/41). Letter to Lord Hankey. CAB 104/227, PRO.

8 Hankey, M. (27/8/41). Letter to J.T.C. Moore-Brabazon. CAB 104/227, PRO.

9 Cherwell, F. (27/8/41). Letter to W.S. Churchill. CAB 104/227, PRO.

10 Churchill, W.S. (1950). *The second world war: the grand alliance*. Cassell, London.

11 Wheeler-Bennett, J.W. (1962). *John Anderson: Viscount Waverley*. Macmillan, London.

12 Scientific Advisory Committee minutes (16/9/41). CAB 90/8, PRO.

13 Chadwick, J. (17/9/41). Letter to Lord Hankey. CAB 104/227, PRO.

14 Pye, D. (19/9/41). Letter to J. Chadwick. CAB 104/227, PRO.

15 Chadwick, J. (20/9/41). Letter to E. Mellanby. CAB 90/8, PRO.

16 Scientific Advisory Committee minutes (19/9/41). CAB 90/8, PRO.

17 Scientific Advisory Committee (22/9/41). Report on Maud Report. CAB 90/8, PRO.

18 Chadwick, J. (17/9/41). Letter to F.E. Simon. CHAD I, 19/8, CAC.

19 Simon, F.E. (19/9/41). Letter to J. Chadwick. CHAD I, 19/8, CAC.

20 Peierls, R.E. (23/9/41). Letter to J. Chadwick. CHAD I, 19/6, CAC.
21 Chadwick, J. (27/9/41). Letter to Lord Hankey. CAB 104/227, PRO.
22 'C', Menzies, S.G. (28/12/41). Letter to Lord Hankey. CAB 104/227, PRO.
23 'C', Menzies, S.G. (2/1/42). Letter to Lord Hankey. CAB 104/227, PRO.
24 'C', Menzies, S.G. (30/1/42). Letter to Lord Hankey. CAB 104/227, PRO.
25 Frisch, O.R. (22/9/41). Letter to J. Chadwick. CHAD IV, 2/10, CAC.
26 Chadwick, J. (6/3/42). Letter to D. Pye. CHAD I, 24/1, CAC.
27 Chadwick, J. (1941). Letters to T. Lunt. CHAD III, 3/1, CAC.
28 Chadwick, J. (1943). Letter to F. Denmark. CHAD III, 3/1, CAC.
29 Chadwick, J. (1942). Letters to Sq. Ldr. J. McGibbon. CHAD I, 24/1, CAC.
30 Chadwick, J. (18/10/41). Letter to J. Thewlis. CHAD I, 12/3, CAC.
31 Chadwick, J. (30/10/41). Letter to E. Appleton. CHAD IV, 10/49, CAC.
32 Oliphant, M.L. (3/11/41). Letter to J. Chadwick. CHAD I, 19/3, CAC.
33 Chadwick, J. (10/11/41). Letter to M.L. Oliphant. CHAD I, 19/3, CAC.
34 Oliphant, M.L. (12/11/41). Letter to J. Chadwick. CHAD I, 19/3, CAC.
35 Chadwick, J. (6/1/41). Letter to G.P. Thomson. CHAD I, 19/7, CAC.
36 Seaborg, G and Perlman, M. (1942). Report on element 94. RG 77, Box 53, National Archives 2.
37 Jones V.C. (1985). *Manhattan: the army and the atomic bomb.* Center of military history United States army, Washington DC.
38 Anderson, J. (30/7/42). Memorandum to W.S. Churchill. PREM3 139/8A, PRO.
39 Anderson, J. (5/8/42). Letters to V. Bush. AB 1/207, PRO.
40 Bush, V. (1/9/42). Letters to J. Anderson. AB 1/207, PRO.
41 Conant, J. (1942). Undated note to V. Bush. RG 227, Box 1A, OSRD S-1 files, National Archives 2, College Park, Maryland.
42 Bush, V. (1/10/42). Letters to J. Anderson. AB 1/207, PRO.
43 Anderson, J. (15/10/42). Memorandum to V. Bush. RG 227, Box 1A, OSRD S-1 files, National Archives 2, College Park, Maryland.
44 Groves, L.R. (1962). *Now it can be told.* Harper and Row, New York.
45 Kinsey, B.B. (20/1/42). Letter to J. Chadwick. CHAD IV, 1/9, CAC.
46 Chadwick, J. (4/3/42). Letter to B.B. Kinsey. CHAD IV, 1/9, CAC.
47 Kinsey, B.B. (8/3/42). Letter to J. Chadwick. CHAD IV, 1/9, CAC.
48 Kinsey, B.B. (10/8/42). Memo to Tube Alloys. CHAD IV, 1/9, CAC.
49 Chadwick, J. (17/8/42). Letter to N. Feather. Feather's papers, CAC.
50 Chadwick, J. (27/11/42). Letter to B.B. Kinsey. CHAD IV, 1/9, CAC.
51 Feather, N. (6/12/42). Letter to J. Chadwick. Feather's papers, CAC.
52 Chadwick, J. (16/12/42). Letter to N. Feather. Feather's papers, CAC.
53 Chadwick, J. (21/10/42). Letter to F. Soddy. CHAD II, 3/1, CAC.
54 Soddy, F. (23/10/42). Letter to J. Chadwick. CHAD II, 3/1, CAC.
55 Chadwick, J. (28/10/42). Letter to F. Soddy. CHAD II, 3/1, CAC.
56 Akers, W. (16/11/42). Letter to M. Perrin. AB 1/128, PRO.

57 Akers, W. (11/12/42). Letter to M. Perrin. AB 1/128, PRO

58 Lawrence, E.O. (23/11/42). Letter to J.B. Conant. RG 77, 85/201, National Archives 2, College Park, Maryland.

59 Chadwick, J. (3/12/42). Letter to R. Fowler. CHAD IV, 1/9, CAC.

60 Akers, W. (11/12/42). Letter to J.B. Conant. AB 1/128, PRO

61 Akers, W. (21/12/42). Letter to M. Perrin. AB 1/128, PRO

62 Conant, J.B. (2/1/43). Letter to C.J. Mackenzie. AB 1/128, PRO.

63 Conant, J.B. (7/1/43). Memo to Tube Alloys. AB 1/128, PRO.

64 Anderson, J. (7/1/43). Memorandum to W.S. Churchill. PREM3 139/8A, PRO.

65 Akers, W. (26/1/43). Minutes of meeting with Conant and Groves. AB 1/128, PRO.

66 Simon, F. (20/1/43). Letter to J. Chadwick. CHAD I, 19/8, CAC.

67 Churchill, W.S. (1950). *The second world war: the grand alliance,* p. 478. Cassell, London.

14 The new world

When he first read Conant's pre-emptive memorandum, Sir John Anderson[1] believed that it was all the work of the US military and could not have been sanctioned by Roosevelt. In fact, Conant had carefully prepared three alternatives for the President which ranged from free exchange of information to a complete cessation, and included the option of limited disclosure, which was ultimately adopted. The President made his decision over Christmas 1942, after talks with Bush and Stimson, and was influenced by his recent discovery that the British had signed a secret agreement with the Soviets some months earlier on the exchange of information about new weapons.[2] The agreement was concluded between Molotov and the British Ambassador in Moscow,[3] who had no knowledge about the Tube Alloys project, and it seems improbable that either side intended to honour it.*

The minute to the Prime Minister from Anderson, dated 11 January 1943 which detailed the new restrictive American attitude, was composed just in time to be carried to Casablanca; Churchill and Roosevelt commenced one of their pivotal, wartime conferences there two days later. Chadwick was assured that 'the Prime Minister went fully briefed on the Tube Alloy problem to the meeting with Roosevelt.'[4] Sir John Anderson's fond hope was that a quiet word between the two leaders would mend the recent, unfortunate rift, and the two Allies would again join forces on an equal footing—in terms of control if not in material commitment—in the quest for the atomic bomb. Churchill came to the Moroccan city with a formidable list of unresolved issues weighing heavily on his mind. On his arrival he cabled to Attlee back in London: 'Conditions most agreeable. I wish I could say the same of problems'.[5] There was unremitting pressure from Stalin to open a second front in France in 1943, whereas the British Chiefs of Staff thought this unrealistic and wanted to prosecute the war in the Mediterranean, once the Axis (German and Italian) forces had been cleared out of Tunisia. The American attitude was almost the reverse with the military favouring a cross-Channel inva-

* The terms of the agreement make astonishing reading in view of later East–West hostility: 'The Government of the USSR and the Government of the UK will furnish to each other on request all information, including any necessary specifications, plans, etc., relating to weapons, devices or processes which at present are, or in future may be, employed by them for the prosecution of the war against the common enemy . . . If either Government considers that in the common interest there would be a disadvantage in giving such information in a particular case, they shall be entitled to withhold the information in question, but in that event they will indicate the reasons which led them to take this view.'

sion in the next year, but unable to convince their Commander-in-Chief. As they grappled with these historic decisions and other matters of high strategy over the next ten days, it seems that there was only the briefest exchange between the two leaders about the vicissitudes of atomic policy. When the subject was broached, Roosevelt's right-hand man, Harry Hopkins, who was truly an innocent abroad on this question, reassured Churchill that the problem could be easily staightened out as soon as the President returned to Washington.[2]

If anything, it was Hopkins who was straightened out on the return to Washington; he was briefed by Bush and Conant on the intricacies of atomic politics, and persuaded that the new limitations on communication were designed to prevent the British from gaining a post-war advantage and should not result in any hampering of the war effort.[2] So as he received telegrams from Churchill over the next few months seeking to break the impasse, Hopkins either ignored them or requested more details. By such devices, he ensured that the subject was effectively kept in abeyance. For the scientists and executive of Tube Alloys, it was a period of complete despondency. Their hopes were not lifted by their own detailed review carried out by Akers and his assistant Michael Perrin, another ICI employee and a chemist by training, which was completed in April. They calculated that the total current expenditure on Tube Alloys in the universities, Government and industry was approximately half a million pounds per annum.[6] To construct even a pilot diffusion plant would cost several million pounds, and for a full scale plant producing 1 kg of ^{235}U per day, the cost would rise to fifty million pounds over several years. Even if they abandoned ^{235}U and made all their atomic eggs plutonium, the bill would still be tens of millions of pounds, and new techniques would have to be devised for producing heavy water in commercial quantities. Aside from the purely financial considerations, such an effort would require hundreds of extra scientists at a time when there were no experienced workers not already engaged on important war projects, and the number of new graduates was minimal. Once the remaining technical questions had been settled by the scientists and engineers, huge quantities of steel, concrete, electricity would be needed in the construction phase, not to mention many thousand able-bodied workers. It was, therefore, for good reason that Akers and Perrin concluded:

> It cannot be emphasised too strongly that it is grossly inefficient for the British and American efforts to be separated if the object is the creation by the United Nations in the shortest possible time of a decisive war weapon. Furthermore we are convinced that the carrying out of this project competitively by Britain and America can only make almost impossible a post-war control problem which from its nature will be difficult enough even if the two Governments are working in the closest co-operation.

Plainly, a working arrangement with the Americans had to be restored if the British were to remain in the field.

The decline in the British position was appreciated as keenly by Chadwick as anyone, both through his senior position in Tube Alloys and because of his experiences as a leading figure in Rutherford's Cavendish Laboratory, in the days when it led the world. In case he had any illusions about the current programme at the Cavendish, Kinsey seemed determined to dispel them in a series of terse letters and an unforgiving annual report in which he wrote:[7]

> We have probably done better than could have been done in peacetime but only because we have not limited our expenditure. Even on the programme of research with which you are content, I can see no excuse whatever for the inefficiency with which we work nor any excuse for continuing to work in these conditions. The greater part of my time here has been spent as a second rate engineer and Jack-of-all-trades. I had enough of this sort of thing in peacetime and I am very reluctant to continue it. It is quite obvious to me that about 80% of our work can be done by relatively unskilled labour. What excuse can there be for using highly trained research men for work which, with a little enthusiasm and better organisation, can be done by laboratory assistants.

Chadwick suffered an endless series of colds, influenza and tonsillitis that winter and for once did not rise to the challenge of attempting to answer Kinsey's criticisms; seemingly exhausted, he made vague mention about providing better facilities.[8] Apart from trying to soothe Kinsey, Chadwick was cast as the peace-maker amongst the various factions in Cambridge. Although Halban departed for the Montreal Laboratory at the end of 1942, the bitterness of his rift with Kowarski lingered on; Bretscher's views on Halban were said to be 'quite unprintable and in general much more lurid than Kowarski at his best'.[9] Bretscher also remained deeply suspicious of Feather, who in turn was finding Kinsey hard to manage. Oliphant, after a visit to the Cavendish, gave his impression with characteristic, Australian, candour:[10] 'Viewed as contributions to the war effort,' he wrote to Chadwick, 'most of what is going on in Cambridge is quite useless'. He thought it would be far better if some scientists there reverted to full-time teaching and released others for research. In the circumstances, it is not surprising that Chadwick was rather lukewarm in his support for Feather's election to Fellowship of the Royal Society. Chadwick was Chairman of the Physics Committee and signed the nomination certificate, confessing to Cockcroft that he had been 'rather disappointed with Feather'.[11] He was, on the other hand, particularly sympathetic towards Kowarski, who was not included in the Montreal heavy water project and remained at the Cavendish as the head of a team reporting to Chadwick. He took a close interest in their research and encouraged Frisch to visit Cambridge to promote the cross-fertilization of ideas, 'putting no great trust in the bogus security which relies on compartmentalizing knowledge'.[12]

The fast neutron work in Liverpool continued regardless of the general despond. With Chadwick's encouragement, Frisch had devised an instrument for analysing samples of uranium and measuring the amount of each isotope present. The principle depended on selectively counting α-particles of a particular energy and Frisch constructed the apparatus with help from ICI.[12] It was the prototype for what became known as the pulse height analyser or kicksorter—a versatile research tool widely used in nuclear physics laboratories. In December 1942, a sample of uranium enriched in the 235 isotope to about 15% by weight was sent to Liverpool from Lawrence's Berkeley Laboratory. Chadwick had requested it some months earlier because, as he wrote to Lawrence,[13] 'so much depends on the fission cross section of 235'. Lawrence agreed about the vital need to carry out careful experimental studies and not to rely entirely on theory, and it says much about his opinion of Chadwick and the Liverpool department that he was prepared to part with the precious sample. One presumes that if the material had become available one month later, Lawrence would have been prevented from sending it by the terms of the Conant memorandum.

In January 1943, Chadwick's good standing in the international world of science led to an approach by a Secret Service agent in Liverpool. He was told that the British authorities wished to smuggle Bohr out of occupied Denmark to join Tube Alloys.[14] No doubt they had calculated that to secure the services of the world's greatest living nuclear physicist would enhance the attractiveness of the British programme to the Americans. Chadwick was prepared to lend his name to this effort, after some details about the proposed escape plan had been explained to him, but as his elliptical message makes clear, he would not presume to put any pressure on Bohr. His primary concern was Bohr's safety.

The University of Liverpool
George Holt Physics Laboratories
25th January 1943.

I have heard in a roundabout way that you have considered coming to this country if the opportunity should offer. I need not tell you how delighted I myself should be to see you again; and I can say to you , there is no scientist in the world who would be more acceptable both to our university people and to the general public. I think you would be very pleased by the warmth you would receive. A factor which may influence you in your decision is that you would work freely in scientific matters. Indeed I have in mind a particular problem in which your assistance would be of the greatest help. Darwin and Appleton are also interested in this problem and I know they too would be very glad to have your help and advice. You will, I hope, appreciate that I cannot be specific in my reference to this work, but I am sure it will interest you. I trust you will not misunderstand my purpose in writing this letter. I have no desire to influence your decision, for you alone can weigh all the different circumstances, and I have implicit faith in your judgement, whatever it should be. All I want to do is to assure you that if you decide to come, you will have a very warm welcome and an opportunity of service in the common cause. With my best wishes for the future and my deepest regards to Mrs Bohr.

Yours sincerely,
J. Chadwick [14]

Chadwick's message was reduced to a microdot, the size of a pinhead, which was concealed in the hollow handle of a doorkey. Bohr was informed in advance of the key arriving in Copenhagen via Stockholm by Captain Volmer Gyth, a member of the Danish General Staff with close connections to the resistance movement. He was responsible for reconstituting the microdot and delivering a legible version of the letter. In his reply,[15] Bohr expressed his gratitude to Chadwick, and indicated that he would not leave Denmark at that time because he had a duty to stay to resist the threat to the freedom of Danish institutions and to protect the exiled scientists in Copenhagen. Seeing the real purpose of Chadwick's message, he replied in a way which showed that he had still not come to accept the imminent, military consequences of nuclear fission, while he was mindful of the problems that lay ahead for international co-operation:

> neither such duties nor even the dangers of retaliation to my collaborators and relatives might carry sufficient weight to detain me here if I felt that I could be of real help in other ways but I do not think this is probable. Above all I have to the best of my judgement convinced myself that in spite of all future prospects any immediate use of the latest marvellous discoveries of atomic physics is impracticable. However, there may, and perhaps in a near future, come a moment where things look different and where I, if not in other ways, might be able modestly to assist in the restoration of international collaboration in human progress. At that moment, whether it will come before or after the cessation of hostilities, I shall make an effort to join my friends and I shall be most thankful for any support they might be able to give me for this purpose.

Bohr handed his reply to Gyth the next day, and he arranged to have the letter reduced to 2×3 mm size. It was wrapped in foil and handed to a courier, who took it to a dentist. The foil package was inserted into a hollow tooth and covered with a filling.[14] Two months later, Bohr heard rumours about large scale production of metallic uranium and heavy water for use in atom bombs; these prompted him to write to Chadwick again to state that he thought the prospect of slow neutron reactions to make an explosion was not promising. He still did not countenance the separation of the uranium isotopes in sufficient amounts to constitute a critical mass—an impossibility that he took for granted.[16]

In May 1943, Churchill travelled to Washington to meet Roosevelt again. The supply of uranium was to become the next point of contention between the Allies. Almost as soon as he arrived, he received a cable from Sir John Anderson telling him that the Americans had virtually cornered the world market in uranium for the next few years. Almost worse was that they had been able to do this by placing orders with the Canadians that would guarantee a monopoly supply from the Eldorado Mine in the Artic Circle. Churchill promptly informed the Canadian Prime Minister that they 'had sold the British Empire down the river'.[6] The Trident Conference lasted almost two weeks and was for the most part indecisive; on the last day, Churchill and Roosevelt held a private meeting about the atom bomb and Anglo-American research. Following this, the Prime Minister sent a reassuring cable[17] to Anderson in London.

> The President agreed that the exchange of information on Tube Alloys should be resumed, and that the enterprise should be considered a joint one, to which both countries would contribute their best endeavours. I understood that his ruling would be based upon the fact that this weapon may well be developed in time for the present war, and it thus falls within the general agreement covering the interchange of research and invention secrets.

Once again this was to prove to be a false dawn. On his own return to London at the beginning of June, Churchill attempted to formalize his agreement with Roosevelt by sending a confirmatory telegram to Harry Hopkins, the 'most faithful and perfect channel of communication'[18] to the President. Hopkins' reply was sufficiently non-committal to worry the increasing number of sceptics in Tube Alloys, and their worst fears were gradually confirmed by the ensuing silence from across the Atlantic. Although no one in London realized it, the occasion for the latest interruption was a second meeting held in Washington on 25 May between Stimson, Bush and Lord Cherwell.[2] The 'Prof.' had begun by attacking American behaviour in the preceding months, and was rebutted by Bush's calm defence of the Conant memorandum. Floundering after his initial thrust had been turned aside so easily, Cherwell stated that the real reason the British wanted full access to the information now was so that they would be able to pursue their own atomic weapons programme after the war. This echoed his opinions to Churchill at the time of the Maud Report about Britain being a dominant world power, and can hardly have been calculated to assuage the Americans. Bush recounted Cherwell's remarks to Roosevelt at the end of June, who found them astounding and observed that Cherwell was 'a rather queer-minded chap'.[2] The President was on the horns of a dilemma—he had given Churchill unequivocal assurances in their private meeting, but now he was worried about the implications of Cherwell's statement. Unable to reconcile the two, he made no decision, and Bush left the White House with the impression that nothing had changed since the policy on limited interchange agreed six months earlier.

Although the Tube Alloys staff had no knowledge of Cherwell's gaffe in Washington, he was nevertheless giving rise to frustration within the organization. Perrin reported to Akers,[19] who was in Canada so that he could travel to the US at a moment's notice, on 12 July:

> To . . . my amazement, Cherwell said that he was under the impression (which he believed was shared by the Prime Minister) that we were actually proceeding at full speed with research and design work on the assumption that a plant was to be put up here by us independently.

That week Bush and Stimson arrived in London and held several meetings with Churchill, where he left them in no doubt about his anger at the continued embargo on Tube Alloys despite several promises from the President. When the question of post-war ambitions was brought up, Churchill replied tersely that 'he did not give a damn about any postwar matter'.[2] Bush had already explained to a representative of Tube Alloys that the President could make no postwar commitments without the approval of Congress.[20] Bush also said that he had 'no doubt that things were now in a complete mess', and blamed the British for going behind his back to the President. The senior British staff by this time had little residual trust or goodwill for Bush and his colleagues: Akers, for example, complained that it was 'a great pity Groves, Bush and Conant gave me three different reasons for the Conant memorandum, and not one of them the real one'.[21]

At a 10 Downing Street meeting with Bush and Stimson on 22 July, Churchill, flanked by Anderson and Cherwell, attempted to repair the diplomatic damage; he offered the concession that 'the President might limit the commercial or industrial uses of Great Britain in such a manner as he considered fair and equitable in view of the large additional expense incurred by the United States'.[2] Bush and Stimson seemed satisfied with this and the other terms set out by Churchill, but Anderson was still concerned that a final agreement might elude them. He wrote to Churchill on 26 July:[22]

> I think we have made some impression on Stimson and Bush and that, if we let Bush down lightly about his bad memory*, we can be reasonably confident that they will go back with the intention of trying to find ways of restoring collaboration. We shall thus have at least one friend in the political and one in the scientific camp.
>
> But we have not touched the Generals. Is there not a danger that General Groves, at any rate, will simply tell Stimson and Bush that, like all Americans who come to our misty island, they have been taken in by our hypocritical cunning and carried away by our brilliant Prime Minister? And that, as the memories of our talk grow dim, he will successfully smother their consciences.

Anderson continued that the 'shipwreck last January' might have been avoided 'if those who knew the reefs and shoals of American politics had been watching the American situation for us'. He suggested that the Government's senior permanent representatives in Washington—the Minister, Colonel J.J. Llewellin and the Head of the British Joint Staff Mission, Field-Marshal Sir John Dill—should be brought into the picture so that they could liaise with the President and his advisers. Anderson added that they would need technical advice and this could be provided by appointing a scientist from the Tube Alloys Directorate to work in Washington.

A few days later, Anderson himself arrived in Washington to negotiate a formal agreement with Roosevelt's advisers, which was concluded by the end of the first week of August. On 19 August 1943, the opening day of their conference in Quebec, Churchill and Roosevelt signed 'Articles of Agreement Governing Collaboration Between the Authorities of the U.S.A. and the U.K. in the Matter of Tube Alloys'—the Quebec Agreement.[6] So eager were the British to make up for lost time that four of their top scientists, Oliphant, Peierls, Simon and Chadwick arrived in Washington in the days before the agreement was finally ratified. They had flown by the Pan American flying boat or Clipper service from Foynes on the west coast of Ireland—incomparably more luxurious than the alternative of riding in a bomber.[23] Akers tried to set up a preliminary meeting for them with Bush, but was told they would have to wait until the ink was dry in Quebec.[24] Bush was, by now, anxious to expedite the collaboration, but determined that it should follow the protocol of the agreement; he was not prepared to pre-empt the Combined Policy Committee (CPC) called for by the Quebec Agreement, but not

---

* When Churchill produced the Conant Memorandum at the Downing Street meeting, Bush appeared shocked and said he had never seen it. The original memorandum, however, had been handed to Akers by Conant and Bush together![6]

yet formed. In a memorandum to the President on 23 August,[25] Bush echoed Anderson's suggestion that 'it would help if a top British scientist, accepted and of sound judgement, could be sent here as chief liaison under Sir John Anderson, to help make arrangements for the committee's work'. He did not want one of the scientists actively working on a single phase of the problem, and explained his residual worries to the President:

> In previous negotiations difficulty was encoutered because the British representative was an industrialist, Mr Akers of International [sic] Chemical Industries. This same man is now here, apparently to make similar arrangements. He recently, and without consulting us, brought four eminent British scientific workers here for interchange. As we cannot use them until the combined committee has laid down the rules, they are likely to think us reluctant to interchange, whereas the exact opposite is true and we are anxious to get appropriate interchange going in an orderly fashion, so that relations will not this time become tangled. Akers is a very able man, but not the one to handle this matter.

The objections to Akers angered Anderson,[26] who had not the 'slightest doubt' about his integrity and ability. Putting this aside, as an experienced diplomat he recognized Akers' continued presence in Washington would further impede progress on the joint project, and recommended to Churchill that Sir Edward Appleton should at once go out to Washington to help to arrange the launching of the CPC.

Bush's fear that the British would be suspicious of any further delay was well founded as far as their scientific contingent was concerned, and Chadwick for one let Akers know that he was 'not too pleased about the speed of progress' so that Akers felt 'a bit guilty at having got him out here so soon'.[27] Chadwick was accompanied by Mark Oliphant, who was by then a seasoned visitor to the US. Like all travellers from wartime England, Chadwick was dazzled by the abundance of items such as eggs and fresh fruit, which were luxuries of distant memory at home. They stayed in New York for a few days, where Chadwick found the heat oppressive. On his insistence late one evening they visited the famous oyster bar at Grand Central Station. In the early hours, Oliphant[28] was awoken by terrible noises from the bathroom of their hotel suite, where he found a distinctly green-faced Chadwick writhing on the floor. Chadwick was able to tell him that he was carrying some medicinal powder in his suitcase that should be mixed with water. Oliphant, deciding that the patient was indeed in a serious condition, liberally exceeded the recommended dose and administered the potion. The next morning they had a meeting at the New York offices of the Manhattan Project with General Groves amongst others. Chadwick strode from the hotel at the appointed hour and was driven to Manhattan, where, according to Oliphant, he put up a forceful performance before returning to their hotel and collapsing into bed. Arriving in Washington a day or so later, where the temperature was in the nineties and it was very humid, Chadwick still looked 'like death'.[29]

On reaching Washington, they found Akers in despair at the lack of any signal from the Americans that they were going to resume co-operation. Chadwick decided to test the water himself, and arranged a dinner with Conant that evening. Initially, Conant told him and Oliphant that he was not at liberty to discuss Tube Alloys, but as the meal went on he opened up and gave his views off the record. He assured his guests that he was entirely in favour of a full restoration of interchange of information, but cautioned the British against forcing the pace. His reason was that 'there was a general feeling prevalent in America that the British always got the best of any deal.'[30] He also mentioned the reservations about Akers as the agent of liaison in view of his connection with ICI. The two visitors came away with the impression that Conant was a reasonable man, who was intent on establishing close collaboration. Chadwick formed the view that Conant was alarmed by the rate of expenditure on the American programmes, and would welcome independent confirmation of their worth or, alternatively, constructive criticism.

Within days of the dinner with Conant, the Americans formally objected to Akers,[30] who stood down with perfect willingness; Llewellin, the Minister in Washington who was now actively engaged in the project, decided that his place should be taken by Chadwick. At last friendly relations were restored, but there was still a snare on the American side which had been affixed by Groves in his quest for impenetrable security. Conant explained to Chadwick[30] that a meeting between all the section heads of the American, British and Canadian sides was impossible because the American heads were not allowed to know anything outside their own sections. There was a hastily arranged meeting of the CPC on the 8 September at the Pentagon, and they established a technical sub-committee with tripartite membership under the chairmanship of General Styer: he was an engineer, the chief of the US Army's supplies service, and had been involved in nuclear matters since 1942.[2] The sub-committee was authorized to act independently on interchange providing there was unanimous agreement between its four members. Chadwick, the British representative, made use of this arrangement at the very first meeting, when he persuaded his colleagues to agree to extensions in the provisions recently set out by General Groves. It was also agreed to set up working groups for fast neutron physics, the diffusion project and plutonium production.

The fast neutron group was scheduled to meet at the Pentagon on the following Monday, 13 September, and represented the first encounter between the leading British and American scientists in the presence of Groves. Groves opened the discussions with an account of his security measures.[31] He explained the system of compartmentalization which had been adopted to prevent any individual, except Tolman, his scientific adviser, and himself, from gaining an overall perspective of the project. Indeed, the members of any scientific or developmental subsection were isolated as much as possible from the construction and operation of the large-scale plant. He made it clear that he alone was responsible for recruitment and expenditure—at which point both Chadwick and Oliphant made a mental note about Groves as dictator. The British learned for the first time about the site under construction at Los Alamos in New Mexico, where there were already 300 scientists (of whom 200 were at least Ph.D.s) working under Oppenheimer. The site, known by the code name 'Y', was under army administration, and although it had proved impossible to cut it off completely from the outside world, communication was limited. The British were amused to learn its alternative name was 'Shangri-La'. Groves repeatedly stressed the prime object was the production of a military weapon in the shortest possible time and 'all considerations of efficiency or of scientific elegance were subordinate to this necessity'. He concluded by suggesting that Chadwick and Oliphant should come to Los Alamos at once since, despite the large number of scientists already engaged at 'Y', there was a dearth of practical experimental physicists.

In his reply to Groves, Chadwick was quietly resolute and began by saying that in Britain security was achieved by different methods, pointing out that the American system just would not work. He urged the necessity for close contact between scientists and engineers, and warned of the grave dangers which attended any separation. This must have caused an inward gasp amongst the American scientific contingent present—Conant, Oppenheimer, Bacher and Tolman—who seemed completely dominated by Groves. Chadwick then moved onto scientific matters, listing a series of experiments that he regarded as crucial to establish the practicability of the bomb. This led to a lively debate with Oppenheimer who believed there was an uncertainty of a factor of about 2 in the theoretical calculations of the critical size, and that direct measurements would not reduce the uncertainty below 30%. Chadwick disagreed and thought 'more precise measurements were both possible and necessary'. Chadwick was also concerned about the necessity of demonstrating that the time interval between absorption of a neutron, fission and emission of neutrons did not exceed about 10^{-8} seconds. In order to produce an explosion, it had been calculated that the fission reaction had to be completed in less than one hundred-millionth of a second—any longer and the energy released would merely blow the mass of fissionable material apart, short-circuiting the full energy release. Oppenheimer agreed it was a crucial point, but felt certain that since the whole process was predicted to take only 10^{-17} seconds on theoretical grounds, they need not worry unduly. Conceding the theoretical expectation, Chadwick nevertheless thought the question was so vital that direct evidence should be obtained, and believed modulation experiments down to 10^{-8} seconds were possible.

During this discussion, Oppenheimer repeated the invitation to Chadwick to come to 'Y', and suggested that his presence would facilitate these difficult experiments. Oppenheimer remembered the sombre Chadwick from a brief and unhappy spell in the Cavendish in 1925, where as a postgraduate his aptitude for experimental research was not strong enough for him to be recruited by Rutherford. Since then he had founded the first internationally recognized school of theoretical physics in the United States; Groves had paid him a visit at Berkeley in October 1942, and Oppenheimer had suggested the need for a central laboratory devoted to bomb work.[32] His family owned a ranch just north of Sante Fe, and he thought the remote, mountainous region would be the perfect site for the new laboratory. He and Groves toured New Mexico together, and settled on Los Alamos, situated at 7,000 feet above sea level on a flat topped, volcanic, rock or mesa.

Once Groves decided that this was the site, a secure Army camp materialized at Los Alamos 'with unbelievable dispatch'. Groves liked Oppenheimer from their first meeting and it was his decision to place him in charge despite questions about his political suitability. While recognizing Oppenheimer's intellectual brilliance, Groves could not follow the mathematical intricacies of his work, and, no doubt, felt uncomfortable at times with the obscurity of his conversation. Now he was seeing Oppenheimer challenged by a straightforward scientist, who presented his arguments with limpid simplicity and retained a reassuring scepticism of theory in the absence of experimental data. As a Nobel Laureate, Chadwick obviously enjoyed universal respect and was regarded as one of the foremost authorities on nuclear physics by everyone Groves had asked. This was a shrewd, poker-faced, operator, who seemed beholden to no one. Who better to keep an eye on those 'crackpots'[33] Oppenheimer was gathering around him at Los Alamos?

When the meeting broke up, Groves took all the participants to lunch at the Mayflower restaurant, 'as he felt the Cosmos Club to be peopled with gossips'[31]—it was the Cosmos Club where Conant had entertained Chadwick and Oliphant to dinner ten days earlier! Groves told Chadwick[34] that the project would stop automatically with the end of the war because funding would cease subject to a lengthy Congressional review. Chadwick, who from his initial scepticism in 1939 had become increasingly confident on the basis 'the more we know the more favourable the relevant factors seem to be',[13] replied it should be possible to start on a big scale at once without Congressional enquiry. This positive response, he noted, 'appeared to impress Groves considerably'.[34]

Chadwick's first impression of Groves was that he was 'the dominant personality in the U.S. group. Although nominally controlled by a committee he appears in effect to be a dictator . . . Groves' idea of collaboration seems to be to incorporate into the U.S. project such sections of the British team as seem likely to promote a speedier and more certain realization of the project'.[34] Chadwick was to have plenty of opportunity to revise his initial opinions of Groves, since the two men met many times over the next ten days in New York, Montreal and then back in Washington. On 14 September, Chadwick held discussions,[34] completely off the record, with Oppenheimer, when he put the British position in general terms and asked Oppenheimer to explain it to Groves, saying that 'we appreciated the U.S. difficulties'. Groves' major worry was the spiralling cost as the Manhattan Project became ever larger and more complex. Initial funding had been approved directly by Roosevelt without Congress' knowledge and was channelled through the Army Corps of Engineers' budget. The sums were becoming so enormous—US$300 million in May 1943, and a projected US$400 million for 1944—that they could not be concealed under general expenditure for much longer.[32] When at the meeting in Groves' office on 16 September, Lawrence, with the support of Chadwick and Oliphant, strongly urged that 3 to 10 bombs should be made and tested before the design could be regarded as established, Groves, mindful of potential opponents in other branches of the services and in Congress, replied firmly that it was the duty of the scientists 'to make certain the first bomb did go off'.[35] According to Chadwick,[34] the meetings were held in an 'excellent spirit', and 'Groves was most friendly and cordial, and he seemed no longer to have any reservations'.

By the time Chadwick returned to Liverpool at the end of September, the arrangements for collaboration had been largely agreed, although they were not finally ratified by the two sides until mid-December. Anderson[36] informed Churchill in mid-October that Chadwick would continue 'to act as immediate Scientific Adviser to the British members of the Combined Policy Committee'. After explaining that his first choice for the position, Akers, had been disqualified 'on grounds which are intelligible in terms of American politics', Anderson suggested that the new arrangement should be accepted, even though Chadwick could not be 'expected to understand all the minutiae of the industrial side of the project', if it meant getting British 'experts in each branch of the project admitted to the United States'. It was clear to Anderson that, in this way, the British would acquire much more information 'than would be possible if the present effort was continued in this country'. Chadwick's diplomatic successes with Conant and Groves were misunderstood by a few in Britain, who suggested that 'he had sold himself to the Americans'.[6] He had, in fact, abandoned any previous chauvinism in the face of abundant evidence of the Americans' overwhelming superiority in resources and technical capacity. He now realized that if Britain wished to preserve any nuclear role in the future, it was better to become involved on almost any terms than to be excluded from the project. He shared Groves' single-minded ambition to produce the weapon to hasten the end of the war, and this together with his palpable lack of personal ambition were strong reasons for Groves to trust him. Apart from his influential position on the technical sub-committee, Chadwick had been appointed as head of the British contingent of scientists and was the only man apart from Groves and Tolman to have access to all the American research and production facilities. In addition it was expected that he would take part in the supervision of experimental work at Los Alamos and bring with him a small team of men that would include Frisch, Rotblat, Bretscher, Titterton and Niels Bohr.

Chadwick had mentioned to Groves during his visit to the US that Bohr might soon be persuaded to leave Denmark. He had suggested bringing Bohr to Los Alamos, and Groves seemed delighted. This impression was confirmed by Oliphant,[37] who had remained in the United States. According to him, Groves was very anxious that Bohr should come at once with Chadwick. The General also wanted Chadwick to come with the authority to deal with him directly. In the same cable, Oliphant gave notice of a threat to the entente so freshly established:

> Blackett is here preaching against the employment of scientists on Tube Alloys and owing to his unique position, he may have some disturbing effect. Have had one talk with him and will see him again.

Patrick Blackett,[38] whose dissenting view on the Maud Committee about the amount of time and money needed to produce a uranium bomb could already be seen to be correct, was steadfastly against the bombing of civilian populations on moral grounds. He did not view the atomic bomb as a revolutionary weapon and disagreed with many aspects of Anglo-American policy.

Bohr had finally left Copenhagen at the end of September 1943, smuggled out by the Danish underground to Sweden.[14] Once the British heard he was there, they sent an unarmed Mosquito, a fast fighter-bomber, to fetch him from Stockholm, and he flew across the North Sea in the unpressurised bomb bay. He was equipped with a parachute, inflatable lifevest and flares so that if the plane were attacked, the bomb doors would be opened, and Bohr released into the North Sea. He failed to follow the instructions about wearing an oxygen mask and was unconscious for some of the trip. When he received no response from his passenger on the intercom, the pilot realized what had happened and completed the journey at low altitude; fortunately, Bohr seemed none the worse when he landed in Scotland. The next day, he flew down to London's Croydon airport where he was met by Chadwick, Aileen and a secret service officer. Bohr was installed in the Savoy Hotel, and Chadwick immediately began to tell him about the Frisch–Peierls Memorandum, the Maud Report, Tube Alloys, the Manhattan Project and General Groves. Bohr was astounded. That evening he had dinner with Sir John Anderson and was asked to join the British Tube Alloys team that would go to the USA. The next day, Aileen took Bohr on a shopping expedition around London to buy clothes and other essentials. By chance, Groves' scientific adviser, Tolman, was in England during October to agree specific working procedures for Anglo-American co-operation with Chadwick and Anderson. Tolman also met Bohr and conveyed an invitation from Groves for him to join the Americans at Los Alamos. Bohr found himself subject to divided loyalties, and although he was paid by the British, it was recognized that his would be a joint appointment. Indeed, Anderson saw some advantage in Bohr's independent status and his unique standing amongst all the scientists on the Manhattan Project, and it was agreed that Bohr should relay his impressions of progress and the level of co-operation to Anderson in London via Lord Halifax, the British Ambassador in Washington. This arrangement would lead to American suspicions about Bohr's neutrality later in the war.

After a week in England, Bohr was joined by his son Aage, who would act as his companion and personal assistant over the next two years. They both visited Liverpool and stayed with the Chadwicks. While in Liverpool, Chadwick had long technical discussions with Bohr over several days about the bomb. They also exchanged news about old friends and colleagues both in America and Europe. Both men were intensely concerned about the activities of the physicists who remained in Germany. Chadwick knew from 'C's' secret service reports that Heisenberg had visited Copenhagen in 1941. He thought that Heisenberg might have used the visit to throw Bohr off the scent, and the reason that Bohr thought there were no military applications of nuclear energy was perhaps because he had been 'sold this idea by Heisenberg'.[39] The two men kept no record of their conversation, but it seems probable that Bohr told Chadwick that as early as September 1941, Heisenberg saw an open road to the atomic bomb and was himself involved with the project. In the spring of 1943, Chadwick[40] had received word from

Cambridge that Heisenberg was now entirely pro-Nazi and working on the German Tube Alloy project. Although this was little more than rumour,* Chadwick had warned the Americans during his recent trip to Washington that Heisenberg, in his opinion, was 'the most dangerous possible German in the field because of his brain power'.[39] After his conversation with Bohr, there seemed no reason to change his assessment.

In what he knew would be a short hiatus before returning to the USA, Chadwick had to plan for the future and divest himself of some old responsibilities. Oliphant was now pushing on an open door, when he wrote to Chadwick at the beginning of October 1943 'the American program is such that it will be impossible for us to be useful there unless we get down to work at once'.[41] Chadwick wanted to involve as many good British scientists in the project as he could muster, subject to their acceptance by Groves. One of the conditions already agreed was that only British citizens would be allowed to come to Los Alamos to work. This presented immediate problems for Chadwick since so much of the work in England had been done by European immigrants. One of his first choices to come was Joe Rotblat,[42] who quite suddenly received a visit from a policeman at his rooms in Abercrombie Square. The policeman started to ask him a lot of personal questions, and Rotblat, assuming that he was in breach of some restriction concerning 'friendly aliens', asked what the purpose of the questions was. The policeman replied that it was in reference to Rotblat's application for British citizenship. When Rotblat informed him somewhat indignantly he had made no such application, the policeman scratched his head and said, 'Well, I don't know, I am only doing my duty.'

Rotbalt went to Chadwick for an explanation, and told him that he was not prepared to give up his Polish citizenship. It had always been his intention to return to Warsaw and now virtually all the physicists in Poland had been killed by the Germans, he felt it was his duty to go back after the war. Chadwick was rather upset about his refusal, and said 'You realise you won't be able to come with us to the States.' Rotblat replied that this was more important than his going to America, and it was left at that. With Frisch, the other Liverpool alien, Chadwick took a direct approach and asked him if he would like to work in America. To his relief, Frisch was instantly enthusiastic and when Chadwick nervously mentioned this would mean becoming a British citizen, Frisch replied 'I would like that even more'. As Frisch[12] recalled:

> After that, things began to happen with bewildering rapidity. Within a few days a policeman appeared and started to take down personal data as well as the names of people who knew me and could vouch for me, explaining that he had been instructed to start naturalization proceedings. He added in an oddly confiding manner, 'You must be a pretty big shot. I have been told to get everything done in a week!'

---

* Chadwick's informant was Kowarski, who, in turn, was reporting on a conversation between Dirac and Victor Goldschmidt, a Norwegian chemistry professor, flown to England in March 1943 by the SIS. For details of Goldschmidt's relationship with the spy Paul Rosbaud, see Kramish, A. (1986). *The griffin*. Houghton Miflin, Boston.

In Birmingham, Peierls' reserved German assistant, Klaus Fuchs, went through a similar security clearance, some months later, in order to obtain a place on the team. He had already been naturalized as a British subject in the summer of 1942. Although there were some unsubstantiated reports of his communist affiliations, these were not taken seriously.[43] To General Groves[44] after the war, it seemed that the British never made any investigation of Fuchs at all.

Chadwick left for the States again at the end of November 1943, having put Rotblat in charge of the Tube Alloys work in Liverpool. He spent the next few weeks criss-crossing the US acquiring a virtually complete picture of the Manhattan Project. The headquarters of the project had now moved from Manhattan to Oak Ridge in rural Tennessee.[32] This was a newly constructed town on a site measuring 16 miles by 7 miles that had been acquired by the army engineers for approximately US$3 500 000 in September 1942. Apart from dwellings for 13 000 people, by the time of Chadwick's visit there were gigantic buildings over a mile long to house Lawrence's electromagnetic separation plants. Chadwick also saw work on the gaseous diffusion plant on a 5000 acre site in one corner of the reservation, and heard about the civil engineering challenges posed by such an enormous yet delicate operation. These two programmes at Oak Ridge were intended to provide sufficient uranium-235 for a bomb, and neither was running smoothly at the time of his visit. The scale of the activities was staggering, and Chadwick now realized that it could never have been undertaken in wartime Britain. He also realized that a place like Oak Ridge would be impossible to camouflage from aerial reconnaissance, and that unless the RAF discovered huge new industrial sites being erected in Germany, the threat of a Nazi atomic bomb was receding.

The US Army and the Du Pont Corporation were also proceeding with plutonium production for a second type of weapon, and the plant for this was the Hanford Engineering Works, an isolated site, 670 square miles in area, in southern central Washington State.[32] There was no major centre of population within a hundred miles, important for safety and security reasons, yet it was well positioned for electrical power and fast flowing water. Even Chadwick was not allowed to visit Hanford. Whereas the separation of isotopes remained the major technical hurdle for the uranium bomb, it seemed that the production of plutonium, while not without its dangers and difficulties, would be less elaborate. A year earlier, Chadwick had raised concerns about the necessary purity of plutonium and the attendant danger of premature detonation. On his first visit to Los Alamos, he was intrigued to learn that Oppenheimer's team were working a novel solution to this problem. Previously it was assumed that the critical mass of fissionable material in a bomb would be assembled by firing two sub-critical masses together by a gun mechanism: this was the proposal put forward in the Maud Report and was still the intended mechanism in the case of ^{235}U. With plutonium, the maximum velocity of approach by this method was probably too slow to prevent predetonation. The new method under investigation was known as *implosion:* a sub-critical mass of plutonium was to be surrounded by explosives inside the bomb case in such a way that when they were triggered, a symmetrical shock wave of immense force would result, instantly squeezing the plutonium into a mass the size of a walnut (which would become critical because of its extremely high density). It had been calculated that implosion offered two related advantages: a lower level of purity of plutonium and a lesser amount of the metal. The theory and practicality of implosion were very exacting and previously untried, but it was such a clever approach that Chadwick felt impelled to discuss it with Oliphant at the first opportunity. Oliphant was about to return to England and while there he briefed Lord Cherwell. The Americans were subsequently dismayed to find that Cherwell knew all about this top secret plan, and its casual mode of transmission worried the security service at Los Alamos.[45]

The dispositions of several senior British scientists were agreed by the CPC in the months after the Quebec Agreement was signed. Simon was involved with the gaseous diffusion project as was Peierls initially; he then moved to Los Alamos and together with Fuchs made important contributions to the implosion problem. Oliphant, with a small research team, went to Berkeley to join forces with Lawrence on electromagnetic separation of uranium isotopes. Chadwick's own first visit to New Mexico brought a memorable example of the deadpan wit, which seems to have been an inseparable accompaniment to his gloomy manner. He was met by a small reception committee from Los Alamos, when his plane landed at Albuquerque. They had arranged for a party of Indians dancing in warpaint as the distinguished visitor descended from his plane. To the disappointment of his hosts, Chadwick ignored the display and made no comment at all. Finally one of them said to him: 'Look, Indians!' to which Chadwick replied: 'Well, this is America isn't it?'[46]

Chadwick believed that the first duty of the British was to expedite the delivery of the atomic bomb in accordance with Groves' plan, but he also wished to bring out as many younger scientists as possible so that they would gain experience that could be drawn on in postwar Britain. He approached General Groves about Rotblat, and convinced him that he was a Polish patriot whose integrity he would personally guarantee. Groves accepted Chadwick's word and gave permission for Rotblat to come to Los Alamos without changing his nationality. Rotblat[47] now accepted the invitation, and it was arranged that Kinsey would return to Liverpool from Cambridge to supervise the continuing Tube Alloys research. Chadwick communicated his intentions to Anderson, by now the Chancellor of the Exchequer, who replied by cable[48] thanking Chadwick for his forcible account of the position and promising to arrange the transfer of more TA experts to the US as soon as possible.

There was of course a limit to the number of British scientists that Groves would accept. He was determined that the bomb would be produced by American workers in time to end the war, and as a deeply proud and loyal soldier, he had a strong reluctance to share this weapon or the ancillary technology with any other country, no matter how close an ally. Chadwick recognized this and decided it would be futile to try to pressure Groves to go further than he wished. It also meant that Chadwick had to select his team with great care. This led to some disappointments amongst those young scientists eager to transfer their Tube Alloy research to the USA. One such was Denys Wilkinson,[46] who was working at the Cavendish. Chadwick had told him that he would be moving to Los Alamos, but changed his mind when he found that Wilkinson had sustained a pneumothorax, after a recent climbing mishap. He told Wilkinson the bad news as though reading from a prepared telegram: 'Los Alamos . . . 7000 feet . . . lungs . . . no good'. Despite entreaties from Wilkinson and his physicians, Chadwick would not relent because he could not risk one of the hand picked team being incapacitated.

The other way to broaden the British nuclear experience was by promoting the Montreal heavy water facility. This project was in jeopardy, amongst other reasons, because of shortage of raw material. Ironically, this was in part because the Americans had acquired the total output of the Canadian uranium mines and refinery; but they had been less successful in obtaining fresh supplies of the much richer ore from the Belgian Congo. Chadwick had received private intelligence that the Belgians would not sell to either country exclusively, but might be prepared to make a joint allocation.[6] He convinced the Minister, Llewellin, that this would be the best way to proceed, and together they approached Groves with the plan. The initial American reaction was cool—they had become used to complete dominance—but they soon realized that their ally did wield influence, since the Belgian Government was in exile in London, and 30% of the equity in Union Minière was in English hands. The Combined Policy Committee charged the sub-committee on which Chadwick was the British representative to devise a plan to secure international control over the supply of uranium ore.[32]

In an ancillary move, Chadwick tried to enlist Groves' support for the Montreal project. This would provide invaluable opportunities for British nuclear scientists and engineers, and, as it was on the same continent, would also probably lead to some worthwhile contact with the research group in Chicago. The two men had a preliminary meeting on the subject on 10 December, when Chadwick[49] put the broad outline to Groves, who was non-committal but not openly hostile. Chadwick decided that the Americans would be more likely to co-operate if the Montreal laboratory were to be run by a well-known British scientist whom they could trust, and suggested John Cockcroft for this role. He kept plugging away at Groves, and arranged that the Canadian team led by Halban should come to Chicago in January 1944 for talks with the workers in the Metallurgical Laboratory there.

The Chicago meeting was chaired jointly by Chadwick and Groves.[50] Two dozen scientists attended, including Compton and Fermi from Chicago, and a ten-strong contingent from Montreal led by Halban and the other Frenchmen, Auger and Goldschmidt. Groves came in late, and was immediately approached by Chadwick, with whom he entered 'a more and more animated dialogue away from the table'. Chadwick then turned to the assembled company and announced bashfully 'that it was never good for the parents to argue in front of the children', upon which he and the General withdrew for a private talk. On their return, half an hour later, they revealed the subject of their dispute, and it was immediately obvious that

Chadwick had been forced to concede. There would, stated Groves, be no collaboration on plutonium chemistry nor on the separation of plutonium from other fission products between the Americans and the Montreal laboratory, since this did not, in his view, fall within the terms of the Quebec Agreement.

Following this meeting, Groves consulted Conant about the Canadian project and was informed that this was not an issue to be decided by him and Chadwick.[49] In an attempt to maintain momentum, Chadwick wrote to Sir Ronald Campbell who had taken Colonel Llewellin's place as the British Minister in Washington and asked him to bring the proposals for the Canadian project before the CPC at its next meeting.

By this time Chadwick was becoming pessimistic about Montreal, which seemed to him 'somewhat disorganised and unhappy';[51] Halban was ill with heart disease and requested sick leave, commenting to Chadwick that 'morale is running low and temperaments are running high'.[52] Chadwick informed London that Cockcroft's presence in Montreal was now essential if it was to succeed. Despite his concerns regarding Canada, he was able to report that 'my relations with Groves continue to be most cordial . . . He is very ready to discuss matters of general policy, to listen to my suggestions, and when convinced, he acts with surprising speed and decision'.[51] To London, this was the most reassuring news of all.

At the next meeting of the CPC on 17 February 1944, a draft agreement was proposed stating that the three governments would collaborate in an attempt to secure control of all known sources of uranium ore around the world.[32] The draft was tentatively approved and it was agreed that a tripartite agency should be created in Washington with its first objective to procure the Congo ore. The CPC moved on to other business, but there was no reference to Montreal, despite Chadwick's letter to Sir Ronald Campbell asking that it should be put on the agenda. Chadwick, therefore, raised the matter himself and pressed for the construction of a large-scale, heavy-water pile in Canada that could be used for plutonium production.[2] Chadwick envisaged that the funds would be provided by Great Britain and Canada, while the United States would supply the heavy water, and all three countries would exercise joint control of the programme. Chadwick mentioned that he and Groves had discussed the subject, and had referred it to Conant. Groves was not at the meeting, but Conant was, and he, at first, 'denied it, and then said "unofficially" he had been consulted.'[50] According to Chadwick, a 'rather warm discussion' ensued with Conant and Bush, 'who seemed to want to throw cold water on the Heavy Water Pile and to be generally obstructive'. The Canadian representative, Mackenzie, said that he required an early decision, one way or the other, about US co-operation. The CPC adjourned without taking a decision and commissioned Groves, Chadwick and Mackenzie to prepare a report on the military advisability, the resources needed and the time to completion.

Mackenzie said that he would support any agreement reached by Chadwick and Groves,[6] and it was decided that Groves' staff would address the problem. It soon became apparent to Groves that the cost of a full-scale pile was beyond the means of the British—a fact which Chadwick readily admitted. The detailed draft report judged that Hanford would produce enough plutonium for the current war and therefore there was no military advantage in building a production facility in Canada. Although there might be some important advantages to a heavy water as opposed to a graphite-moderated pile, the recommendation was firmly against any new construction in Canada. When Groves presented the draft to Chadwick at the end of March, he expressed strong disagreement with the final conclusion, if not with the body of evidence presented.[6] Groves invited him to redraft the report, which he did, making the case for a modest-sized, pilot plant so that research into the heavy water reactor could continue. Groves acceded, and the concept was approved unanimously by the CPC on 13 April. Groves and Chadwick then spent another five weeks hammering out a set of detailed ground rules governing interchange: these were broadly drawn, but more interesting for what they excluded—details on Hanford, plutonium separation and chemistry. Chadwick again dissented, and as one final concession, Groves permitted a limited amount of irradiated uranium to go to Montreal so that the scientists there could work independently on plutonium separation and purification.[49]

To Chadwick's chagrin, his painstaking efforts to revive the moribund Canadian project, culminating in the CPC approval, which stretched the boundaries of the Quebec Agreement, were received grudgingly in London.[6] He knew that there was little in the scheme for the Americans, since it was essentially going to be a postwar undertaking, and that Groves had gone against his natural inclination to have nothing to do with a team of foreign scientists—especially when so many of them were French. Without Groves' support, which the General would soon regret, the Montreal programme would have been finished. The Tube Alloys staff in London were by now hopelessly mistrustful of negotiations with the Americans, and being unaware of the complex technical details, could not understand why Groves and Chadwick took so many weeks to reach a final, working, agreement. The Montreal scientists were in 'abysmally low spirits'[6] and it took them some time to regain a sense of purpose. In fact, as the scale of Chadwick's triumph sunk in, the Cavendish and the defence authorities agreed to release Cockcroft from his radar work; he was flown in a bomber to Montreal at the end of April to take over the directorship from Halban, and to restore vigour and morale.

References

1 Anderson, J. (11/1/43). Minute to W.S. Churchill. PREM3 139/8A, PRO.

2 Hewlett, R.G. and Anderson, O.E. (1990). An uneasy partnership. In *A history of the United States Atomic Energy Commission: the new world*, pp. 255–88. University of California Press.

3 Corell Barnes (1944). Minute on 1942 Anglo-Russian agreement. CAB 126/30, PRO.

4 Perrin, M. (27/1/43). Letter to J. Chadwick. CHAD IV, 11/52B, CAC.

5 Lamb, R. (1991). North Africa and Casablanca. In *Churchill as war leader*, pp. 207–26. Carroll and Graf, New York.

6 Gowing, M. (1964). *Britain and atomic energy, 1939–1945*. Macmillan, London.

7 Kinsey, B.B. (1943). Annual progress report on Tube Alloys at the Cavendish Laboratory. CHAD IV, 1/9, CAC.

8 Chadwick, J. (16/3/43). Letter to B.B. Kinsey. CHAD IV, 1/9, CAC.

9 Perrin, M. (11/8/43). Letter to W. Akers. AB 1/376, PRO.

10 Oliphant, M.L. (14/10/43). Letter to J. Chadwick. CHAD I, 19/3, CAC.

11 Chadwick, J. (26/10/43). Letter to J. Cockcroft. CHAD III, 3/5, CAC.

12 Frisch, O.R. (1979). Liverpool 1940–3. In *What little I remember*, pp. 133–47. Cambridge University Press.

13 Chadwick, J. (24/5/42). Letter to E.O. Lawrence. The Bancroft Library, University of California, Berkeley.

14 Pais, A. (1991). Bohr, pioneer of 'glasnost'. In *Niels Bohr's times*, pp. 473–508. Oxford University Press.

15 Bohr, N. (1943*a*). Letter to J. Chadwick. Niels Bohr Archive, Copenhagen.

16 Bohr, N. (1943*b*). Letter to J. Chadwick. Niels Bohr Archive, Copenhagen.

17 Churchill, W.S. (1951). *The second world war: the hinge of fate*, p. 723. Cassell, London.

18 Churchill, W.S. (1950). *The second world war: the grand alliance*, p. 24. Cassell, London.

19 Perrin, M. (12/7/43). Letter to W. Akers. AB 1/376, PRO.

20 Perrin, M. (19/7/43). Minute of conversation with V. Bush at the Royal Society. AB 1/376, PRO.

21 Akers, W. (29/7/43). Letter to M. Perrin. AB 1/376, PRO.

22 Anderson, J. (26/7/43). Memo. to W.S. Churchill. PREM3 139/8A, PRO.

23 Peierls, R.E. (1985). War. In *Bird of passage*, pp. 145–81. Princeton University Press.

24 Akers, W. (13/1/43). Memo Tube Alloys project. AB 1/129, PRO.

25 Bush, V. (23/8/43). Memorandum to F.D. Roosevelt. PREM3/139/8B, PRO.

26 Anderson, J. (28/8/43). Memo to W.S. Churchill. PREM3/139/8A, PRO.

27 Akers, W. (31/8/43). Letter to J. Munro. AB 1/376, PRO.

28 Oliphant, M.L. (1991). Speech at Chadwick Centenary, Gonville and Caius College, Cambridge.

29 Akers, W. (13/9/43). Letter to M. Perrin. AB 1/376, PRO.

30 Akers, W. (13/9/43). Memo. Tube Alloy project: negotiations with the Americans after the signing of the Quebec Agreement. AB 1/129, PRO.

31 Oliphant, M.L. (1943*a*). Notes on conversations with Americans in Washington, Monday, September 13, 1943. AB 1/376, PRO.

32 Jones V.C. (1985). *Manhattan: the army and the atomic bomb*. Center of military history United States Army, Washington DC.

33 Fermi, L. (1954). *Atoms in the family*, p. 231. University of Chicago Press.

34 Chadwick, J. (24/9/43). Continuation of Aker's memo on Tube Alloys project (30). AB 1/129, PRO.

35 Oliphant, M.L. (1943*b*). Notes on meeting with General Groves, 16/9/43. AB 1/376, PRO.

36 Anderson, J. (15/10/43). Memo to W.S. Churchill. PREM3/139/8A, PRO.

37 Oliphant, M.L. (21/11/43). Letter to J. Chadwick. CHAD IV, 3/16, CAC.

38 Lovell, B. (1975). Patrick Maynard Stuart Blackett, Baron Blackett, of Chelsea, 1897–1974. *Biographical memoirs of fellows of the Royal Society*, **21**, 1–115.

39 Notes on meeting of sub-committee September 10, 1943 (R.C.T.). RG 77, File 334, Box 60. NA2.
40 Kowarski, L. (12/4/43). Letter to J. Chadwick. CHAD IV, 1/9, CAC.
41 Oliphant, M.L. (2/10/43). Letter to J. Chadwick. CHAD I, 19/3, CAC.
42 Brown, A.P. (1994). Interview with Joseph Rotblat, London.
43 Gowing, M. (1974). *Independence and deterrence*, Vol. 2, p. 146. Macmillan, London.
44 Groves, L.R. (1962). *Now it can be told.* Harper and Row, New York.
45 de Silva, P. (22/3/44). Memo. re British mission to J. Lansdale. RG 77, file 000.71, NA2.
46 Wilkinson, D. (1995). Letter to the author.
47 Rotbalt, J. (22/12/43). Letter to J. Chadwick. CHAD IV, 1/2, CAC.
48 Anderson, J. (20/1/44). Cable to J. Chadwick. AB 1/376, PRO.
49 Chadwick, J. (1944). Proposals for collaboration on large scale heavy water programme. AB 1/197, PRO.
50 Goldschmidt, B. (1990). The calm returns. In *Atomic rivals*, pp. 191–201. Rutgers University Press.
51 Chadwick, J. (5/2/44). Letter to M. Perrin. AB 1/376, PRO.
52 Halban, H. (9/2/44). Letter to J. Chadwick. AB 1/129, PRO.

15 The scientist–diplomat

It is hard to imagine a greater contrast between the dismal, leaden, skies of Liverpool and the breathtaking, clear, air of New Mexico with its wide horizons, arid canyons, and brilliant colours. When James and Aileen Chadwick arrived to take up residence at Los Alamos early in 1944, the distant Sangre de Cristo Mountains had their peaks capped with snow. Where just over a year earlier, there had been a small collection of log and stone buildings of the Los Alamos Ranch School for Boys, the top of the mesa—locally known as the Hill—was now a military base housing 3500 people. Most of them lived in newly erected, prefabricated barracks or dormitories, and the provision of family accommodation was deliberately restricted to limit pressure on essential supplies and to minimize the need for extra community services such as schooling.[1] James and Aileen had made a Canadian detour before travelling to Los Alamos to see their daughters in Halifax. James had seen them once during his first transatlantic visit a few months earlier. On that occasion it seemed to the twins that their father was as nervous about seeing them as they felt about meeting him again. Aileen had not seen them since the summer of 1940, when they had left Liverpool as dainty, thirteen-year-old, English schoolgirls. Now they were nearly seventeen, spoke with Canadian accents, wore make-up and had become used to a degree of personal independence that their mother had not imagined. Reeling from the transformation, Aileen berated her husband, who, if he had noticed the changes, had failed to tell her about them.

General Groves had allocated the Chadwicks one of the most desirable residences on the Hill—a larger log cabin, with two bedrooms, at the rear of the main school building.[2] Many of the less prestigious accommodations lacked private amenities, but this cabin was part of 'bathtub row'. As soon as the Chadwicks arrived, James flew back to the East Coast in order to take charge of negotiations regarding the Montreal laboratory. During the first two months of 1944, Chadwick's itinerary was so full that he found himself spending a great deal of money on travel and hotels, which he was unable to recover because he had no time to submit expenses nor was there a practical way to reimburse him. His agreed salary of seventeen hundred pounds per annum,[3] although the highest of any British scientist on the Manhattan Project, was patently not going to be enough. Eventually he mentioned this to the British Central Scientific Office in Washington, and the senior civil servant wrote a pleading letter to London on his behalf. He pointed out that 'for us [Chadwick] is as much of a key figure as is Groves on the American side, and in principle, he ought to cover the whole territory of the project . . . He will certainly not be able to approach this task without wearing himself to death, unless he has a very real personal relief from financial worries.'[4] The Treasury accepted the weight of the argument and in March 1944 agreed that in view of Chadwick's special responsibilities, he would receive an extra five hundred pounds per annum, backdated to November.[3]

In the year or so that the Chadwicks were to spend at Los Alamos, James was often there for only a few days at a time, and the longest stay of a few weeks was for convalescence from a debilitating attack of shingles. Aileen was rather lonely, at least initially, and her inherent aloofness did not make it easy for her to join what was a lively and egalitarian community. Despite her American grandfather, she had not inherited the gene that codes for frontier spirit—she never saw the attraction in roughing it. To help keep the cabin clean, the Chadwicks had the services of a Pueblo Indian, called Appolonia, who would come to work wearing a native blanket and beautiful turquoise jewellery. On one occasion, Aileen invited a number of other wives to English high tea, and caused some offence amongst the Americans by attacking 'the primitive nature of life in the United States'.[2]

At the end of February, Joe Rotblat arrived from England and the Chadwicks invited him to stay with them, as they had a spare room. Following his own initiation by high tea, when he first arrived in Liverpool not speaking any English, and his occasional time with the Chadwicks at their holiday cottage in North Wales, Rotblat had become a close family friend. He provides an affectionate reminiscence of the domestic scene in the Chadwicks' log cabin:[5]

> Most of the time he was very happy just to read a book. If we were sitting around in the evening, I would chat with Aileen, he would just sit with his book near the fireplace and read it. This was most happy for him. From time to time, he would put in a word or two; in other words, he listened but did not actively take part in the conversation.
>
> On the whole we talked usually about a few things from the lab which had to be done. I tried to bring in the political element at that time because of my worries about the bomb throughout the years. He did not agree with me, you know, in this respect we were quite different. He respected my views, he never tried to put me down or criticise me, although from time to time we would argue. Of course we would always talk about the progress of the war, this was a natural topic of conversation.

On the rare occasions that Chadwick did have a few spare hours at Los Alamos, there was nothing he liked better than to go trout fishing in the fast running rivers. In order to do this, he naturally had to acquire a fishing licence. Kenneth Bainbridge, the Harvard experimental physicist, who knew Chadwick from a spell at the Cavendish and also as a result of a wartime visit to London (when he had attended one of the Maud Committee meetings), took him to Espanola, New Mexico, in order to obtain the necessary permit.[6] The status of members of the British Mission outside Los Alamos was nebulous—they were not supposed to exist. Some, most famously Niels Bohr, were given an alias;* others had invented their

* Bohr was given the name Nicholas Baker and his son, Jim. The disguise was not entirely convincing,

own disguises. Chadwick had not, and when the town official looked at him and asked what name to put on the fishing licence, Chadwick turned to Bainbridge, hesitated, and asked 'What is my name?'[7]

Bohr was a regular visitor to the Chadwick's log cabin, and struck up a warm friendship with Rotblat. Rotblat[5] had bought a short wave radio on which he and Bohr would often listen to the BBC World Service. General Groves would also visit frequently, when he was in Los Alamos. On one occasion he came to dinner, and in front of Rotblat made a remark to the effect that the real purpose in building the bomb was to subdue the Soviets. This shocked Rotblat—not because he had any illusions about Stalin's regime, which through its pact with Hitler had allowed the Germans to invade his native Poland—but because it was said at a time when thousands of Russian soldiers were dying every day on the Eastern Front in the battle against a common mortal enemy. Rotblat[8] felt deeply 'the sense of betrayal of an ally', and it left a lasting impression on him. Whether Groves was anticipating the Cold War or whether he deliberately wished to provoke Rotblat, whom he may have regarded as a security risk, is impossible to know. Chadwick, who always tried to avoid political discussion, made no comment.

Now the frightening prospect of war with Russia using atomic weapons had been planted in Rotblat's mind, his conversations with Bohr took on a new intensity. Bohr had already forseen the possiblity of an East–West nuclear arms race and thought it might have dire consequences for the whole future of mankind. Chadwick took a less cataclysmic view, but also listened very intently to Bohr's ideas. It seemed to him that Bohr 'thought almost as deeply about politics and other human matters as he did about physics'.[9] Chadwick never doubted that the international control of nuclear weapons would be one of the most urgent and difficult problems to be faced in the postwar world, nor did he imagine that the ability to construct such weapons would remain an Anglo-American prerogative. He knew French, Russian and German scientists who could all solve the inherent scientific problems—beyond that it was merely a question of national resources and the political will to become a nuclear power.

Another innately conservative man who listened very carefully to Bohr's warnings was the British Chancellor of the Exchequer, Sir John Anderson. He also took counsel from Chadwick, and as early as March 1944 was broaching the subject of international control with Churchill. The expectation was that the allies would have a bomb ready by the first half of 1945. Anderson informed the Prime Minister[10] that all the available evidence suggested that the Germans were not working seriously on a bomb, but that the Russians might be soon. He said that some of Roosevelt's advisers seemed ready to move on this matter, and thought that the Allies would face a choice between international control and 'a particularly vicious form of armaments race with, for a time, a precarious and uneasy advantage'. In Anderson's opinion, as soon as discussions with the Americans began, 'one point

since he arrived in New York with a suitcase emblazoned with the name Niels Bohr, but he was known affectionately as 'Uncle Nick' for many years by those who knew him at Los Alamos.

which we shall have to settle quickly is whether, and if so when, we should jointly say anything to the Russians'. He continued:

> If we jointly decide to work for international control, there is much to be said for communicating to the Russians in the near future the bare fact that we expect, by a given date, to have this devastating weapon and for inviting them to *collaborate* with us in preparing a scheme for international control. If we tell them nothing, they will learn sooner or later what is afoot and may then be less disposed to co-operate. At the same time, there would seem to be little risk of the Russians, if they choose to be unco-operative, being assisted in their own plans by a communication of the kind suggested.

On the original copy of this memorandum, Churchill circled the word 'collaborate' and wrote 'On no account' in the margin next to it; adjacent to the line about 'little risk' he penned the single word 'No'. Churchill had just been briefed on atomic matters again by Cherwell[11] who told him that plutonium 'has quite incalculable industrial and indeed defence possibilities and will probably be the biggest step forward since the introduction of steam power'. A month later, Anderson[12] again tried to interest Churchill in discussing international control with the President, and received the following rebuff: 'I do not think any such telegram is necessary; nor do I wish to widen the circle who are informed.'

With the departure of so many of the leading nuclear physicists for the US, those left behind in Cambridge and Liverpool naturally felt insecure about their futures and began to question the relevance of what they were doing. They were not reassured by Oliphant, who, as we have seen, was the first British scientist to grasp the essential importance of collaboration with the Americans, and who held a fairly low opinion of the performance of the Cavendish nuclear group in particular. A short visit to Liverpool by him in March 1944 created what amounted to a crisis in laboratory morale, and the sense of gloom soon reached their erstwhile colleague, Rotblat, in Los Alamos. He wrote to Chadwick,[13] who was in Washington, to tell him that he had received a desperate letter from Liverpool saying the work there had completely stopped. Rotblat took it on himself to try to boost their spirits by writing an open letter[14] to the whole laboratory which began with the traditional Rutherford greeting, 'Dear Boys'. He regretted that Oliphant's visit 'instead of cheering you up, has caused the deepening of your sense of frustration.' He held out the prospect of new work being assigned to them:

> I have been trying the whole time to convince the Professor about these things. I have not succeeded completely, but he agreed for enriched samples [of uranium] to be sent to you, which should enable you to carry out some experiments. I will do my best to arrange with the Prof. that you should get more information about the progress of the work over here.
>
> Cheer up and show that one can do in England as much as in America.

Apart from the discontent stirred up by Oliphant in Liverpool, he had also caused trouble at the Cavendish, as Chadwick now learned from the beleaguered Norman Feather. Feather had been ill with influenza and had been on sick leave;

Oliphant's visit did nothing to bolster him, as the following extract from his letter to Chadwick[15] shows.

> I need hardly say that Oliphant took a very clear-cut (and probably very much over-simplified) line in discussing with me both the past and the future—and stuck to it . . . It was nothing less than the assertion that both he and you had come to the conclusion that I was not, and never had been really 100% interested in T.A. work, except with the academic curiosity which might be expected of a worker in allied fields. If I had been 100% interested, the argument ran, I should have seen to it myself that I was released from all college and university duties and I should have built up in Cambridge a team of workers who were similarly untrammelled by other duties. The chief sin concerning the lack of output of T.A. work from the Cavendish was at my door . . . It was no real sign of effective interest in the project that I should have produced some dozens of individual 'screeds' nor have been a perpetual worry to you by my frequent correspondence.
>
> Well, I will not elaborate the argument, but I hope you do not subscribe to the whole of it.

The truth was that Chadwick did, at least in part, subscribe to Oliphant's argument, but would not have confronted Feather in this way. He sent the following apology:[16]

> I told Oliphant that I had for some time in the past been very disappointed with your side of the work in Cambridge, but that I had come to realise that I had expected too much and that I had indeed been unjust. My last two visits to Cambridge had convinced me that it was not possible for you to give more than fleeting thought to T.A. problems for you were overburdened with university and college work . . . I confess again that it is only in the last year that I have sufficiently understood the difficulties of your position.

He was concerned about Feather's health and warned, 'You are getting near the edge and should be careful'. Within days another letter followed from Feather[17] complaining about Kinsey being unable to understand directions, and seeking to convince Chadwick that he had not abdicated the Cavendish. By chance, Chadwick had just received a forceful and detailed account from Bernard Kinsey,[18] who had taken over control of the Liverpool cyclotron after Rotblat's departure and was still working half-time in Cambridge. The first three pages of Kinsey's letter consist of a report, which is often critical, of Tube Alloys work in progress in Liverpool; it included reference to joint research with Powell in Bristol using his photographic plate method. Under the heading of 'future work', Kinsey amplified the concerns that Rotblat had already mentioned about the uncertainties facing those left behind in Britain.

> Oliphant expressed his opinion that any further nuclear research here is entirely futile. As you know, I have felt for a long time that the scale on which we have been working has been quite inadequate and now that Oliphant's view is known to everyone, it is difficult to carry on.
>
> We do not know your present views on this matter. The last of Rotblat's letters to Liverpool indicated that at that time (February 25) you still wanted Liverpool to carry on. Oliphant gave us the opposite impression. I made it clear to

Perrin, before Rotblat left, that I was prepared to carry on in Liverpool if, and only if, D.S.I.R. would provide me adequate equipment and full facilities . . . I doubt now the wisdom of carrying on, even with proper facilities. There are two reasons for this. Firstly, the work is likely to remain uncoordinated with that going on in the United States, and I cannot see any way of effecting this coordination without myself making periodical visits to the United States. Secondly, I think that Pickavance, Holt and Rowlands . . . need a change and I think they have been there too long already . . . In my view, if Tube Alloys cannot take any decision for some time to come, it would be better for Liverpool to close down completely and let its personnel take up other employment.

Whatever the merits of Kinsey's arguments, his presumption was too much for Chadwick, who began his five page reply[19] with a sharp rebuke:

I welcome the information but not many of your comments. Some of these show a lack of sense of proportion and even of understanding, while others verge on rudeness. I assume that this is unintentional and due to your natural incontinence of expression: if not, let me know.

As a reminder to Kinsey of his relatively junior position, Chadwick continued:

On the general question of future work on the nuclear side I shall write more fully to Feather, who will no doubt discuss my letter with you.

Chadwick then proceeded to answer Kinsey's scientific criticisms, point by point, repeatedly scolding him for apparently hasty judgements.

You state very positively that the plates were exposed in unsatisfactory conditions. It would be well if you would think a little more before making such a statement. As you learn more about the experiment you will realise that one has to make a compromise between intensity and geometry. I think the compromise we decided upon was on the whole satisfactory.

He closed by giving Kinsey some 'general views' on future work.

(a) Not a great deal can be done in England since we have neither the men nor the material, but some useful and interesting work still remains.

(b) I am not going to sacrifice my laboratory where this work started and where some useful information has been obtained in order to keep the Cambridge cyclotron in action. The facilities in Liverpool are better than in Cambridge provided you take some assistants with you.

Before I left England I discussed with you the future arrangements for nuclear work with the cyclotron, and you agreed to transfer to Liverpool, taking with you certain technical assistants and scientific staff, and some apparatus. You have not carried out your promise to me and I am extremely annoyed.

(c) I do not like your remarks about the 'wisdom of carrying on' and about the men in Liverpool. I do not consider that your limited experience entitles you to make such remarks, which betray a lack of judgement and of understanding.

Still fuming about Kinsey's insolence, Chadwick wrote to Feather again on 6 May:[20]

I received a few days ago a long letter from Kinsey giving me some data about

> the fission spectrum and also some information about the course of experiments in Liverpool and Cambridge. In this letter he gives the impression that he, Kinsey, is in complete charge of all nuclear work and that he is exercising the power of life and death over all and sundry. I hope this is a very great exaggeration for it will lead to rapid collapse of such work as can still be carried on. It is quite evident from his remarks that he has not taken the trouble to try to understand what has been done, for example in connection with the fission spectrum, or what is planned, as for example in the delay experiments.

Chadwick admitted to being hurt by Kinsey's behaviour since he had taken him into Tube Alloys. He asked Feather to 'find time and energy to pay a visit to Liverpool to talk to my boys and to re-assure them, and to discuss with them future experiments'. For his part, Kinsey wrote to Chadwick again on 2 June,[21] displaying remarkable insouciance:

> Dear Professor,
>
> Many thanks for your interesting letter of 2nd May. I am sorry you have taken offence at some of my remarks and criticisms; I had intended none.

He then presented a further progress report before offering an explanation of why he had not transferred full-time to Liverpool, as had been originally agreed. In essence, it seemed to make for a more efficient use of resources to leave the Cambridge workers where they were, and Kinsey had really been treading water until the political decisions regarding Tube Alloys work were settled. Then:

> on March 3rd, Oliphant arrived in Cambridge and expressed his view that it was a waste of time to carry on any research in this country in view of the far better equipment available in America. He had come, he said, with a proposal that I should go to Berkeley, but he then said that it was likely that this move would be changed after decisions had been made as to whether slow neutron work should proceed in this country or in Canada. On March 8th I met him in Liverpool, where he expressed this view again, this time in the presence of Moore, Rowlands, Holt and Pickavance. He made it clear that there was no point at all in continuing any experiments on nuclear physics . . . He gave me the impression, and he gave others the same impression, that he was speaking as your representative. He said also that we might expect to hear within a fortnight at which place in the U.S. or Canada to which we should be assigned. We have been waiting eagerly for news ever since.

He pointed out to Chadwick that until his letter arrived in May, 'there was no question at all' on the part of Tube Alloys or others such as Cockcroft that any nuclear research should be continued in England. He pointed out that all decisions were made on scanty reports from America, and his own supposed lack of understanding was due to the absence of technical information at his disposal. He apologized for not fulfilling the original promise he had made Chadwick, but in the circumstances did not see how he could have taken any other course. Chadwick had been so consumed by his policy negotiations with Groves that he had probably not realized the confusion that the lack of clear directives was causing back in England; even now the problem had been brought to his attention there was little he could do to rectify matters because of the delicate state of his arrangements with Groves. He had not completely forgotten the Liverpool laboratory, and in the middle of April had formally asked Groves[22] for 100 mg of enriched uranium to be sent there. He justified this request by saying the Liverpool workers would provide 'independent confirmation of work being done at Y [Los Alamos], so that we can feel more confident of some fundamental constants'. He could not tell Kinsey that such important work was in prospect, but did write[23] to absolve him from all blame and explained that he could not go into the matters in detail. Clearly there had been 'a complete misunderstanding' on all sides, and Chadwick warned that 'all our troubles are not quite over, but it should not be long before a clear decision is reached'.

To the British in Washington there was 'a natural but increasing sense of American self-confidence and buoyancy as to their ability to carry through this project'.[24] With this conviction came a growing recognition of the corresponding responsibilities, both domestically and with respect to international affairs. The project was secretly consuming hundreds of millions of dollars, which in itself was a powerful force driving towards the ultimate use of the bomb, and Groves knew there would be a day of reckoning with Congress, once the war was over. The unprecedented power that they would soon acquire and the terrible threat that it would represent reinforced the tendency for Groves and others to want to keep atomic energy and weapons as an American preserve. 'The real difficulty', according to Chadwick,[25] 'was the reluctance of the Americans to have anything to do with Montreal. They have no confidence in the team as a whole and they have a distrust of certain members.' Chadwick admitted to Appleton that collaboration would have been easier if there had been 'a really sound British team' in Canada, and if the project were not so one-sided in terms of human and material resources.

At a practical level, Groves was still unwilling to provide Montreal with information on plutonium chemistry and separation so that he could tell Congress that some secrets had been withheld from the British scientists. Although the UK was implicitly allowed to possess enriched uranium and even atomic weapons under the Quebec Agreement, Groves now suggested that Britain itself was too vunerable to attack and such items should be housed in Canada. When this was first suggested at a meeting in Washington, it was Chadwick who said such a restriction would be unacceptable to the United Kingdom Government. Further he warned the British to expect a similar argument based on security considerations when it became plain that work was to start on full-scale plant in the UK.[26] The Ambassador reported back to Anderson in London, who was sufficiently disturbed by the attitude of the Americans to send a message to Chadwick on 2 June 1944:[27]

> This refusal seems to me to be inconsistent with the assurances obtained at the meeting of April 13th of the Combined Policy Committee and indeed with Clause 5 (c) of the Quebec Agreement. Kindly report on developments with a view , if necessary, to personal action on my part.

Chadwick had left Washington at the end of May to spend a few days at Los Alamos. He had conceived a solution to the latest disagreement, which would preserve the formal non-disclosure of information to Montreal, while allowing the chemists there to receive small amounts of plutonium with which to work. He travelled to Chicago for a meeting with Groves, Cockcroft and the Canadian MacKenzie on 8 June, two days after D-Day. Groves eventually agreed to Chadwick's initiative and agreed to send some irradiated uranium rods to Canada so that the workers in the Montreal Laboratory could extract their own plutonium.[28] Whenever such a working compromise was reached, Chadwick would smile a little and in an exaggerated Lancashire brogue announce that 'It's all jam and kippers'. MacKenzie wrote in his diary for 9 June: 'Both Chadwick and I feel that for the first time we have everything on the rails so far as the negotiations are concerned. His difficulty now is to try to keep Sir John Anderson on the rails.'[28]

Later in June, Chadwick seemed close to despair as he complained to Akers about 'the complacency, almost amounting to indifference'[29] with which the whole project seemed to be treated in London. Two months later, Chadwick was able to express his concerns directly, when he was recalled to London to brief Anderson on all aspects of the Manhattan Project and developments in Canada. Before leaving he wrote to Cockcroft[30] mentioning that he felt 'very low just now' and needed a rest. Cockcroft needed to expand his staff in Montreal now that work there was moving ahead, and Chadwick offered to recruit men for him in England; he thought it would be a 'grave mistake' for Cockcroft to leave Montreal since no further delays in their programme could be afforded. In Chadwick's opinion, Groves was already disappointed with progress in Canada and it was important from the view of future collaboration to demonstrate a drive and efficiency that matched the US efforts. While in London, Chadwick met Frédéric Joliot, who just two weeks earlier had led his staff and students in fierce battles against the last German tanks on the streets of Paris, hurling Molotov cocktails made in his laboratories.[31] His immense personal courage and inspirational role in the Resistance added to the lustre of his reputation, so that he had become the natural successor to his mother-in-law as France's premier scientist. His authority was formalized by the Provisional Government, which appointed him Director of the CNRS, the national scientific research council.

Chadwick and Joliot met on 7 September 1944—the first time they had seen each other since collecting their Nobel Prizes in 1935. It was a strained reunion between two men who had never felt at ease in each other's company. The Americans had already interrogated Joliot in Paris about wartime activity in his laboratory and had been relieved by his assessment that the Germans were 'not remotely close to making an atomic bomb'.[32] Chadwick[33] felt constrained by the terms of the Quebec Agreement so that he could not 'communicate any information about Tube Alloys to third parties' without the consent of the Americans, which he certainly did not have. Joliot knew that several of his former colleagues like Halban and Kowarski had left England to work in North America, but Chadwick formed the impression that Joliot believed their efforts were directed towards slow neutron research and that Joliot harboured 'no idea about a military weapon'.[34] Joliot had always been left wing in his political views, and he now told Chadwick quite openly that he was a member of the Communist Party. Plainly this would make Joliot *persona non grata* with the Americans, especially Groves. Chadwick tried to convince Joliot that it would be very much against the French interest for him to make vigorous enquiries at this time: it would have been more pertinent to state that it would have been against Anglo-American interests.

Those interests seemed to be inextricably sealed in September 1944 following a private conversation between Churchill and Roosevelt at the latter's family estate, Hyde Park. An *aide-mémoire*[35] of their discussion gives a succinct summary of the leaders' perceptions and intentions in September 1944.

> 1. The suggestion that the world should be informed regarding Tube Alloys, with a view to an international agreement regarding its control and use, is not accepted. The matter should continue to be regarded as of the utmost secrecy; but when a 'bomb' is finally available, it might perhaps after mature consideration* be used against the Japanese, who should be warned that this bombardment will be repeated until they surrender.
>
> 2. Full collaboration between the United States and the British Government in developing Tube Alloys for military and commercial purposes should continue after the defeat of Japan unless and until terminated by joint agreement.
>
> 3. Enquiries should be made regarding the activities of Professor Bohr and steps taken to ensure that he is responsible for no leakage of information, particularly to the Russians.

Such high-level concern about Bohr stemmed largely from Churchill, who had been persuaded to meet the soft spoken and diffident Dane in London to hear his views on the need for international control of atomic weapons. It was a historic non-meeting of two of the great minds of the twentieth century; as Bohr remarked at the time 'It was terrible. He scolded us [Bohr and Cherwell] like two schoolboys.'[36] Bohr had subsequently had a much more sympathetic audience with Roosevelt, to whom he had been introduced by a Supreme Court Justice, Felix Frankfurter. At the Hyde Park meeting, Churchill must have a mounted a vigorous denunciation of Bohr, and the intensity of his animosity is plain from a telegram[37] he sent to Lord Cherwell, as he sailed back to Britain two days later.

> The President and I are much worried about Professor Bohr. How did he come into this business? He is a great advocate of publicity. He made an unauthorized disclosure to Chief Justice Frankfurter who startled the President by telling him he knew all the details. He says he is in close correspondence with a Russian professor, an old friend of his in Russia to whom he has written about the matter and may be writing still. The Russian professor has urged him to go to Russia in order to discuss matters. What is all this about? It seems to me Bohr ought to be confined or at any rate made to see that he is very near the edge of mortal crimes. I had not visualized any of this before, though I did not like the man when you showed him to me, with his hair all over his head, at Downing Street. Let me have by return your views about this man. I do not like it at all.

---

* The phrase 'might perhaps after mature consideration' appears in Churchill's handwriting on a draft of the Hyde Park aide-mémoire in place of the original, single, word 'should'.

Cherwell's reply[38] was waiting when Churchill arrived back in London, and he strongly defended Bohr, pointing out that the possibilities of 'a super weapon on Tube Alloy lines have been publicly discussed for at least six or seven years'. It was decided that Bohr should be interviewed in Washington before being allowed back to Los Alamos,[39] and the task fell to General Groves and Chadwick, who must have found any attack on Bohr, from whatever quarter, distasteful. The Russian professor referred to was Peter Kapitza, who had sent a letter via the Russian Embassy in London when word had reached him of Bohr's departure from Copenhagen. Bohr had behaved scrupulously, clearing every response he made with British secret services. He now explained his communications with Frankfurter, whom he had first met on a visit to Washington in 1939. Frankfurter had entertained him to lunch at the Supreme Court in February 1944, and indicated to Bohr that he knew about the Manhattan Project. Again Bohr seemed to have behaved impeccably and with more discretion than the Chief Justice; Chadwick and, more importantly, Groves were completely satisfied with his explanation of events and the matter appeared to be settled.

His nomadic existence meant that Chadwick hardly saw his daughters, who had arrived in Los Alamos in early summer. In contrast to their mother, they loved the surroundings and opportunities for outdoor life. Rotblat had moved out of the cabin and into the bachelor quarters known as the Big House. He still visited the Chadwick family and on weekends would take Aileen and the twins out for hikes onto the wooded slopes of the Jemez Mountains. There was a security fence around the perimeter of the camp which was patrolled by armed Military Police in Jeeps. These were men who had volunteered for a top secret mission and found themselves undertaking such dangerous assignments as baby-sitting for eccentric European professors. On one hot Sunday afternoon, Rotblat and the Chadwick ladies were returning thirsty and exhausted from a long walk. They would have to walk an extra mile or so to the main security gate, where they were supposed to sign in, and then another mile back to the cabin. Rotblat noticed at a point about one hundred yards from their ultimate destination inside the fence that somebody had lifted the bottom wire so that one could wriggle underneath it. He suggested, 'half in jest', to Aileen that perhaps they should take this shortcut. To his surprise she did not rebuke him, but, egged on by her daughters, agreed. The girls and Rotblat slid under, followed more slowly by Aileen. It was the only time that Rotblat ever saw her lower her dignity. A minute or so after she stood up, a Jeep came along. Rotblat amused himself with the notion of the diplomatic incident that would have followed if the wife of the leader of the British team had been shot in the rump, while crawling under the security fence by a US soldier starved of action.[5]

Aileen's husband, meanwhile, found himself at the centre of a real diplomatic dispute that once again threatened Anglo-American relations. At the eye of the storm were the five Frenchmen working in Montreal—Auger, Halban, Kowarski, Goldschmidt and Guéron. The first four of them were on the British payroll, but Guéron had enlisted in de Gaulle's Free French Forces in London in 1940 and his foremost loyalty was plainly to the new French Provisional Government. When de Gaulle visited Montreal in July 1944, he was secretly briefed by Guéron about the nuclear weapon that was being developed.[28] In October, Guéron made a visit to France, a natural enough journey, but one which alarmed the Americans who were desperate to confine nuclear technology to the North American continent. The issue was raised at a meeting in October between Chadwick, Groves and his security advisers.[34] Chadwick indicated that 'he was beginning to have some concern about the foreigners who are on the project, in as much as it appears that they might for the most part be loyal to their respective Governments and would therefore present an increasingly acute security problem.' He told the assembled group about Guéron's visit to Paris and although 'he had every confidence in Guéron's discretion, the latter was very much interested in the new French Government'. Chadwick stated that Lord Cherwell (and therefore Churchill) was very much against Guéron's going to Paris at all, but Chadwick had not been in any position to prevent him. Chadwick was convinced that de Gaulle 'must have wind of the project'. He had asked British Intelligence to ascertain Guéron's contacts in Paris, but they had refused; General Groves and his staff immediately offered to undertake this surveillance. Chadwick advised that Joliot 'must at all costs' be prevented from coming to North America because it would be 'impossible to keep him from obtaining a vast amount of information about the project'. It was concluded that every inducement should be made to keep the French scientists working in Canada for the duration of the war.

The fundamental tension that was now emerging reflected a lack of congruence between the bipartisan terms of the Quebec Agreement and the reality that a third nation, France, had made a significant contribution to the nuclear 'boiler' project.[40] The Chancellor of the Exchequer, Sir John Anderson, who had drafted the Quebec Agreement, held the view that it would be natural for France to receive some benefit for the pioneering contribution of her nuclear scientists in any postwar arrangement. Although he had told Bush[41] about the French patents by letter in August 1942, neither man had explicitly raised the issue when the Quebec Agreement was signed a year later. The Agreement thus did not mention France, and instead indicated that the United States would be expected to recoup the lion's share of commercial gains from nuclear power after the war. Anderson now injudiciously threw fuel onto the spluttering fire in the form of an *aide-mémoire* to Winant,[42] the American Ambassador in London. He pointed out to the Ambassador that 'the nature of the agreements with Halban and Kowarski and the proposals concerning the French patents, were made known to and understood by the U.S. authorities, who raised no objection to them.' Winant forwarded the document to Washington where it was read by an astonished and furious General Groves, who had no prior knowledge of any foreign patents or obligations to the French.

Worse was to follow since the French contingent in Canada were lined up like swallows on a telephone wire, waiting to migrate. Early in October, Halban had visited Chadwick in Washington and casually remarked that he wished to visit London and to see Joliot, in order to discuss the patent arrangements.[43] Chadwick agreed that Halban had a responsibility to Joliot to discharge, but did not think the matter urgent and counselled some delay. He heard nothing more until the 27 October when he received a letter from Cockcroft saying that Halban was about to leave Montreal for London. Chadwick, knowing that Groves was already deeply upset, phoned Cockcroft and urged him to delay Halban's departure. He also spoke to Lord Cherwell on the matter, and word reached Churchill. Churchill sent a worried minute to Anderson:[44]

> This is a very alarming state of affairs. Has the President been informed? I do not see what we can do until we get Lord Cherwell back, except on one pretext or another, to stop other Frenchmen coming over. We must discuss this together.

Anderson meanwhile had been petitioned by Halban,[45] who complained that his planned trip had been postponed on Chadwick's instructions, and that this interference implied a lack of trust. After telling Churchill that he must be left free to handle the situation, Anderson sent two cables: one to Cockcroft asking him not to delay Halban further, and a second to Chadwick stating the postponement of Halban's visit was not in accord with his wishes.[46] Chadwick's inclinations were much closer to Churchill's, and as soon as Halban set sail for England on the *Queen Mary*, he summoned the next Montreal scientist wishing to visit France, Bertrand Goldschmidt. Goldschmidt was anxious to lay the foundations of a university career in post-war France. Chadwick, while sympathetic to the young man's aspirations, explained his concern about the delicate state of Anglo-American relations, which were being jeopardized, in his view, by the premature visits to France. Much to his relief, Goldschmidt proved more tractable and consented to a delay, 'disappointed, but not anxious to get caught up in a maelstrom of orders and counterorders'.[28]

Halban arrived in London on 9 November; he immediately started to press his case for being allowed to travel to Paris, specifically to meet Joliot. He convinced Akers of his right to such a meeting and together they drafted a memorandum[47] to Anderson to support the visit and to delineate those topics he might be permitted to discuss with Joliot. It was suggested that Halban should travel as a member of DSIR, and the Americans should be informed accordingly. Anderson was in broad agreement with the plan and expressed his support in a minute to Winant, the US Ambassador, whose job it was to inform Groves. The General,[48] scarcely able to believe the treachery of the British, sought to impose two conditions on Halban's activities. The first was that any talks he held with French scientists in London should be monitored; the second was simply that Halban should not be allowed to visit France. To reinforce this advice he dispatched his chief security adviser, Colonel Lansdale to London, to point out that the Americans could see no possible advantage and many probable disadvantages in allowing Halban and Joliot to meet. John Winant, the Ambassador, was given a 'pretty sticky message from Washington'.[49] Oblivious to the American concerns, Halban flew to Paris on 24 November. Over the next ten days, he had several lengthy conversations with Joliot, who felt that 'no arrangement on the patents could be envisioned outside a general agreement for Franco-British collaboration in atomic energy'. Joliot also pointed out that any negotiations would have to be authorized by General de Gaulle. On his return to London, Halban[50] prepared a report in which he concluded that Joliot was clearly intending to start an atomic energy project in France and the only way he might be dissuaded would be to include France in a partnership with Britain and the United States. If not granted this opportunity, Halban warned, the French might turn to the Russians as partners.

Before flying to Paris, Halban had submitted for Anderson's approval a self-imposed list of scientific and technical items that he might discuss with Joliot. In his report, Halban indicated that he had kept within these guidelines, although there was some scepticism even amongst the British that he would be able to forbear from at least dropping hints to Joliot about the extraordinary scale of the Manhattan Project or what he knew of Los Alamos. For his part, Groves found it inconceivable that Joliot did not receive more information than Halban admitted to giving. Frustrated that Anderson seemed intent on ignoring the American viewpoint and indeed the terms of the Quebec Agreement, Groves countered by attacking the Canadian representative on the CPC, the Minister of Munitions, C.D. Howe. He protested in the strongest terms that Halban, a scientist from Montreal should have travelled to Paris and disclosed highly secret information about the US part of the Manhattan Project. Groves' tactics were astute; Howe[51] was at once embarrassed and angry that he had not been consulted about the visit and sent a cable to Anderson registering his displeasure, and informing him that Halban would not be allowed back into the Montreal Laboratory.

At last the magnitude of the gaffe was beginning to be appreciated in London. Howe's cable was forwarded to Anderson with an official's perspicacious comment: 'I am afraid it looks as if we are in for trouble.'[52] Only two days before, Anderson with apparent confidence that he was in control of an increasingly volatile situation had sent Chadwick a cable[53] asking him to lift his ban on Goldschmidt. Blinded by his own condescension, Anderson assured Chadwick that Lansdale, Groves' security representative, had come round to the British view and his previous doubts about the visits to France were 'largely due to misunderstanding and ignorance of past history'. He told Chadwick that the problems of the French relationship to atomic matters would have to be faced sooner or later, and that 'if we treat them as "prisoners" [the] only result will be that the question will be raised prematurely and in an acute form from the French side'. This was too much for Chadwick who was not dealing with Groves at arm's length, but had been close enough throughout to feel the heat of his temper. He replied to Anderson:[54]

> The reason for my desire that French members of the Montreal team should not visit Paris at this juncture was to give you time to consider relations with France and agree with U.S. on a common policy, at least as regards immediate future. It seemed to me and seems still, that no French member can visit Paris without giving away some information on the project. All, and especially Halban, have considerable knowledge of U.S. projects, about which any leakage whatever would be taken most seriously here.

> ... There was no intention of treating the French as prisoners nor was such an interpretation possible in the face of the explanation I gave. Gueron and Goldschmidt fully appreciate the difficulties and share my views on the need for caution. There is complete understanding between us. Goldschmidt agreed readily and without pressure to postpone his visit for a time, although he was naturally disappointed.
>
> Halban's visit to Paris produced a violent reaction here and Lansdale's return has not helped to allay it. Lansdale appreciates the motives underlying the attitude in London to this problem but disagrees emphatically with the actions which have been taken. The Americans are seriously alarmed about the possible implications and the position, difficult already, may become grave.
>
> A visit to Paris by Goldschmidt at this point would precipitate matters, for it would be looked upon by the Americans as an open disregard of their views.

Chadwick also warned Anderson that the Canadians were furious that there had been no prior consultation with them and the whole imbroglio revived American distrust of Halban in particular and of foreigners in general. On Christmas Eve, Chadwick and the British Minister in Washington, Sir Ronald Campbell, met Groves and listened to his concerns.[33] It transpired that he had little or nothing in the way of specific allegations about leaks from Halban to Joliot, but was seriously alarmed about hints which might have been given, intentionally or not. On 28 December, Anderson, who by now was acutely aware of the trouble his policy towards the French had caused, sent three separate telegrams[55] to Campbell in Washington. He now acknowledged that he was the target of American anger, and shifted his ground with the poise of a seasoned politician. Whereas in his previous statements on the subject, he had painted himself as the honourable broker anxious to see the French obtain their just deserts, he now claimed 'my whole policy so far has been to play for time'. He also requested that Chadwick should pay a flying visit to see him in London, suggesting that 'if he went both ways by bomber he should not need to be out of Washington for more than a few days'. Campbell refused the request saying 'Chadwick's continuous presence in D.C. essential both because of his personal relations with Groves and because of his ability to answer on the spot any technical questions from the American side and to deal with matters connected with the Montreal team.'

During the days after Christmas, Halban was sailing back across the Atlantic, blissfully unaware of the furore surrounding his visit to Paris. With typical black humour, Chadwick, who was exhausted by the strain of the affair, confessed to Goldschmidt that 'in spite of being a good patriot, he found himself wishing the *Queen Mary* had been torpedoed on the way and sunk with everyone on board'[28] since this would have placated Groves. The General himself was stoking the fire of discord as hard as he could and attended a long meeting on the subject at the White House with Roosevelt and Secretary of War, Stimson. Stimson told the President how Anderson had 'hoodwinked poor old John'[56] (Ambassador Winant), and Groves warned that he was not to be trusted in any circumstances. Assailed on all sides and engulfed by a torrent of critical cables from Washington, Anderson made an unconvincing attempt to shift some blame for the row onto the Canadian minister, Howe, for exaggerating Groves' complaints;[57] when this ploy fooled no one, he eventually apologized to all and sundry for the misunderstandings and lack of communication. It was left to Chadwick and Campbell to try to re-establish an atmosphere of trust in Washington.

A contemporary impression of the disruption caused comes from Mark Oliphant, who arrived in Washington in January 1945 to find that all the hotels were full because of Roosevelt's fourth inauguration so that he stayed with Chadwick. He wrote to Akers in London:[58]

> You must be aware of the terrific uproar created by the French business. Halban's visit and what he might have revealed has made a stir of the first magnitude. Chadwick is unable to sleep and is very upset over the whole business. The direct result is that he has time for nothing else, and will have none till this business has quieted down.

After deploring that 'the preoccupation with these nonessentials' should eclipse the main objective of the Manhattan Project, Oliphant went on to castigate Lansdale, the security chief, for his 'smiling urbanity', calling him 'a barefaced liar' in some matters. He also reported Ernest Lawrence's view that 'high level Anglo-American accord is at a low ebb' due to what Groves termed 'British rascality'. Chadwick[59] thought that there was not the slightest accusation of unreliability against Guéron, Goldschmidt or Kowarski, and that all the Americans, including Groves, were willing to accept his opinion that they were all entirely trustworthy. The postion of Halban, despite the lack of definite evidence against him, was very different. Chadwick had come to share the loss of confidence, but still thought he must be treated fairly and reasonably, and was wary of making him the scapegoat. On the central question of Anglo-American understanding, Chadwick felt impelled to give his own views to Sir John Anderson:[60]

> In my opinion, the fundamental question is whether we wish to collaborate fully with the U.S.A. after the war or not.
>
> If not, then my presence here has no longer any meaning to me and I should be unwilling to remain.
>
> But this answer does not make sense. I have always believed that your object was the fullest collaboration and most intimate understanding on this very far-reaching project, and all I have done or tried to do here has aimed at building up a basis for future collaboration.
>
> Our participation in this project represents a very small fraction indeed of the total effort and it will not be immediately obvious to Congress and the American public that we have taken any significant part. We shall have to rely very much on the U.S. authorities, and especially on Groves, to work with us towards collaboration, not merely to accept collaboration if it is forced on them.
>
> I think that Bush, Conant, and Groves are now quite a long way on the path we want to tread, but it would not be difficult to scare them off it. It seems to me that we must keep our interests identified with theirs as far as we can, and that we must certainly contribute our utmost towards the realisation of the project. We shall have to work our passage home, by doing everything in our power to show that this project has been a co-operative effort as far as our resources and other

> war commitments permitted. And I think this means doing nothing to the contrary, such as taking men home whom the Americans say are useful and needed for the production of the weapon, such as diverting men and effort to starting up plants in England which can have no significance for the war, and so on.
>
> I firmly believe that we can be satisfied with the progress we have made in gaining the confidence of the Americans and in working towards a good post-war understanding, but I also believe that we dare not reduce our efforts or relax our vigilance.

Perhaps the most enduring consequence of the fracas over the French scientists resulted from Anderson's loss of face and therefore diminished influence at the highest levels. He had always been attracted by Bohr's philosophy of openness regarding the international regulation of atomic weapons and power. This was a risky and somewhat nebulous policy, which would have been difficult to sell to Churchill or Groves under any circumstances. Churchill's determination for secrecy was if anything firmer than ever: he did agree in January 1945 to inform his own Chiefs of Staff about the Manhattan Project, but this was about as far as he was prepared to go.

When Anderson attempted to launch further negotiations with the French early in 1945, Churchill's response was to suggest that if Joliot had acquired more information than he was entitled to, he 'should be forcibly but comfortably detained for some months'.[61] Nor was the Prime Minister's distrust of scientists confined to the foreign variety—in a minute to the Chief of Staff, General Ismay,[62] on 19 April, he repeated that he wanted to keep the secret as long as possible and warned that 'for every one of those scientists who is informed there is a little group around him who also hear the news'. He criticized those around him for not appreciating the lead obtained by the US, as a result of their huge expenditure, and pointed out 'there is all the difference between having certain paper formulae and having a mighty plant in existence'. In a crisp minute to the Foreign Secretary, Churchill encapsulated his views:[63]

> Our policy should be to keep the matter so far as we can control it in American and British hands and leave the French and Russians to do what they can. The Chancellor [Anderson] said that the Frenchmen whom he had interviewed would never betray the secret to de Gaulle, and he vouched for their good behaviour. Now we are threatened that the Russians will be told . . . Once you tell them they will ask for the very latest news, and to see the plants. This will speed them up by two years at least. You may be quite sure that any power that gets hold of the secret will try to make the article and that this touches the existence of human society.

Churchill was clearly exasperated by Anderson's advocacy for the French, confessing that he was 'getting rather tired of all the different kinds of things that we must do or not do lest Anglo-French relations suffer'. He commented archly that he had 'never made the slightest agreement with France or with any Frenchman'. For his part, Chadwick was still attempting to negotiate a settlement of the problem with Groves. He had told Groves[64] in January that on returning home and joining

Joliot's CNRS, the French scientists could not be expected 'not to make use of the knowledge they have acquired', and they should be asked only to confine the dissemination of information about the Manhattan Project to 'official and government circles'.

References

1 Jones V.C. (1985). The atomic communities in New Mexico. In *Manhattan: the army and the atomic bomb*, pp. 465–81. Center of military history United States Army, Washington DC.

2 Szasz, F.M. (1992). The British mission at Los Alamos: the social dimension. In *British scientists and the Manhattan project*, pp. 32–45. St. Martin's Press, New York.

3 Tube Alloys salary file (1943–44). AB 1/267, PRO.

4 Webster, W. (24/2/44). Letter to E. Appleton. CHAD IV, 3/16, CAC.

5 Brown, A.P. (1994). Interview with Joseph Rotblat, London.

6 Bainbridge, K. (29/1/74). Letter to J. Chadwick. CHAD IV, 12/65, CAC.

7 Peierls, R.E. (1994). Recollections of James Chadwick. *Notes and records of the Royal Society London*, **48**, (1), 135–41.

8 Rotblat, J. (1985). Leaving the bomb project. *Bulletin of the Atomic Scientists*, Aug., 16–19.

9 Weiner, C. (1969). Sir James Chadwick, oral history. American Institute of Physics, College Park, Maryland.

10 Anderson, J. (21/3/44). Minute to W.S. Churchill. PREM3/ 139/2, PRO.

11 Cherwell, F. (March 1944). Minute to W.S. Churchill. PREM3/ 139/2, PRO.

12 Anderson, J. (April 1944). Minute to W.S. Churchill. PREM3/ 139/2, PRO.

13 Rotblat, J. (10/4/44). Letter to J. Chadwick. CHAD IV, 1/2, CAC.

14 Rotblat, J. (22/4/44). Letter to physics department, Liverpool University. CHAD IV, 1/2, CAC.

15 Feather, N. (12/3/44). Letter to J. Chadwick. CHAD I, 25/1, CAC.

16 Chadwick, J. (9/4/44). Letter to N. Feather. CHAD I, 25/1, CAC.

17 Feather, N. (16/4/44). Letter to J. Chadwick. CHAD IV, 1/8, CAC.

18 Kinsey, B.B. (11/4/44). Letter to J. Chadwick. CHAD IV, 1/9, CAC.

19 Chadwick, J. (2/5/44). Letter to B. Kinsey. CHAD IV, 1/9, CAC.

20 Chadwick, J. (6/5/44). Letter to N. Feather. CHAD I, 25/1, CAC.

21 Kinsey, B.B. (2/6/44). Letter to J. Chadwick. CHAD IV, 1/9, CAC.

22 Chadwick, J. (15/4/44). Letter to L. Groves. RG 77, file 201, NA2.

23 Chadwick, J. (26/6/44). Letter to B. Kinsey. CHAD IV, 1/9, CAC.

24 Gowing, M. (1964). *Britain and atomic energy, 1939–1945*, p. 340. Macmillan, London.

25 Chadwick, J. (17/4/44). Letter to E. Appleton. AB 1/561, PRO.

26 Campbell, R. (31/5/44). Cable to J. Anderson. PREM3/139/11A, PRO.

27 Anderson, J. (2/6/44). Cable to J. Chadwick. AB 1/197, PRO.

28 Goldschmidt, B. (1990). *Atomic rivals*. Rutgers University Press.

29 Chadwick, J. (24/6/44). Letter to W. Akers. AB 1/615, PRO.

30 Chadwick, J. (5/8/44). Letter to J. Cockcroft. AB 1/197, PRO.

31 Goldsmith, M. 1976). Resistance. In *Frédéric Joliot-Curie*, pp. 97–124. Lawrence and Wishart, London.

32 Groves, L.R. (1962). *Now it can be told*, p. 212. Harper and Row, New York.

33 Relations with France. CAB 126/30, PRO.

34 Lansdale, J. (23/10/44). Memo: conference with Dr Chadwick. RG 77, Box 85, file 201, NA2.

35 Aide-mémoire of Hyde Park agreement, (18/9/44). PREM3/139/8A, PRO.

36 Pais, A. (1991). Bohr, pioneer of 'glasnost'. In *Niels Bohr's times*, pp. 473–508. Oxford University Press.

37 Churchill, W.S. (20/9/44). Cable to Lord Cherwell. PREM3/139/8A, PRO.

38 Cherwell, F. (23/9/44). Letter to W.S. Churchill. CAB 126/39, PRO.

39 Professor Bohr (October 1943–December 1944). CAB 126/39, PRO.

40 Barnes, C. (29/10/42). Letter to M. Perrin. CAB 126/30, PRO.

41 Anderson, J. (5/8/42). Letters to V. Bush. AB 1/207, PRO.

42 Anderson, J. (23/10/44). Aide-mémoire to J. Winant. CAB 126/30, PRO.

43 Chadwick, J. (31/12/44). Cable to J. Anderson. CAB 126/30, PRO.

44 Churchill, W.S. (29/10/44). Letter to J. Anderson. CAB 126/30, PRO.

45 Halban, H. (27/10/44). Letter to J. Anderson. CAB 126/30, PRO.

46 Barnes, C. (30/10/44). Cable to J. Chadwick. CAB 126/30, PRO.

47 Akers, W. (13/11/44). Memo re Halban to J. Anderson. CAB 126/30, PRO.

48 Groves, L. (17/11/44). Cable to US embassy, London. CAB 126/30, PRO.

49 Winnant, J. (29/11/44). Message to Cabinet Office. CAB 126/30, PRO.

50 Halban, H. (1944). Report on visit to Paris. CAB 126/30, PRO.

51 Howe, C.D. (30/12/44). Cable to J. Anderson. CAB 126/30, PRO.

52 Barnes, C. (21/12/44). Note to J. Anderson. CAB 126/30, PRO.

53 Anderson, J. (19/12/44). Cable to J. Chadwick. CAB 126/30, PRO.

54 Chadwick, J. (22/12/44). Cable to J. Anderson. CAB 126/30, PRO.

55 Anderson, J. (28/12/44). Cables to R. Campbell. CAB 126/30, PRO.

56 Hewlett, R.G. and Anderson, O.E. (1990). The quest for postwar planning. In *A history of the United States Atomic Energy Commission: the new world*, pp. 322–46. University of California Press

57 Anderson, J. (31/12/44). Cable to R. Campbell. CAB 126/30, PRO.

58 Oliphant, M.L. (21/1/45). Letter to W. Akers. CAB 126/30, PRO.

59 Chadwick, J. (29/1/45). Letter to M. MacDonald. CHAD IV, 6/25B, CAC.

60 Chadwick, J. (22/2/45). Letter to J. Anderson. CHAD IV, 3/16, CAC.

61 Churchill, W.S. (1/2/45). Minute to A. Eden. CAB 126/30, PRO.

62 Churchill, W.S. (19/4/45). Minute to H. Ismay. PREM3/139/11A, PRO.

63 Churchill, W.S. (25/3/45). Minute to A. Eden. CAB 126/30, PRO.

64 Chadwick, J. (7/1/45). Minute to L. Groves. RG 77, Box 85, file 201, NA2.

16 A different world

There was one joyous piece of news for Chadwick while he was struggling with the wilful crisis of the French scientists: he was knighted in the New Year Honours List for 1945. With characteristic modesty, he regarded this as recognition of the efforts made by the Tube Alloys team first in England and now in the United States. Amongst the messages he received, one which brought him particular pleasure was from Holt, Pickavance and others in Liverpool that combined economy of expression, of which he always approved, with obvious affection, which he found touching. It simply read: 'CONGRATULATIONS DEAR SIR'. Aileen, now in her element as Lady Chadwick, had decided that since her husband seemed to be an indispensable and permanent fixture in Washington, there was no point in her and the girls remaining isolated in Los Alamos. She left the Hill with no regrets; in April the family found a house to rent near Dupont Circle in DC, and they were properly reunited for the first time since the summer of 1940.

Chadwick continued to work long hours, seven days a week, running the British atomic energy organization in Washington. The recent alarms over the French had left their mark, and as Sir Ronald Campbell, the British Minister, wrote 'the salad is heaped in a bowl permanently smeared with the garlic of suspicion'[1] as far as the Americans were concerned. The British might resent the American attitude and ridicule it in private, but in Campbell's view it was 'a part of the scenery'. It could not be disregarded and should not be exacerbated by ill-judged behaviour. Given this background, Campbell thought it would be the greatest possible mistake to attempt to conceal from the Americans plans for an atomic research establishment in postwar England. Chadwick had attended a meeting of the Tube Alloys Consultative Council during his visit to London in September 1944, when it was agreed that atomic bombs should be produced in the UK as soon as possible and the 'construction of a Gaseous Diffusion Plant, to produce 600 grams per day of two-fold enriched material, should be undertaken at once'.[2] Chadwick agreed with this conclusion on technical grounds, but made it clear that he had significant reservations to make regarding the overall policy. On his return to Washington, Chadwick had held a two day meeting[3] on the same subject with other senior British scientists—Peierls, Oliphant and Cockcroft. At Chadwick's suggestion, in order to give their meeting a definite objective and in the hope that positive recommendations would emerge, the group assumed that the main policy objective would be the fabrication of British atomic weapons as quickly as possible.

By default this became the unofficial policy in London for the next few months. Early in 1945, the British aims were broadened to include all aspects of atomic energy, and it was suggested to Churchill that he should try to reach some general understanding with Roosevelt about continuing Tube Alloys work in the UK.[4]

Mark Oliphant had been pressing, with characteristic enthusiasm, since early 1944 for a start to be made on the British organization. In Chadwick's view, Oliphant had been premature in his recommendations and was too optimistic about the electromagnetic method of isotope separation that had been invented at Berkeley and too negative about the gaseous diffusion method.[5] Chadwick thought a decision about which methods to adopt in England should be deferred until these and other technical questions were more settled. By pointing out that he was the only British scientist with access to all phases of the Manhattan Project, and since his recommendations were economically convenient, Chadwick's arguments for delay held sway with the Chancellor of the Exchequer, Sir John Anderson, and the committees. Oliphant continued to chafe and complained to Akers,[6] the Director of Tube Alloys in London, in February 1945: 'it is becoming increasingly clear to me that if a British effort on any effective scale is to be made within the next five years or so the background organisation must be worked out now, and the immediate steps must be taken to build and equip the basic laboratories and plants.' Unable to contain his frustration at Chadwick's apparent inertia on this subject, he continued:

> Chadwick's other duties evidently prevent him from even guessing at a date for the meeting proposed by Anderson in London. It is evident that the magnitude of the American effort in this sphere, the dominance of the Americans in the politics of the subject, the British desire for compromise in this as in all other questions demanding a clear-cut decision, and the feeling that by lying low we are getting our knowledge for nothing, all combine to prevent any serious consideration of a policy of our own.

Akers thought Oliphant's points quite logical and persuasive. Since Chadwick had been sent a copy of the letter, Akers asked him for his comments. They arrived by cable and were unyielding:

> I can see no reasoned argument in Oliphant's letter of February 8 and I cannot accept many of the statements he makes. Nothing, in my opinion, could be more ill-judged than his proposal for immediate action on EM [electromagnetic] plant, although I can understand his anxiety to be doing something at home. In interpreting Olpihant's letters you must remember that he can see only part of the picture and that he is much troubled in mind over personal matters.[7]

The previous summer, Chadwick had written to Wallace Akers[8] complaining about 'the complacency, almost amounting to indifference', with which it appeared to him that Anderson's Consultative Council were viewing the propect of a British post-war, nuclear programme. Although he agreed strongly with Oliphant that decisions needed to be made on the organization of the British Experimental Establishment, he still thought it was too soon to select the exact technology to be used. Akers informed Sir John Anderson[9] that it seemed impossible to reconcile the statements in Chadwick's cable with those contained in Oliphant's letters. It appeared to him that Chadwick had decided that there should be no atomic energy developments in England, other than to agree in principle that there would be an experimental establishment at some future date. Anderson[10] wanted Chadwick, Oliphant, Cockcroft and Peierls to return to London before the end of March 1945 to meet his Consultative Council, and suggested that Chadwick should be the one to tell Groves about the reason for the visit. Akers[9] feared that 'Groves will be hostile to any suggestion of T.A. development in this country' and would oppose any visit by the British scientists at this juncture. He suggested to Anderson that the British Ambassador in Washington, Lord Halifax, be requested to impress on Chadwick 'the need to take a strong line with General Groves' in the matter.

Over the preceding months, Chadwick[11] had already discussed postwar policy with Groves, whom he found well disposed towards the continuance of Anglo-American collaboration. Chadwick explained to the General that his guiding principle when considering present difficulties was to aim for 'orderly and stable arrangements' which could be carried over to atomic policy in the postwar era. Even with good intentions, Chadwick thought that the alternative to full collaboration could easily be competition, which might have disastrous results. He found Groves took the same view and was particularly concerned about how the current relationship might determine post-war nuclear politics. He was sure that Congress would examine the wartime collaboration on the Manhattan Project and thought that candour about future intentions was the best policy to forestall any suspicion of British motives. Now that England was not under direct threat from Germany, he warned that the British should 'be very careful to avoid giving the impression that [they] were not very interested in the Japanese war.' Significantly, Groves admitted to a long-standing feeling, reinforced by the French situation, that Anderson had a strong interest in the commercial aspects of atomic energy. Chadwick commented that 'any mention of patents immediately reminds them of past difficultes and causes some mistrust of our future intentions'.

The British Ambassador, Lord Halifax, sent Anderson a cable[12] dissuading him from recalling the senior British scientists to London for the policy review meeting. He explained that Peierls was overseeing a group of theoreticians working on the implosion problem and that 'his absence might very well delay the date on which the weapon would be ready.' Halifax had just taken over from Campbell as the British representative on the Combined Policy Committee and was unwilling to lose Chadwick's advice, while he was 'still fresh to this business'. It also seemed to Halifax that Chadwick's absence could cloud his personal relations with Groves, which were the linchpin of Anglo-American co-operation at this stage. Anderson quickly abandoned his plans for a meeting. Chadwick himself wrote a letter[13] apologizing for his inability to travel by bomber, since he was 'not in very good shape'. 'It may seem somewhat careless of me' he continued, 'not to have adjusted my state of health more conveniently, but it was hardly possible in face of the events and of the changes of the past few months.'

Instead of travelling to London at the end of March, Chadwick sent Anderson a seven-page memorandum[14] giving his views on future Tube Alloys policy. He told Anderson that Bohr, who was visiting London, could supplement the written account because he, unlike Cockcroft or Oliphant, had an overview of the whole position. Chadwick passed on General Groves' opinion that it would be 'a grave mistake' for the British to begin a large-scale independent programme now. Anderson was told to expect 'a very powerful military weapon' to be ready by August. Chadwick thought the technical problems of civil nuclear power could be overcome, and although it was impossible to say how it would work economically, 'both aspects, the military and the industrial, of TA development have such great potentialities that we cannot afford to neglect either of them'. He did not believe that this meant matching the enormous expenditure of the Americans, 'but any worthwhile effort must be a big one. We cannot expect to get milk without feeding the cow, and this is a very big cow.' Chadwick continued to believe that from a scientific point of view, a delay in the British programme was not a bad thing because several outstanding questions would be answered in the next few months. 'But', Chadwick wrote, 'the chief consideration in my mind is political rather than technical, and concerns our relation with the U.S.' By collaboration, the two countries could 'establish a predominating position in this field', and so put themselves into a strong military position from which he hoped they could ensure 'the wise and just use of a very dangerous power.' He urged that the British continue 'to throw ourselves heart and soul into the U.S. effort' because this would help future relations. The next few months, in his view, would be the crucial period as the culmination of the project was approached. 'It is as certain as such things can be,' he continued, 'that a weapon will be ready by August, and then the whole project will come to the test. It seems to me essential that we should be associated with the project to the fullest possible extent, right up to its culmination and beyond.' He recognized that this might demand some sacrifice not only from individuals working in the US, but more broadly from the universities, who would be deprived of well qualified staff, and from commercial organizations, notably ICI. Chadwick realized that in advocating this policy he lay himself 'open to the charge of neglecting or forgetting the interest of my country.' This he regarded 'as a superficial and short sighted view.'

In terms of immediate action regarding an experimental establishment in Britain, Chadwick[15] proposed that potential staff should be earmarked, raw materials such as graphite and uranium oxide acquired, and some equipment such as a cyclotron ordered. There should be long term planning for a controlling organization and for integration with the current programme in Canada, and 'the possibility of a far-reaching agreement with U.S. for post-war collaboration' should be kept 'ever in view'. So concerned was Chadwick not to alarm the Americans that any inkling of independent activity in England would cause him to send a note of warning to Sir John Anderson. Thus on 2 April, he feared 'Cockcroft may be somewhat too ready to get very active on detailed design and construction'.[16] Two days later, 'some remarks of Oliphant have given me cause to think that the situation in Montreal might be incompletely appreciated in some quarters and that the promise of the undertaking and the advantages to us were being underestimated.'[15] He reminded Anderson of the substantial contribution of the US to the Montreal programme—they provided information, heavy water and uranium. He was anxious, too, to avoid over-optimism, and wished to counter 'the natural rush of enthusiasm' with 'simple fundamental facts'.

Oliphant was becoming increasingly unhappy and was missing his wife and family, who did not accompany him to Berkeley. He announced to Groves that he intended to leave Berkeley at the end of March 1945 in order to return to England. He had felt for some time that his team were approaching the end of their useful work and would be more valuable back in the UK, where they would be in position to revitalize their university science departments as soon as the war was over, not to mention being in on the ground floor of any independent British nuclear programmes. In Oliphant's eyes, what the British 'are doing now gives only the impression that we are trying to muscle in on a racket we have been too dumb to develop ourselves.'[17] Always plain spoken, he had not flinched from open criticism of Groves in the past, as when referring to 'the clumsiness of the Army organisation, which neither controls nor checks the operation except in a very desultory and inefficient manner'.[18] Apart from the nature of the remaining work, Oliphant was finding it increasingly difficult to maintain the necessary subservient role to Ernest Lawrence, who did his best to make adjustments, but who in Chadwick's words,[17] although 'a delightful friend and a very able man', could only run a laboratory as a dictator. Chadwick was concerned that the relationship between Lawrence and Oliphant had reached breaking point, and wished to avoid an open disagreement between them which could spill over onto the wider canvas of scientific co-operation. As he told Anderson, 'it is impossible to restrain Oliphant from a distance of 3000 miles' so that his policy became 'to keep things straight with Groves, who understands Oliphant's little ways.' It was great credit to Chadwick's tact and openness that Oliphant left Berkeley in March 1945 with Groves' blessing and with no ill-feeling on Lawrence's part.

There had been one earlier return to England in which Chadwick had again been closely involved. Joe Rotblat[19] had decided that since it was taking the Americans so long to build the bomb, his 'fear of the Germans being first was groundless'. He found, therefore, 'the whole purpose of my being in Los Alamos ceased to be', and he asked Chadwick for permission to return home. When Chadwick passed on the request to the Army authorities at Los Alamos, he had been shown a thick dossier on Rotblat containing highly incriminating evidence which pointed to him being a spy for the Russians. It appeared that he had informed a young woman of his acquaintance in Santa Fe that he intended to return to England, join the RAF to parachute out over Russia or occupied Poland, and then to establish contact with the Russians in order to betray his considerable knowledge of the Manhattan Project. In Rotblat's words:

> The trouble was that within this load of rubbish was a grain of truth. I did indeed meet and converse with a person during my trips to Sante Fe. It was for a purely altruistic purpose, nothing to do with the project, and I had Chadwick's permission

> for the visits. Nevertheless, it contravened a security regulation and it made me vulnerable. Fortunately for me, in their zeal the vigilant agents had included in their reports details of conversations with dates, which were quite easy to refute and to expose as complete fabrications. The chief of intelligence was rather embarrassed by all this and conceded that the dossier was worthless. Nevertheless, he insisted that I not talk to anybody about my reason for leaving the project. We agreed with Chadwick that the ostensible reason would be a purely personal one: that I was worried about my wife whom I had left in Poland.

Chadwick[20] could see other serious, political difficulties from Rotblat's continued presence on the Hill as the bomb project came to fruition. He thought that if Rotblat's family were still alive, they would probably be living in Russian-occupied territory after the war, raising the prospect of blackmail. He discussed this with Groves, who agreed that Rotblat should leave immediately. Chadwick also saw advantages in Rotblat's return in that he had acquired such a close knowledge of all branches of nuclear physics which should prove useful in the postwar period. Rotblat travelled to the East Coast and stayed with Chadwick in Washington for a few days. This was just before Christmas 1944, when there was such uproar about Halban's visit to France. Before leaving Los Alamos, Rotblat[19] had packed all his research notes and personal documents into a wooden box made by his laboratory assistant. Chadwick accompanied Rotblat to Union Station, where he boarded the train for New York and together they carried the box to the train. When Rotblat arrived in New York, a few hours later, the box had disappeared and he had to sail back to Liverpool without it. The box was never recovered and it seems most likely that it was confiscated by US Army Intelligence, who still harboured suspicions about Rotblat.

The Americans were also dissatisfied with the security vetting of non-British scientists working in Montreal,[21] and continued to press John Cockcroft for their biographical details. This put him in a difficult position because the British security forces were in the habit of issuing clearance notifications with no further details. Chadwick could not see why 'a definite statement that foreigners have been thoroughly investigated by our Intelligence people'[22] should not suffice, but attempted to placate General Groves by occasionally giving a brief sketch of the 'more interesting cases in private conversation'. Groves[23] found that he was 'most punctilious in informing me of the slightest question of background, including that of German blood'. There was one case where Chadwick badly misjudged one of the British scientists. With remarkable prescience but no real suspicion, Groves[24] had become concerned about the amount of time that Alan Nunn May, who was based in Montreal, had been spending at the experimental nuclear pile at Argonne, near Chicago. Chadwick interviewed May, who informed him that he had not picked up much information while working at Argonne.[25] Chadwick knew May both as a research student at the Cavendish in the early thirties, and more recently from Tube Alloy research carried out jointly between Bristol and Liverpool, early in the war. May had been Powell's chief assistant in Bristol, and had been one of the co-authors with Chadwick of the letter to *Nature* in 1941; Chadwick reassured Groves that May 'was exceptionally reliable and close-mouthed'.[25] Despite his own special status, even Chadwick was not above suspicion apparently: in January 1945 he was unable to visit Canada because of strong objections on the grounds of security from the Americans, which he thought 'it would be most impolitic to ignore'.[26]

The divergence of opinion between Chadwick and Oliphant[27] on the development of a nuclear research establishment in England rumbled on for a while. Chadwick's constant calls for restraint plainly frustrated both Oliphant and Cockcroft at times, but in fact he became increasingly proactive as the prospect of the atomic age became ever clearer. He sent Anderson[28] a detailed minute on the UK proposals at the end of May. Commenting on a suggestion of Cockcroft's that the research station should be sited so that there was easy access to Oxford or Cambridge, Chadwick could see that there were 'obvious advantages to be gained by promoting the interchange of visits with university laboratories, but we must avoid degrading the Establishment to a mere house of pleasure for visiting scientists'. He favoured a site in the Midlands, in proximity to an industrial centre with its manufacturing facilities, materials and supply of skilled labour. By this time Chadwick urged some immediate action to:

1. find an airfield which is reasonably suitable as regards facilities and reasonably convenient;
2. get it;
3. get on with the job of providing equipment;
4. begin to offer posts to staff.

He did not foresee any difficulty in recruiting young scientists, but thought there would be considerable competition for the 'few senior men' required. By this stage, Sir Edward Appleton, the Director of DSIR, had agreed with Sir John Anderson that the first chief of the new organization should be John Cockcroft. Appleton contacted Chadwick asking for his endorsement of this choice, which was granted but with some apparent reservation:[29]

> He has many virtues which would contribute to the smooth running of the Establishment, and he would keep his hand on all the strings. In common with most scientists he also has many faults and those we must recognise from the beginning so that we can supplement him by a suitable choice of assistants. For example his knowledge is wide but it is not at all profound; his views are of rather a dull, everyday hue. On the other hand, his temper is so equable and his patience and persistence so inexhaustible that we can put in lively and relatively irresponsible men who have the real feeling for research without fear of upsetting the balance.

Chadwick himself had not had the opportunity to indulge his own love of basic research for some years. This omission contributed to the disenchantment he sometimes felt for his new status in life. In a letter to Oppenheimer, written in April 1945, he expressed some melancholy sentiments about his present duties for many of which 'I am fitted neither by temperament, training nor ability'.[30] For a while after his permanent move to Washington, Chadwick continued to supervise administrative matters concerning the British group at Los Alamos. These would include dealing with requests from scientists who wished to return to England, and with letters from their mothers in England asking what had become of them. There were also the usual wrangles over promotion and pay, and for example when Philip Moon was given an increase in salary, Chadwick[31] asked him not 'to breathe this over the loudspeaker system' at Los Alamos. He soon realized that the arrangement was inefficient and asked Rudolf Peierls to take charge of the Los Alamos affairs. He also asked Peierls to keep him abreast of scientific progress, which Peierls did by writing a series of detailed letters, sometimes in conjunction with Moon.

Moon joined the team headed by Kenneth Bainbridge, the Harvard physicist, which was to test the implosion model of the plutonium bomb in Project Trinity. Bainbridge had chosen as a proving ground for the test an area of flat scrub land in the Jornado del Muerto valley in southern New Mexico.[32] The site was about 200 miles south of Los Alamos and formed the northern limit of the huge Alamogordo Army Air Field. As the name 'Journey of Death' suggests, it was an intensely inhospitable place—fiercely hot and replete with poisonous snakes, spiders, and scorpions. As a rehearsal for the actual Trinity test, Bainbridge's group detonated a 100 ton charge of high explosive mixed with a small amount of radioactive material on 7 May. Apart from serving to test functions like transportation and communication, the 100 ton shot provided the opportunity to calibrate instruments that would be used at the definitive bomb test to measure the magnitude of the explosion and the distribution of radioactive fission products. Peierls and Moon[33] soon conveyed the lessons learned to Chadwick. They described the difficulties encountered in measuring the blast waves in air and seismic earth shock due to the 100 ton explosion, but thought that most instances of equipment failure had obvious causes. According to them, 'the only surprising thing was that some mice that had been placed at about 400 yards distance were killed by lung haemorrhage. The pressure at that distance was believed to be something like 1 pound per sq. inch, which British mice withstand successfully.'

The letters continued at a rate of about one a week through May and June, and apart from including reports on contributions made by British scientists, they also gave details on the new methods available for analysing implosions (the technique developed to detonate the plutonium bomb). As a result Chadwick was conspicuously well-informed of all developments—to such an extent that Groves started to feel inadequately briefed by comparison. When he found out that Peierls was the source of Chadwick's inside information, he sent word to Peierls asking for copies of the letters to be sent to him too. Peierls[34] had been concerned that by sending the letters, he was in contravention of the strict rules governing the communication of information from Los Alamos; Groves' flattering request brought him both relief and amusement.

Two letters written at the end of May 1945[35] reveal how many intricate problems relating to the fusing of a plutonium bomb remained to be solved. The methodology of implosion (whereby the central mass of plutonium would be squeezed into a critical form) required the science of explosions to be developed at a more detailed level than ever before. The theoretical solution had been provided by the Hungarian mathematician von Neumann a year earlier. The greatest English expert on fluid mechanics and the theory of explosions was Geoffrey Taylor from Cambridge—in happier times, one of Rutherford's golf partners in the Trinity College Sunday morning foursome. Groves was so keen to have Taylor at Los Alamos that Chadwick wrote to Appleton that 'anything short of kidnapping would be justified'.[36] Another key ordnance expert whom Chadwick had prised away from London was William Penney, a mathematician from Imperial College, who had worked on the effects of high explosives for the Home Office during the blitz. Taylor and Penney both advised on the device being tested at Los Alamos, termed the lens system, which depended essentially on shaped high explosive charges (the lenses) surrounding the plutonium core of the bomb. There were two forms of explosive used with different speeds so that after detonation the pressure wave would be shaped inwards and focused onto a sub-critical arrangement of plutonium, which would experience a compression of tremendous force and be squeezed down into a solid, super-critical mass. In his letter of 23 May, Peierls[35] wrote that the production of full-scale lenses was just beginning. One practical problem concerned the presence of minute flaws in the mass of explosive:

> The question of the effects of gaps or cracks in the H.[igh]E.[xplosive] is being given much attention . . . The most spectacular effect is an acceleration of the detonation wave which is observed if the detonation wave runs along a crack of finite length. The explanation is probably that the crack acts as a cavity charge and a jet of some power runs down the crack and initiates the H.E. at the end ahead of time. This effect has been observed with gaps as small as 1/84th inch.

This gave some concern about the integrity of the explosive lenses following assembly and transportation, and Peierls mentioned that they would be lacquered and reinforced to prevent powdering or chipping.

The other crucial challenge facing the Los Alamos scientists was to initiate the nuclear chain reaction in the plutonium at the split-microsecond that it reached its maximum density. A stray cosmic neutron could not be relied upon to provide the necessary stimulus, and it was decided to incorporate a source of neutrons at the centre of the plutonium core. The device to be used was the same combination of materials, polonium and beryllium, that Chadwick had employed in his modest apparatus at the Cavendish in 1932, when he had identified the neutron for the first time. To achieve a successful initiation, the α-particles from the polonium had to be prevented from coming into contact with the beryllium metal until the instant that the chain reaction would be triggered. Peierls[35] was able to report that the chemists had made remarkable progess by plating the polonium with a microscopic coating of nickel which would arrest the α-particles until disrupted by the detonation wave. Two British workers were evidently making significant contributions to this work:

> In connection with the theory of initiators Fuchs has worked a great deal on the

> theory of the penetration of Munroe jets. An existing theory by Mott does not seem quite adequate for our purposes. Considerable help in this will come from a recent experiment by Tuck*, who succeeded in photographing with X-rays the penetration of a jet of wax into a block of wax having a sheet of lead over its surface. Tuck has a series of photographs at different times showing up the mechanism most beautifully. This apparently is the first time anyone has succeeded in photographing the penetration of a jet, although it has often been tried.

Another point which only occurred to the physicists at this late stage was that some neutrons from the polonium-beryllium mixture could be scattered back from the explosive shell as slow neutrons and 'if they could return into the active material they would considerably decrease the critical mass'. Peierls[35] was led to conclude that:

> pre-detonation is determined not by the number of background neutrons during the time interval in which the gadget will be critical, but during the interval from which one of the off-spring of such a neutron has a chance of surviving until the gadget does become critical. This point is very similar to the point made by Feather a long time ago in connection with delayed neutrons. All this makes it necessary to absorb as many of the slow neutrons as possible and there are essentially three proposals.**

The first nuclear explosion at the Trinity test site had been planned for 4 July, but on 13 June Peierls[37] mentioned to Chadwick that there would be a delay of a week or two because the fabrication of the explosive lenses had started later than scheduled. At the end of June, refinements were still being made in the detonators; there had been problems with the insulation of electrical wiring which was now solved by covering the leads with nylon sleeves. Tuck was starting a new programme of experiments 'to observe detonation waves and their immediate effects in a way that will allow one to calculate detonation waves and other data directly'. There had been discrepancies between the measured velocity of shock waves and the compression produced by them, and the theoretical shock wave equations. Moon[38] reported to Chadwick that these discrepancies were 'beyond experimental error (in most cases something like 20%) and . . . beyond the error of the theory, supposing that the physical constant of the explosive and of the tamper have been correctly chosen and that no wrong assumption has been made by the theorists'. There were other major unresolved issues such as what effect the intense electromagnetic radiation generated in the chain reaction would have on the later stages of the nuclear reaction. Edward Teller told Moon 'that these effects could conceivably increase the efficiency [of the bomb] manyfold'.[38]

As the huge scientific team now assembled on the project stretched their collective ingenuity to ensure that the uranium-235 and plutonium bombs would work as intended, there were a few individuals who stepped aside from the juggernaut and asked where it might be leading the world. Foremost amongst them was always

---

* James Tuck had begun his scientific career as an assistant to Lord Cherwell at Oxford.

** The three proposals remain classified information.

Niels Bohr, who was regarded as the philosopher king by the scientific community, but whose ideas on sharing nuclear technology were so roundly scorned by Churchill. Bohr was still refining his proposals, which were expressed in rather abstract and circumlocutionary terms, comprehensible to theoretical physicists, but not likely to find favour with a sceptical audience of generals and politicians. In the autumn of 1944, Sir Henry Dale,[39] the President of the Royal Society, had discussed with both Bohr and Chadwick their deep anxieties about the future uses to which the nuclear weapons they were helping to create might be put. By June 1945, various memoranda from Bohr were circulating in Washington and were passed to Groves and Stimson, the Secretary for War. These writings were characterized by Makins,[40] a senior Foreign Office man in the Washington Embassy, in a memo to the War Cabinet Office in London, in the following terms:

> As Bohr's papers deal in ideas, and as he always avoids or evades questions on the actual working of control, inspection, etc., he has really shot his bolt and can, in the opinion of Chadwick and myself, do no more good at the moment over here either on the political or scientific side. On the other hand, his continued presence may cause some prejudice to our relations with the Americans for a general and special reason.
>
> The general reason is that the Americans are, as you know, still apt to be suspicious and alarmed on security grounds and Bohr is, of course, regarded as a British agent. As he is always popping round, some of the Americans naturally wonder what he is really up to.
>
> The special reason is that he has got himself involved with [Justice] Frankfurter who, as you know, having more or less accidently obtained knowledge of T.A., is taking a keen interest in it, much to the embarrassment of most of the people involved on the American side. You can readily understand how the activities of Frankfurter and Bohr together could muddy the waters at a moment when we particularly want to keep them clear.

Apart from Bohr's voice of conscience, there was a report from six scientists working on the project in Chicago on the social and political implications of the nuclear bomb. This group was led by James Franck,[41] who had shared the 1925 Nobel Prize in Physics and who had previously shown the stamp of moral courage when resigning from his Chair at the University of Göttingen in 1933 as a personal protest against Hitler's Nazi regime. His report regarding the use of the bomb sounded a warning against any action by the US in the present conflict, which might undermine its future standing in the international community and make the international control of nuclear arms more problematic. In particular, the scientists counselled against dropping bombs on Japan without prior warning. The Franck Report surfaced in Washington in June 1945,[42] and at the request of the US War Office was considered by a scientific panel comprising Oppenheimer, Fermi, Lawrence and Compton, the head of the Chicago Metallurgical Lab. It was Arthur Compton who had commissioned Franck to write the document only two weeks earlier. At the end of the panel's deliberations, Oppenheimer drafted a statement[43] which effectively dismissed the idea of a preliminary, non-combat use of the bomb: 'We recognize our obligation to our nation to use the weapons to help save American lives', he wrote, adding: 'We can see no acceptable alternative to military use . . . We can propose no technical demonstration likely to bring an end to the war.' Oppenheimer's panel did, however, endorse the Franck Report's recommendation that the Soviets should be informed before the bomb was dropped on Japan, on the basis that if the secret was shared with them, they would feel less threatened and therefore be less likely to enter a nuclear arms race with the US.

Although the decision would finally rest with the President, Oppenheimer's panel at least had some indirect influence. President Truman had only learned about the existence of the Manhattan Project following Roosevelt's death in April. He was briefed both by his Secretary for War, Henry Stimson, and General Groves on the technical aspects of the Project, its military potential, and also the challenge of international control. Truman agreed to Stimson setting up the Interim Committee—a body of civilians which included Bush and Conant—to consider these matters until they came into the public domain after the first bomb was used. The committee, in its short existence, came to be dominated by James Byrnes, who would later become a dominant figure in American nuclear politics as Truman's Secretary of State. The Interim Committee[44] had little real influence on the decision to use the atomic bomb (the notion that it did was derided by Groves as 'plain bunk').

Chadwick learned of the Franck Report, probably from Groves, and he was well aware of fundamental concerns that were troubling a few scientists such as Bohr and Rotblat. He was less inclined to disclose information to the Soviets at this stage. Although not overtly political, Chadwick certainly did not share Rotblat's socialist sympathies, and his close relationship with General Groves had made him acutely aware of the future military threat posed by the Soviets. Philosophically, he believed that there should be international co-operation to avoid potentially catastrophic competition, but he could not see how the necessary safeguards could be enforced and was constitutionally opposed to excessive risk taking in such a dangerous sphere.

In the early summer of 1945, there was still no guarantee that the plutonium bomb would work because of the untested implosion method of detonation, and there was insufficient of the 235 isotope to manufacture a uranium bomb. These concerns weighed heavily with Chadwick on the question of whether there should be a preliminary demonstration of the bomb before it was used in earnest. In his opinion, there was not 'enough material to play around with'; more subtly, even if it could be made to work, he was concerned there was no readily identifiable group who could be assembled at will, to whom the bomb could be demonstrated, and who would then be able to ensure that Japan surrendered.[45]

On 4 July, Chadwick was invited with Halifax, the British Ambassador, to a meeting of the Combined Policy Committee in the Pentagon.[42] Groves also attended, but the formal business was conducted between Stimson, for the US, and Field-Marshal Sir Henry 'Jumbo' Wilson, for the UK. The prime purpose was to satisfy the conditions of the Quebec Agreement in that the Americans needed to obtain the consent of the British before using the bomb or disclosing its existence. Agreement to use the bomb against Japan was readily given by Field-

Marshal Wilson, but there were reservations both on the narrow issue of whether Stalin should be told about the bomb at the forthcoming Potsdam Conference and on the broader question of a public scientific report to be issued once the bomb had been used. It was decided that a sub-committee comprising Bush, Chadwick and Groves should draft guidelines governing the general release of information. Chadwick's opinion carried no particular weight, although it would be of interest to Groves who had come to have such a high regard for his judgement. The two men would meet several times a week, usually in Groves' office, and would talk privately for 15 or 30 minutes before others were admitted to the room. Their mutual confidence was now so great that it was natural for Chadwick to confide in him, although the brusque General would still on occasion startle him. Once he sought advice from Groves, when an off-hand remark about the treatment of captured German scientists had been transmitted directly to London as Chadwick's serious policy suggestion. It appears he had stated that the scientists should be kept apart from the French and the Russians, and sent to the Ruhr Valley to teach. By this time, the likes of Anderson and Akers in London tended to pay attention to any of Chadwick's opinions. Groves' response 'that the German scientists should all be shot as they were undoubtedly as great war criminals as anyone else in Germany' was 'not relished by Dr Chadwick'.[46]

On the same Independence Day that the CPC met in Washington, Peierls[47] wrote to Chadwick from Los Alamos. There would be no more evening colloquia until after the Trinity test, he said, 'since people can hardly be expected to have the detached attitude necessary for that kind of discussion', such was the level of anticipation. At the last meeting in the series, Hans Bethe had reviewed the theoretical basis for their predictions of the bomb's efficiency. In Peierls' view the estimates were reasonable, providing 'nothing goes wrong with the detonators, that the initiator works, and there is no predetonation'. Although the newly insulated detonator wires seemed much more reliable, and the discrepancies concerning the shock waves from the lenses and the compressiblity of the core material were being narrowed, there was still an abundance of reasons for Chadwick to remain sceptical about how all this would work in practice.

While the dramatic events in New Mexico were progressing rapidly to their historic climax, Chadwick wanted to concentrate on them to the exclusion of everything else, and his correspondence shows he was increasingly unwilling or unable to give his mind to other matters. Rotblat wrote to him regularly about the state of the Liverpool department. He was very troubled by the shortage of staff, the status of the honours course, and the future of the research programme, which Chadwick had so carefully nurtured in the late 1930s. The university were contemplating the purchase of a new site for a bigger cyclotron after the war, and desperately wanted Chadwick's advice. For once, Chadwick did not reply to these enquiries and Rotblat was left in limbo. The lack of communication fed the growing belief that Chadwick would not return to Liverpool after the war.[48]

Another voice that was uncharacteristically ignored by Chadwick during this period was Mark Oliphant's. He had returned to Birmingham, with Chadwick's help, but was still agitated about other British scientists being kept on at Berkeley and not being allowed to return to their universities in Britain, where, in his opinion, they were desperately needed. He saw this as a political ploy on General Groves' part to keep them tied up, 'wasting their time on trivialities'. He told Philip Moon[49] that he thought he had persuaded Chadwick that this was the true state of affairs, and expected him to arrange their return as quickly as possible. Oliphant confessed to Moon he felt unhappy that he 'found it necessary to press [Chadwick] so hard.' Oliphant continued: 'He has a thankless task to perform, and I know of no-one else who could have carried on through all the vicissitudes, and despite the indignities he has had to suffer.' Regardless of this, Oliphant felt that '*our* task is now here and our loyalty is to Britain—not to General Groves'. In truth, by this stage, the momentum of the Manhattan Project was irresistible, and Chadwick would have found it impossible to dissociate himself from Groves, even if he had wished to do so. Instead Chadwick[46] continued to strive for the placement of British personnel in key positions so that they would gain as much valuable information as possible. He wanted to include at least one British scientist in the base party for the forthcoming Trinity test, in addition to himself. Ernest Titterton had proved himself to be such a master of electronics that he was entrusted with the task of generating the electronic pulse which would fire the plutonium bomb. Although Groves was reluctant to include another English scientist, Oppenheimer had recommended Otto Frisch, who Chadwick did not think would be a good choice because he was too inclined to make unnecessary, last minute, scientific changes. His preference was for Penney, who had been advising on the placement of bunkers and shelters for observers at the test. Penney was also to be responsible for certain critical measurements such as the ignition of structural materials from the radiant heat emitted by the blast.*

The revised date for the Trinity test was Monday, 16 July. A hill called Compañia, 20 miles northwest of ground zero was selected as the viewing point for some of the most senior scientists, including Chadwick. He had been driven there across the Jornada del Muerto with Edwin McMillan, a physicist from Berkeley. To pass the time on the journey, they discussed the question of international control of atomic power and specifically whether the Soviets should be informed about the newest and most devastating weapon in the history of warfare. Chadwick dismissed the notion out of hand: 'Why should we? They don't give us anything.'[50] At Compañia, they joined McMillan's boss, Ernest Lawrence, as well as Hans Bethe and Edward Teller.

It has been written that Chadwick came to the Trinity site 'to see what his neutron was capable of'.[51] The most devout of Rutherford's followers, Chadwick would have instantly recoiled from any suggestion of proprietary rights to the neutron. At the Cavendish, subatomic particles remained the property of God;[52] even Rutherford was only the midwife to the birth of the α-particle. Nevertheless, it is true that

---

* In the event, Frisch attended the Trinity test and wrote a classic description of the explosion, and Penney, who was meant to observe from a bomber, was grounded by bad weather at Albuquerque Air Base.

Chadwick's discovery of the neutron marked the unintentional first step towards man's loss of innocence in the field of nuclear weaponry. There had been other crucial but unwitting contributions, most notably the elucidation of nuclear fission, as we have seen. With the outbreak of a conflict which threatened civilization in Europe, leading physical scientists recognized that the Nazi regime could not be allowed to secure a lead in this deadly race and had deliberately set themselves the task of constructing a bomb of unprecedented force. Chadwick had been in the forefront of this effort, first in England, where he tirelessly co-ordinated work at several universities despite all the difficulties in communication imposed by the war; apart from his accustomed role as scientific mentor, he smoothed personal and academic rivalries, and drafted the crucial Maud Report, which convinced not only the British leaders but more importantly those in the United States that a nuclear weapon was feasible in the present war. When the Americans eventually took over the endeavour and turned it into the mightiest military engineering project the world had ever seen, Chadwick stayed at the centre of activities, lending unswerving support to General Groves and in return ensuring a continuing stake for Britain as a nuclear power. Now all these years of relentless strain and responsibility were to be put to the most dramatic test.

The weather was appalling—tropical thunderstorms, lashing rain and high winds. They waited together in the cold, black, desert night. Chadwick, like Groves, was concerned that the weather would mean a postponement in the the realization of 'the boldest and certainly the most expensive experiment in scientific history'.[53] Groves, who was under immense pressure to conclude the test while President Truman was with Stalin at the Potsdam Conference, reluctantly put his faith in the meteorologist who predicted the storms would abate at dawn, and ordered the detonation for 5.30 am. Chadwick[53] described the birth of the atomic age:

> In the faint light before dawn, the last few minutes went slowly by as we prepared ourselves for the critical moment. I sat on the ground holding a piece of dark welder's glass before my eyes, shielding the sides with my hands. Suddenly I saw an intense pinpoint of light which grew rapidly to a great ball. Looking sideways, I could see that the hills and desert around us were bathed in radiance, as if the sun had been turned on by a switch. The light began to diminish but, peeping round my dark glass, the ball of fire was still blindingly bright . . . The ball had then turned through orange to red and was surrounded by a purple luminosity. It was connected to the ground by a short grey stem, resembling a mushroom.
>
> Meanwhile, about 100 seconds after the appearance of the pinpoint of light, had come the sound of the explosion, sudden and sharp as if the skies had cracked. This was followed by a long rumbling noise—the echoes thrown back by the surrounding hills—as if huge noisy wagons were running around.
>
> Although I had lived through this moment in my imagination many times during the past few years and everything happened almost as I had pictured it, the reality was shattering. What had begun in a few simple ideas had at last been put to the test of experiment, and the awe-inspiring nature of the outcome quite overwhelmed me.

Now that the test had been successful enough to ensure an effective weapon,

Chadwick thought the next logical step was to use it in combat as soon as possible. That is not to say he did not harbour serious doubts about its use. Until the Trinity test he may have retained a quiet, slight, hope that it would not work, although his scientific intellect had long led him to expect it would. Once the Trinity test was over, Chadwick returned to Washington, where for the first time in many months he could attempt to turn his mind to other things. But he found himself shattered by what he had witnessed at Alamagordo, whenever he thought about it; for the first time in his life, he was unable to concentrate at will. He even found himself being indecisive about plans for the Liverpool department, and in a letter written to Rotblat[54] on 26 July, he was almost apologetic:

> I wish to add that it is very difficult for me to give proper consideration to such matters as these under the present circumstances. I have really very little time to spare for them and no one to discuss them with, and, what is worse, I am now quite out touch with conditions in the laboratory. I am therefore content to leave a great deal, almost everything in fact, to the judgement of Roberts and yourself, etc. This does not mean that I am comparatively uninterested. Far from it. But I am living in a different world from which I cannot easily return.

The world became a truly different place on 6 August, when a single uranium-235 bomb exploded in the morning sky over Hiroshima. Three days later, an adaptation of Chadwick's signature, polonium–beryllium, device was used to generate neutrons to initiate the fission process in an imploded mass of plutonium; and Nagasaki was devastated. On 10 August, Japan, with her navy and air force already decimated, agreed terms for surrender, and the war finally ended on 14 August. America had unleashed a weapon of unimaginable terror, her military were cock-a-hoop and already her politicians were plotting to keep the technology the exclusive property of Uncle Sam. Great Britain, which still had imperial pretensions, were quietly determined to produce their own bombs, and unknown to the West, the Russians had already made a start by acquiring vital information from at least two scientists, Fuchs and Nunn May, working on the North American continent.

Even before detailed assessments of the damage caused by the bombs at Hiroshima and Nagasaki were available, Chadwick thought it was a mistake to regard them, as some had suggested, as 'merely a bigger and better bomb'. In his opinion, the two bombs already went 'so far beyond any other weapon of destruction as to constitute a menace to civilisation'.[55] Despite this chilling judgement, he never recanted the use of the bombs against the Japanese. He believed the decision was justified in an attempt to save the lives of American and Allied servicemen, as well as prisoners of war who would die in their thousands if Japan were slowly strangled by conventional methods of bombardment. On the moral question of tens of thousands of innocent Japanese civilians being killed at a stroke, he saw no difference between their fate and similar numbers who had perished in the fire bombing of Tokyo. Six years of war had irredeemably altered the public tolerance for atrocities and brutality. Chadwick found some grounds for hope in the way the bomb had been born out of this historic global struggle:[55]

> It is, I think, fortunate that this weapon was developed and produced during the war, firstly because its development in time of peace would have occurred more or less concurrently in different countries and competition would thereby have been inevitable, and secondly because the sufferings and havoc of the present war have branded into our minds the merciless nature of war and have made us long for peace as never before.

On 12 August, it was decided to draw attention to the British contribution by arranging for Chadwick to give a press conference.[56] He opened with a heartfelt tribute to General Groves, without whom 'the atom bomb would not have been ready on this day'. He thought the industrial applications of atomic energy would be realized in about a decade. His next prediction was much less encouraging for the domestic audience: in answer to a question about how far ahead the US were in the nuclear field, Chadwick answered that 'any nation could learn the secret in about five years of experimentation, assuming it had access to the necessary raw materials'. Although he knew that there could be no lasting monopoly in nuclear weaponry, he did not believe in the immediate disclosure of knowledge gained on the Manhattan Project. Indeed he was if anything even more conservative than General Groves on this issue. Anticipating difficulties with scientists from the project wishing to explain their work to the public, in April 1944 Groves[57] had commissioned Henry D. Smyth, a physicist from Princeton, to write a report which would be the official account and establish the boundaries of what was permissible in terms of national security. The heavily vetted Smyth Report was given a final review at a meeting in Secretary Stimson's office on 2 August, where Chadwick saw it for the first time. After the meeting broke up, he read the entire document and was concerned by its contents. He recorded that his first impression was 'disquiet and almost alarm that it should be considered necessary or desirable to release the amount of information given in the report'. At a subsequent meeting, Groves and Tolman successfully sought to allay his worries, and Chadwick[57] sent the following message to Groves:

> I am now convinced that the very special circumstances arising from the nature of the project, and of its organisation, demand special treatment, and that a report of this kind may well be necessary in order to maintain security of the really essential facts of the project. To judge how far one must go in meeting the thirst of the general public for information, and the itch of those with knowledge to give it away, so as to preserve security secretly on vital matters is indeed difficult, but so far as I am competent to express an opinion, I find myself broadly in agreement with my United States colleagues.

On reflection, Chadwick thought the assistance to a potential competitor given by the report was more apparent than real, and his estimate of the time that it would save them might be three months or so.

Two months later, he expanded his views on the complex problems of international nuclear control in a private letter to an American scientist who had worked at Oak Ridge:[58]

> There rests upon us a grave responsibility to do our best to ensure that this great

advance in the application of physical sciences shall be used to the benefit of all humanity and that it does not become a menace to civilisation. At the same time, while hoping and striving for international understanding and a harmonious relationship between nations, we must not endanger or prejudice the security of our own countries. I mean by this, for example, that before any important technical information is divulged to other parties we should require concessions which will guarantee, as far as possible, common safety against aggression or unscrupulous use. No international agreement which will give any satisfactory form of control is possible without far-reaching concessions from all partners. I believe that the world in general is satisfied that the United States and Great Britain are peace-loving nations and will not act aggressively; we must be assured of the goodwill of other nations before they are admitted to full knowledge and to our fellowship.

I think that our statesmen have a great, perhaps a unique, opportunity to remove some of the suspicion in international relations and to replace it by frankness. But it will need good will, understanding and above all a firm and purposeful policy.

There are serious difficulties in any form of international control which has so far suggested itself to me.

While atomic energy presented the possiblity of enrichment of human existence as an alternative way of supplying large amounts of energy, Chadwick[55] was in no doubt 'the development of the atomic bomb has created a crisis in human affairs'. The remedy to the crisis did not lie in the field of science. 'It is a problem of statesmanship, of human relations, and of peaceful collaboration between nations the solution of which cannot be found by scientists alone, but to which they can render valuable services.'

References

I was permitted to read the file CAB 126/259 (*Proposed construction of an atomic energy plant in England)* at the Cabinet Office archives in London, but it should be at the Public Record Office in Kew by now.

The long, detailed letters from Peierls (and Moon) in Los Alamos to Chadwick in Washington, DC have never been declassified in England; although they are available in part in the National Archives 2 in College Park, Maryland, they have been re-censored because of security concerns as recently as November 1994.

1 Campbell, R. (29/1/45). Cable to C. Barnes. CAB 126/259, PRO.
2 Tube Alloy minutes for meeting, 11/9/44. CAB 126/259, PRO.
3 Minutes of meeting in DC, 11–12/11/44. CHAD IV, 2/11, CAC.
4 Akers, W., and Perrin, M. (1945). Memo Tube Alloys, January. CAB 126/259, PRO.
5 Chadwick, J. (17/4/44). Letter to W. Akers. AB 1/566, PRO.
6 Oliphant, M.L. (8/2/45). Letter to W. Akers. CAB 126/259, PRO.
7 Chadwick, J. (23/2/45). Cable to W. Akers. CAB 126/259, PRO.
8 Chadwick, J. (24/6/44). Letter to W. Akers. AB 1/615, PRO.
9 Akers, W. (24/2/45). Letter to J. Anderson. CAB 126/259, PRO.

10 Anderson, J. (23/2/45). Cable to Lord Halifax. CAB 126/259, PRO.

11 Chadwick, J. (1/2/45). Memo of conversation with General Groves, 1/2/45. CHAD IV, 2/11, CAC.

12 Lord Halifax (15/3/45). Cable to J. Anderson. CAB 126/259, PRO.

13 Chadwick, J. (23/3/45). Letter to J. Anderson. CAB 126/259, PRO.

14 Chadwick, J. (23/3/45). Memo: future Tube Alloys policy and programme. CAB 126/259, PRO.

15 Chadwick, J. (4/4/45). Letter to J. Anderson. CHAD IV, 6/25B, CAC.

16 Chadwick, J. (2/4/45). Letter to J. Anderson. CHAD IV, 2/11, CAC.

17 Chadwick, J. (24/3/45). Letter to J. Anderson. CHAD IV, 3/16, CAC.

18 Oliphant, M.L. (2/11/44). Letter to J. Chadwick. RG 77, file 201, NA2.

19 Rotblat, J. (1985). Leaving the bomb project. *Bulletin of the Atomic Scientists*, Aug., 16–19.

20 Chadwick, J. (18/12/44). Letter to W. Akers. CHAD IV, 3/15, CAC.

21 Minutes of meeting, 8/1/45, in Chadwick's office. RG 77, Box 60, file 334, NA2.

22 Chadwick, J. (27/11/44). Letter to J. Cockcroft. AB 1/197, PRO.

23 Groves, L.R. (1962). *Now it can be told*, p. 144. Harper and Row, New York.

24 Groves, L.R. (16/12/44). Letter to J. Chadwick. RG 77, Box 85, file 201, NA2.

25 Groves, L.R. (3/2/45). Notes on meeting with J. Chadwick, 20/1/45. RG 77, Box 85, file 201, NA2.

26 Chadwick, J. (29/1/45). Letter to T. Allibone. CHAD IV, 3/16, CAC.

27 Chadwick, J. (21/5/45). Letter to M.L. Oliphant. CHAD IV, 2/11, CAC.

28 Chadwick, J. (24/5/45). Letter to J. Anderson. AB 1/611, PRO.

29 Chadwick, J. (2/6/45). Letter to E. Appleton. CHAD IV, 2/11, CAC.

30 Gowing, M. (1974). *Independence and deterrence*, Vol. 1, p. 47. Macmillan, London.

31 Chadwick, J. (27/2/45). Letter to P. Moon. CHAD IV, 3/15, CAC.

32 Jones V.C. (1985). The atomic communities in New Mexico. In *Manhattan: the army and the atomic bomb*, pp. 465–81. Center of military history United States army, Washington DC.

33 Peierls, R.E. and Moon, P. (16/5/45). Letter to J. Chadwick. RG 77, Box 85, file 201. NA2.

34 Peierls, R.E. (1985). In *Bird of passage*, p. 201. Princeton University Press.

35 Peierls, R.E. (23&30/5/45). Letters to J. Chadwick. RG 77, Box 85, file 201, NA2.

36 Chadwick, J. (21/3/45). Letter to E. Appleton. AB 1/615, PRO.

37 Peierls, R.E. (13/6/45). Letter to J. Chadwick. RG 77, Box 85, file 201, NA2.

38 Moon, P. (20/6/45). Letter to J. Chadwick. RG 77, Box 85, file 201, NA2.

39 Dale, H. (6/10/44). Letter to J. Anderson. CAB 126/259, PRO.

40 Makins, R. (23/6/45). Memo to D. Rickett. CHAD IV, 12/65, CAC.

41 James Franck, 1882–1964. In *Nobel lectures: Physics, 1922–41*, pp. 109–11. Elsevier, Amsterdam.

42 Hewlett, R.G. and Anderson, O.E. (1990). Terrible swift sword. In *A history of the United States Atomic Energy Commission: the new world*, pp. 347–407. University of California
Press

43 Bernstein, B.J. (1988). Four physicists and the bomb: the early years. *Historical Studies in the Physical and Biological Sciences*, **18**, (2), 231–64.

44 Alperovitz, G. (1995). The interim committee. In *The decision to use the bomb*, pp. 155–72. Knopf, New York.

45 Weiner, C. (1969). Sir James Chadwick, oral history. American Institute of Physics, College Park, Maryland.

46 Groves, L.R. (1945). Notes on meeting with Chadwick, 1/6/45. RG 77, Box 33, file 001, NA2.

47 Peierls, R.E. (4/7/45). Letter to J. Chadwick. RG 77, *85*, 201, NA2.

48 Rotblat, J. (1945). Letters to J. Chadwick. CHAD IV, 1/2, CAC.

49 Oliphant, M.L. (25/6/45). Letter to P. Moon. CHAD IV, 3/16, CAC.

50 Szasz, F.M. (1992). *British scientists and the Manhattan project*, p. 76. St. Martin's Press, New York.

51 Rhodes, R. (1986). *The making of the atomic bomb*, p. 668. Penguin, London.

52 Oliphant, M.L. (1972). *Rutherford: recollections of the Cambridge days*, p. 145. Elsevier, Amsterdam.

53 Chadwick, J. (1946). The atom bomb. *Liverpool Daily Post*, 4 March.

54 Chadwick, J. (26/7/45). Letter to J. Rotblat. CHAD IV, 1/2, CAC.

55 Chadwick, J. Undated description of first atomic bomb explosion on 16/7/45. CHAD I, 19/6, CAC.

56 *New York Times*, 13 August 1945.

57 Groves, L.R. (1962). *Now it can be told*, pp. 348–50. Harper and Row, New York.

58 Chadwick, J. (5/10/45). Letter to J. Balderstone. CHAD IV, 11/56, CAC.

17 Sir Atom

In the middle of the Potsdam Conference, which took place between the Trinity test and the Hiroshima bomb, there was a changing of the guard in the British contingent. A general election had been held on 5 July 1945, and many of the votes were cast by servicemen still overseas so that three weeks had been allowed for the counting of ballots. When the results were announced on 26 July, the Labour Party, led by the mild and middle-class Clement Attlee, had scored a notable victory. Churchill resigned at once, and Attlee returned to Potsdam, as Prime Minister, with the new Foreign Secretary, Ernest Bevin. Deputy PM in the wartime coalition since 1940, Attlee had been initially invited to Potsdam as 'my good friend' by Churchill. Despite his senior position and unquestioned loyalty in the wartime coalition government, Attlee had no detailed knowledge until now about the atomic bomb. In this regard, his position was analogous to Truman's on the death of Roosevelt, but Attlee was plunged immediately into a crucial international summit, with no opportunity for briefing on the momentous issues at stake. Any communication between Truman and Attlee regarding the use of the bomb against Japan was brief and perfunctory, since the decision to go ahead had already been made. Nor does it seem likely that Attlee would have disagreed with that decision, since he was convinced that it would hasten the end of the war and save the lives of many Allied servicemen.[1]

As one of his last tasks before stepping down from the premiership, Churchill drafted a statement for future public release which detailed the history of the project, expecting that an atomic bomb would be dropped on the Japanese within a matter weeks. There was particular emphasis on the contribution made by British scientists, and it was probably Sir John Anderson who ensured a special mention for Chadwick. Churchill[2] praised the smoothness with which the arrangements for co-operation between Great Britain and America, made in 1943, had been carried into effect. He thought this 'a happy augury for our future relations and reflects great credit on all concerned—on the members of the Combined Policy Committee . . . on the enthusiasm with which our scientists and technicians gave of their best—particularly Sir James Chadwick who gave up his work at Liverpool to serve as technical adviser to the United Kingdom members of the Policy Committee and spared no effort'. In the same vein, Mark Oliphant paid his friend a generous and sincere tribute in his 1946 Rutherford Memorial Lecture:

> Chadwick, to whom Rutherford's leadership in nuclear physics had naturally

> descended, played a part in the development of atomic energy which will be fully appreciated only when the whole story is told. In his handling of men who worked with him, in his delicate task of relationship with the U.S.A., his legacy from Rutherford was clearly apparent. Perhaps he alone of the British scientists who served in the war preserved throughout that clarity of physical insight and feeling for fundamentals which was characteristic of Rutherford. Most of us found ourselves greatly affected by the scientific compromise which seemed essential to progress and as a result emerged from the war more interested in gadgetry and quick results than in the intellectual side of science.

Oliphant[3] also recalled the involvement of students of Rutherford's, like Cockcroft, Dee and others, who had shown surprising adaptability, enabling them to switch from nuclear physics to radar research. According to him, they all acknowledged their reliance on methods 'which they had learnt from Rutherford and Chadwick in the Cavendish Laboratory'.

Chadwick's integrity was unquestioned amongst his fellow scientists, the world over. It was, therefore, fitting that he should be involved in discussions about the future of captured German scientists, who were being held at Farm Hall in England. Their conversations there, including their incredulous reaction to the Hiroshima bomb, were recorded and translated by British agents and immediately sent to General Groves. He showed the reports to Chadwick, who found them both interesting and amusing.[4] While not susbscribing to Groves' preliminary assessment that they were all war criminals, Chadwick[5] did not believe that any of them could be assumed to be reliable when it came to nuclear secrets. He thought conditions for them in postwar Germany should be eased sufficiently that they were not driven into the arms of the Soviets. At a meeting on 14 September 1945, Groves made it clear to Chadwick and Field Marshal Wilson that he did not want the Germans in the USA. Chadwick recorded[6] that it was agreed between the three of them that:

> most of the men are of no particular use to us, but we do not agree that they would be of little value to any other country, nor that they are unlikely to go to another country if asked. On the contrary, the records of their conversations leads us to believe that, in view of probable conditions in Germany, they would go almost anywhere where they would be given facilities for work, and where they and their families would enjoy a more comfortable existence. We have little doubt that many of them would eventually go to Russia and that they would appreciably assist progress on Tube Alloys.

His own preferred solution was that they should be released from Farm Hall, but forced to live in Bonn with their families. There they should find university work 'with adequate but not lavish means of subsistence'. The scientists were released to the British controlled zone of West Germany at the beginning of 1946. Although they were kept under close supervision, the British authorities took an enlightened view, regarding them as leaders who had an influential role to play in the revival of German technology. Otto Hahn, the radiochemist and co-discoverer of nuclear fission, was to be President of the Kaiser Wilhelm Gesellschaft, but his appointment was called into question by US newspaper reports that he was a Nazi. Hahn had been a friend for 30 years, and had gone out of his way to show Chadwick small kindnesses when he had been interned in Ruhleben. During his first visit to Washington in 1943, Chadwick had been sounded out about German scientists and had gone out of his way to dispel any doubts about Hahn, stating that he 'was, is, and always will be anti-Nazi and is doing no war work'.[7] Incensed by the fresh allegation that he was a Nazi, Chadwick[8] wrote to Sir John Anderson:

> The report of Hahn's Nazi leanings was attributed to Dr. Lise Meitner, who worked with Hahn for many years before she fled Germany. I can assure you that Dr. Meitner, with whom I talked only two days ago, strongly resents the American statement and has contradicted it in a Press conference.
>
> It seems to me most important for the establishment of a healthy spirit in German universities and institutions that men of known liberal sympathies, such as Hahn and von Laue, should occupy positions of honour and trust especially also when they are the natural leaders in their field. It is in our own general interest, and it will also contribute towards mutual understanding, to give support to such men whenever opportunity offers, as it does now in Hahn's case.

Lise Meitner had come to Washington in February 1946 to receive the 'Woman of the Year' award from the Women's National Press Club. She spent an evening with Chadwick, when they discussed the charges levelled against Hahn: Meitner told her nephew Otto Frisch[9] that she was taken aback by Chadwick's 'sharp opinion' that even the most decent Germans shared responsibility for the crimes of the Third Reich. There was plainly a distinction in Chadwick's mind between Nazi evil and collective guilt. Further mischief emerged in the US press the following month, when rumours appeared about maltreatment of the German scientists during their internment in Britain. Chadwick thought the charges should be refuted, but Groves did not consider any action necessary. In a rare display of dissension, Chadwick wrote to London that he did not think 'Groves' opinion should control any action that we may think desirable in England'.[10]

While Chadwick was eager to see scientists restored to a meaningful role in postwar German society, his primary concern remained the opportunities and conditions of physicists in Britain. Their prospects would be enhanced, in his view, the more they could bring the wealth of new knowledge acquired on the Manhattan Project back with them. Such exchange of information was permitted under the Quebec Agreement and perpetuated by the Hyde Park *aide-mémoire*[11] signed by Churchill and Roosevelt in September 1944. Roosevelt suggested:

> Full collaboration between the United States and the British Government in developing tube alloys for military and commercial purposes should continue after the defeat of Japan unless and until terminated by joint agreement.

In this spirit, Chadwick wrote to Philip Moon[12] in September 1945: 'We cannot leave all our information behind and go home with empty hands'. Chadwick was going to check with Groves that it was permissible for the British scientists from Los Alamos and other sites to take their laboratory notebooks as well as technical reports back with them. There was now the opportunity to publish some wartime research which did not have any direct military relevance. Chadwick's last original paper, written jointly with Powell, Nunn May and Pickavance, arose from the collaboration between Bristol and Liverpool, combining Powell's photographic techniques with high-energy particles from the cyclotron. In August 1945, Powell offered to put Chadwick's name on another joint paper; Chadwick declined[13] on the grounds that Powell had done nearly all the work, and urged him to go ahead and publish on his own.

While contact with fellow physicists was mostly through letter, Chadwick would occasionally receive a colleague in his Washington office. On one such occasion, the visitor was Denys Wilkinson, the young nuclear physicist whom he had denied permission to come to Los Alamos because of the concern about his lungs. Wilkinson had instead spent the rest of the war in the Cavendish 'measuring more and more fast neutron cross sections with only little idea as to how, or if at all, they fitted into the great design'. Now Wilkinson was on his way to join the Anglo-Canadian programme in Montreal, and made a detour to Washington to present Chadwick with his measurements on the isotope uranium-233, which showed similar fissionable properties to uranium-235. Unlike ^{235}U, the newly discovered ^{233}U could be manufactured in a nuclear pile by bombarding the widely available element thorium with neutrons. In the summer of 1945, uranium-233 was highly favoured as a potential material for future bombs. Passing Chadwick the hand-drawn graph of the cross section as a function of neutron energy, Wilkinson waited for his reaction with a mixture of pride and anticipation:

> He, behind his desk, looked at it for a time and then let it fall, whether into an open drawer or into a waste-paper basket I could not tell, saying: 'It's wrong'.

It is perhaps surprising that Chadwick's habit of snubbing young colleagues in this way never seems to have led to any lasting resentment. They seemed to sense that there was no malice intended, although it happened so often that one suspects that it was a deliberate ploy to test their mettle. Wilkinson contented himself with the reflection that 'Chadwick was a man of few words'.

Peierls and Moon continued their informal reports to Chadwick about events at Los Alamos. The military urgency driving the research had dissipated, and time was now available for tidying up some of the loose ends which had been left because they were not absolutely vital to the production of the first bombs; attempts were also being made at long term planning for scientific teams, which had operated from one week to the next since their inception. One loose end concerned the efficiency of the first explosion at Trinity. There was a marked discrepancy between an assessment based on an analysis of the fission products of the explosion, which gave an estimated efficiency of 17%, and a theoretical calculation from the photographed rate of growth of the initial fireball, which had given a figure of 36%. At the beginning of October 1945, Peierls and Moon told Chadwick 'this difficulty has now been solved by Fuchs'.[14] In a piece of work which combined uncanny physical insight with his great mathematical fluency, Fuchs showed how the presence of heavy materials such as steel at the initial stages of the explosion would affect the expansion of the shock waves at different times over the ensuing microsecond intervals, at first slowing and then hastening the process. He recalculated from the photographic evidence that the bomb's efficiency was indeed about 15%—in very close agreement with the fission products data. In their next letter, on 29 October, Peierls and Moon mentioned[15] that 'Fuchs is in no hurry to fix a date for his departure but he is very keen to get back as soon as arrangements for the Establishment [in England] have reached a stage where he can be useful there'.

Fuchs' success seemed to be the one positive piece of news from Los Alamos where 'the spirit of the place is very low indeed and is still deteriorating'. According to Peierls and Moon:

> Most people feel that the development of weapons is not a satisfactory field to work on in peacetime. Oppenheimer at one time tried to persuade them that the possession of weapons was an extremely valuable factor to be turned to the benefit of the world and to lead to a strong international organisation. However, most people have now become very disillusioned, and do not believe the right use will be made of this. The general impression is that Oppenheimer is beginning to share this view.

Peierls had already sent Chadwick his own penetrating analysis of the global threats posed by the bomb, and there was also a memorandum[16] prepared by the British scientists, which they intended Prime Minister Attlee to study. They concluded that complete international co-operation was essential, especially on the sensitive issues of the inspection of nuclear materials and facilities, 'if there is to be any hope of avoiding the catastrophe of a war waged with atomic bombs'. Thus the thorny concept of *verification* of nuclear stockpiles was clearly identified, just two months after the Hiroshima bomb and while the US was the sole atomic power. The memorandum also argued for the free movement of scientific information and personnel among all countries. Chadwick did not add his name to the signatories, but forwarded the document to London with the comment that it was unanimously agreed amongst the British staff at Los Alamos and Berkeley.

British and American scientists were especially concerned about a piece of legislation then before the US Congress—the May–Johnson bill—which made provision for an Atomic Energy Commission (AEC) of far reaching authority and autonomy.[17] The proposals in the bill were largely sponsored by the War Department, who intended that the Commission would be dominated by members of the Armed Forces. The emphasis was on the military rather than the potential civil applications of nuclear power; in the eyes of its most notable opponents from Los Alamos and Chicago, the May–Johnson bill threatened future research by unreasonable security provisions, backed by Draconian penalties; it would also undercut efforts towards international control through the fledgling United Nations. Although President Truman 'was singularly unimpressed by the clamor among the scientists',[17] he in turn came to oppose the bill because of the proposed domination of the commission by the military, and the fact that the commission would be 'virtually independent of Executive control'.

Across the Atlantic, Attlee took a close personal interest in atomic matters from the first days of his premiership. He had noticed at Potsdam that some diplomats seemed grounded in the nineteenth century, and was surprised that 'people still talked of the line of the Oder–Neisse although rivers as strategic frontiers have been obsolete since the advent of air power'.[1] He recognized that the emergence of the bomb had already 'rendered much of our post-war planning out of date' and was deeply concerned by the lack of any accompanying policy to limit its threat to international security. Attlee[18] did not believe the whole cabinet to be a suitable forum to consider matters of broad policy because individual ministers were too hidebound by their departmental responsibilities, and he preferred to rely on small groups of senior colleagues. He formed an 'Atom Bomb Committee' (known in the blander parlance of Whitehall as 'Gen 75'),[19] whose other members were, like Attlee, men born in the late Victorian age. They did not, however, all share his ability to come to terms with the realities of the modern world. Within a month of the Labour government taking office, President Truman abruptly ended the Lend–Lease arrangement which had financed the British war effort since 1941. Thus, having just scaled the heights of power, the new Labour ministers were to find their country was bankrupt, and the Empire they had grown up with was becoming unsustainable. There was, however, suddenly the tantalising promise of an abundant, new, supply of energy, which could perhaps mitigate the inefficiencies of the coal industry, and an atomic weapon, the possession of which could enhance Britain's prospects of remaining a world power. Equally, by threatening the coal industry, nuclear power could undermine their own political strongholds, and the notion of America as the world's sole possessor of the atomic bomb was enough to trouble any socialist's heart. Gen 75 clearly needed expert advice on these questions immediately, and the only people qualified to give it were the rump Tube Alloys organization.

Lacking any politician in their own party with sufficient scientific training, they agreed that Sir John Anderson, nominally the Independent MP for the Scottish Universities, should be asked to head the Advisory Committee on Atomic Energy (ACAE).[19] Attlee suggested this arrangement on the day that Japan capitulated. Letters were rapidly exchanged between the Prime Minister and Sir John, and the Committee was established; Attlee appointed Chadwick to serve as a member on 17 August 1945. He joined fellow scientists Appleton (the secretary of DSIR), Dale (the President of the Royal Society), Blackett and G.P. Thomson, as well as Sir Alan Brooke, the Chief of the Imperial General Staff, Sir Alexander Cadogan, Under-Secretary of State at the Foreign Office and Sir Alan Barlow, Second Secretary at the Treasury.

When Attlee announced the formation of the committee to the House of Commons,[20] his own backbenchers pressed him on the question of whether the United States intended to retain exclusive possession of the nuclear 'secret', a notion which they deplored. The Leader of the Opposition, Churchill, asked him to make it quite clear that the committee would just deal with technical issues and 'is not concerned with the main policy to be adopted by this or other countries'. Attlee replied blandly that while policy had to be decided by the Government, 'this committee will advise us both with regard to the scientific progress and the general background of the whole subject'.

The secretary to the ACAE lost no time in writing to Chadwick[21] in Washington, sending minutes from the first meeting and asking him to come over and assist. In particular, the Committee wished to receive from him damage assessments for the bombs dropped on Japan, and his ideas for the proposed UK research establishment. Chadwick sent first a cable[22] commenting on the defence issues raised in the ACAE minutes:

> It seems obvious that the Military applications make a production plant of our own essential for the defence of the United Kingdom and the British Commonwealth. I believe that there can only be one opinion on this question, but I expect there will be differences about the way in which we should proceed and the type of plant to be erected.

The ACAE was clearly adopting a broad remit, and Chadwick for one had no hesitation in launching into areas of strategic policy. What seemed obvious to him regarding the need for Britain to become an independent nuclear power was generally assumed by Attlee and his circle of advisers, without being explicitly stated.* Although Churchill might not have liked the idea of scientists recommending national defence policy, he would certainly have been at one with Chadwick's opinion. Indeed, Churchill made one of the few public utterances on this monumental issue,[23] when he spoke in the House of Commons on 7 November 1945 and said: 'This I take it is already agreed, we should make atomic bombs.'

Whereas Chadwick had consistently argued during the last year of the war that lack of knowledge and technical uncertainties made the choice between uranium-235 and plutonium premature, he was now convinced, after studying early reports on Hiroshima and Nagasaki, that plutonium was the superior material for weapons. William Penney, the plain spoken mathematician whose knowledge of blast waves had made him indispensable at Los Alamos and earned him a seat on the observation plane over Nagasaki on 9 August 1945, had walked through the two Japanese cities that autumn, making detailed observations on the blast damage. By studying 'small things such as squashed tin cans, dished metal plates, bent or snapped poles', Penney[24] came to realize that 'the Nagasaki bomb was, in fact more powerful than the Hiroshima bomb.' This was even though Hiroshima appeared 'as a vast expanse of desolation' and nothing as vivid was left in his memory of Nagasaki. In hard numbers, which were Penney's forte, the equivalent amount of TNT to cause the same blast damage in Hiroshima was 5000 tons, and

---

* The official decision for Britain to produce atomic weapons independently was taken by an ad hoc committee, headed by Attlee, in January 1947. Ernest Bevin, the Foreign Secretary, was a notable proponent, and in reference to the Baruch plan under discussion at the UN was reported to have commented 'Let's forget the Baroosh and get on with making the fissile'. The January 1947 decision was not announced to Parliament until May 1948, and during the five years of Attlee's administration, there was not a single debate about atomic policies.[23]

20 000 tons in Nagasaki. Based on Penney's findings, Chadwick now believed the UK should move ahead with a production plant for their own plutonium, and alternative plants for uranium isotope separation should be ignored, because 'we cannot afford to shoot all over the target'.[22] To manufacture 250 to 300 grams of plutonium per day, a large graphite nuclear pile would be required, and Chadwick thought it should be built in Canada. While there were some physical reasons for this (the availability of a prepared site and an ample water supply, for example), his prime consideration was political: Canada would be more acceptable to the Americans.

There was only one informed voice which seriously questioned the tendency for Great Britain towards making its own atom bombs in the immediate post-war period, and that belonged to Patrick Blackett.[23] He did not begin from a position of absolute pacifism, nor even from a primary moral objection to nuclear weapons, but rather asked the question whether the military benefit that would be acquired by Britain was commensurate with the investments made in terms of money and skilled manpower. His own analysis of the defence and economic implications of the issue led him to be 'appalled by the incompetence and sheer stupidity' of the political and administrative history of the Tube Alloys project. In private, he was also very critical of Chadwick's continuing role in Washington, where he thought Chadwick was an unreliable representative, inclined to side with Groves against his own country. These unflattering opinions were conveyed to Chadwick by Oliphant,[25] along with scarcely more encouraging views on Chadwick's latest recommendations to the ACAE, of which Blackett was a member.

> He [Blackett] regards as dangerous and ridiculous the proposals put forward by you, that, as Plutonium is the obvious choice at the moment for a military weapon, effort should be concentrated on it, and that the plant should be in Canada. He takes the point of view that the British effort should be devoted primarily to research and development into the industrial and other possibilities, as well as the weapon, and that any manufacture of weapons should be subsidiary to development. He argues that we must have such supplies of all fissile elements as may be required to secure full development in every direction.

This, of course, ran completely counter to Chadwick's belief that the British could not afford to shoot all over the target, and his conviction that the independent production of nuclear weapons was essential for British defence. Over the next two years, Blackett's wide-ranging arguments were considered and rejected at the highest levels, and he was deemed too left wing by Attlee's socialist government;[23] Chadwick's more circumscribed and conservative opinions continued to have great influence.

Oliphant's friendship with Chadwick was clearly strong for him to feel able to write such a candid account of Blackett's criticisms. Indeed, he even went as far as to say that he was inclined to agree with Blackett—if his version of Chadwick's cables and letters were correct, and not, as Oliphant suspected, exaggerated or incomplete. It does not seem that Blackett misrepresented Chadwick's position, but neither does it appear that Chadwick felt it necessary to justify himself to his old

Cavendish colleagues. In a generous gesture, Oliphant had concluded his letter by offering to come to Washington for a month to give Chadwick a much needed holiday. 'Although we do not always agree about policy our aims are the same', Oliphant wrote and added, 'for a time I would cheerfully submit to your instruction'. The offer was never taken up even though Chadwick[26] had not had 'an hour of freedom', and had been so unwell that 'it was exceedingly painful to do any work at all'. He did manage, nevertheless, to answer the stream of enquiries forwarded from Rickett, the secretary to the ACAE in London. Chadwick's considered replies were often not encouraging. It seemed to him that the plans for the research establishment in England were getting very confused.[27] They seemed to ignore the desperate national shortage of skilled workers: 'We have not an infinite supply of engineers, draughtsmen, etc.' With the same hard realism, he continued:

> The estimate for the total capital cost of the plants mentioned in the Recommendations seems to me fantastically low. In fact I would say it is quite impossible even to build a plutonium plant to produce one kilo per day for £25 million.
>
> I am afraid the Technical Committee does not appreciate the magnitude of the effort which would be required to carry out its Recommendations and the strain this would put on our industrial resources.

There were several reasons Chadwick could not let Oliphant deputize for him in Washington, and the most pressing was hinted at in a letter to London on 14 September 1945:[26] 'a new trouble has arisen in Canada which may have serious repercussions'. This was the unmasking of Alan Nunn May of the Montreal laboratory as the first atomic spy.[28] A young Russian cipher clerk, named Igor Gouzenko, left the Soviet embassy in Ottawa on the evening of 5 September with one hundred secret documents from the military attaché's office stuffed inside his shirt. When translated, the documents revealed a whole network of Soviet agents working in Canada, including one connected with atomic research. The FBI were alerted, and the following week J. Edgar Hoover informed the White House that the Gouzenko papers had revealed that Nunn May, a British scientist working on the atomic bomb project in Canada, had been passing secrets to the Soviets. Apart from technical secrets, in February 1945, Nunn May[29] had also sent details on the organization of the Manhattan Project and information about the highly secret work going on at Oak Ridge, Hanford, Argonne and Los Alamos. On the day of the Nagasaki bomb, he had stolen a platinum foil coated with the fissionable isotope uranium-233 from the Montreal laboratory, and it had been flown directly to Moscow.

A week after Gouzenko's defection, the head of the Montreal Laboratory, John Cockcroft,[30] supplied the British Office in Washington with an assessment of Nunn May's access to information and materials. It was decided that the Ambassador, Lord Halifax, should be the man to inform General Groves, and the top priority was to limit any further damage to Anglo-American relations. Groves was, it can be imagined, incandescent with rage. He had personally laid down the rules governing the exchange of information between men working at the University of Chicago and those at the Montreal laboratory (code name Evergreen) in May

1944. He had become concerned from routine reports that Nunn May was making repeated visits to Chicago, and in October 1944 had banned further trips because he was concerned about how much knowledge Nunn May was acquiring about the latest in nuclear pile technology.[28] Now his worst fears were confirmed. On 18 September his own Intelligence Division provided him with their estimate of the Evergreen scientists' knowledge about operations at the Chicago Metallurgical Laboratory, and whether this knowledge had been gained within or outside the Groves' specific rules. They reported[31] that 'the case of Dr Alan Nunn May is an outstanding example of this extraneous information.' Following Nunn May's prolonged, last visit to Argonne during October 1944, the nuclear pile expert, Dr Walter Zinn, pointed out that such an extended stay 'inevitably permits the visitor to become familiar with all of the activities of the laboratory'. Zinn thought that the breadth of Nunn May's access was such that no single person at Argonne could give a detailed account of all the subjects discussed with him nor of the experiments he had observed. Even though Nunn May had been shown no classified reports or plans, there had plainly been a breakdown of Groves' cherished compartmentalization, on a grand scale. The intelligence report to Groves concluded that 'the British and Canadian employed scientists have, in the past three years, had an opportunity, if they had been so directed and if they were so inclined, to acquire a fairly comprehensive picture in significant detail of the operation of the Metallurgical Project at the University of Chicago'.

We know that Groves discussed his concerns about Nunn May with Chadwick at the end of 1944, and that Chadwick had interviewed him in Montreal before assuring the General he was 'exceptionally reliable and close mouthed'. Nunn May had told Chadwick[32] that he had not picked up any information about the heavy water pile while working at Argonne—a bare-faced lie. Chadwick must have felt personally betrayed by Nunn May, regardless of the damage done to Anglo-American trust, which he had been tirelessly building over the past few years. As a student at Cambridge, Nunn May was 'a retiring, dreamy-looking character . . . who managed to convey the distant impression of always being spiritually in his laboratory'.[33] He had completed a Ph.D. at the Cavendish, under Chadwick's supervision, where the whole research ethos depended on honesty and personal integrity. It was natural when Nunn May was applying for the Readership in Physics at King's College, London in the summer of 1944 that he should ask Chadwick to be a referee.[34] With the support of Chadwick and Cockcroft, he was appointed, and King's College gave him a starting date of 1 September 1945, just a few days before Gouzenko's defection.

Another member of the British Embassy staff in Washington was equally stunned by the exposure of Nunn May, but for very different reasons. Donald Maclean, the First Secretary and family friend of Lord Halifax, had been a contemporary and close friend of Nunn May's at Trinity Hall.* Nunn May had arrived at

---

* A small Cambridge college, not to be confused with Trinity College (J.J. Thomson, Master)—which had its own communist cell in the mid-1930s that included Guy Burgess and Anthony Blunt.

Cambridge a committed communist, perhaps as a result of seeing his father, a small industrialist in Birmingham, suffer financial ruin in the Depression of the late 1920s. Although Nunn May did not talk about his politics inside the Cavendish, he and Maclean openly attended communist meetings at Cambridge.[28] They were both recruited as Soviet agents soon after leaving the university. Now the nuclear scientist had been found out in Canada, would there be a connection made to the equally brilliant, but psychologically taut diplomat in DC? Maclean could not belive his luck when it was decided that Nunn May would not be apprehended immediately. Partly, this was to avoid embarrassing the Soviets, at a time when it was hoped that they could be conjoined in international agreement. The other reason was Gouzenko's papers revealed plans already made for Nunn May to rendezvous with a Soviet agent outside the British Museum in February 1946.[35] He was to be carrying a copy of *The Times* under his left arm and would be approached in a certain way: this seemed like a golden opportunity to the security services to penetrate the espionage ring further, and so nothing would be done to alert the stooge.

Just two weeks after the unwelcome news about Canadian spies, President Truman was in receipt of a remarkable letter from Attlee[36] probing him about his intentions regarding the international regulation of nuclear weapons and suggesting an early meeting to discuss the question. In Attlee's estimation, unless the political relationships between countries were changed, 'sooner or later these bombs will be used for mutual annihilation'. He saw the present position in the following terms, which echo Chadwick's remarks to the press, six weeks earlier, in the immediate aftermath of the Japan bombings:

> ... whilst the fundamental scientific discoveries which made possible the production of the atomic bomb are now common knowledge, the experience of the actual processes of manufacture and knowledge of the solutions which were found to the many technical problems which arose, are confined to our two countries and the actual capacity for production exists only in the United States. But the very speed and completeness of our joint achievement seems to indicate that any other country possessing the necessary scientific and industrial resources scould also produce atomic bombs within a few years if it decided now to make the effort.

With great prescience, Attlee also recognized the Gordian knot inextricably linking the civil and military uses of nuclear energy, when he wrote that 'the successful manufacture of bombs from plutonium shows that the harnessing of atomic energy as a source of power cannot be achieved without the simultaneous production of material capable of being used in a bomb'. This letter did serve to remind Truman of the obligations resting on the US as a result of the 1943 Quebec Agreement, and he amended his speech to Congress on 3 October 1945 to say that discussions on international control would be held 'first with our associates in this discovery, Great Britain and Canada, and then with other nations'.[17] Other than this brief recognition, there was no sign from Truman that he had heeded Attlee's message, and in particular there was no invitation to talks. Attlee sent a cable to Truman,[36] and his persistence was rewarded when it was agreed that a meeting would be held in Washington on 9 November, to which the Canadian Prime Minister, Mackenzie King, would also be invited.

In the weeks before the meeting, perhaps as a result of the Nunn May case and also because of briefings from officials, Attlee's position became less idealistic regarding the notion of national security and more cautious particularly with respect to the Soviet Union. He flew to Washington with Sir John Anderson, bearing no set agenda but intending to cover both international control and the future of Anglo-American relations.[36] In the event, the latter subject was discussed not by the political leaders, but by a quartet consisting of Anderson and Roger Makins, from the Washington Embassy, for the British, and the Secretary of War, Patterson, together with General Groves for the Americans.[17] One wise and experienced counsel who was missing on the British side was Chadwick, who had flown the Atlantic in the opposite direction with his wife, to receive his knighthood from the King. The investiture was held at Buckingham Palace on 12 November and was marked in one Fleet Street newspaper with the headline: *The King Knights British 'Mr. Atom'.*

Despite Chadwick's absence, the Washington summit seemed to have reached a satisfactory outcome with a declaration from the three heads of government, and the conclusion of the parallel 'Groves–Anderson' memorandum[36] on Anglo-American relations. The declaration recognized and supported the free exchange of basic scientific knowledge, between the US, UK and Canada; the sharing of detailed information on the practical applications of atomic energy would not be permitted until 'effective enforceable safeguards against its use for destructive purposes can be devised'. The three premiers called on the United Nations to set up a Commission, without delay, to make recommendations on nuclear matters. The confidential 'Groves–Anderson' memorandum suggested that the still secret 1943 Quebec Agreement should be superseded by new arrangements to be agreed by the Combined Policy Committee. The memorandum also called for full co-operation at the basic level of scientific research, but recognized that applied knowledge should be more closely guarded.

While the Truman administration would obviously bring enormous influence to bear on the shaping of any future international agreements, the President's most pressing need in the fall of 1945 was to wrest control of the United States' nuclear programme from his own military. He needed help from Congress to defeat or radically redraft the May–Johnson bill. Senator Brien McMahon, a Democrat from Connecticut, established himself as the Senate's leading expert on nuclear affairs, after a few preliminary, ineffectual skirmishes in debates, when he repeatedly said that the bombing of Hiroshima was the greatest event in world history since the birth of Jesus Christ.[17] It was agreed that there should be a special Senate committee on atomic power, and McMahon managed to outmanoeuvre more senior men like Johnson to become its chairman. This was both a heavy responsiblity and a remarkable opportunity for a freshman senator. Shortly after the conclusion of the international meeting with Attlee and King, Truman joined forces with the

McMahon in order to bring the Pentagon to heel. At the President's behest, Senator McMahon agreed to bring an amendment to the May–Johnson bill to provide for civilian supremacy in the control of fissionable materials. In a significant move, McMahon intertwined domestic and international policy by publicly stating:'we must find a way of controlling the destructive power of atomic energy on a world level'.[17]

Acting defiantly and clumsily, Groves tried to resist the growing influence of the whipper-snapper from Connecticut. When asked to provide him with classified information for educational purposes, the General refused point blank.[17] Denying a senator in wartime was something Groves had managed to do with impunity, but in peacetime it was much harder to withhold information which was known to be available to foreign-born scientists, still on American soil. The other disastrous mistake made was for the US Army to confiscate three cyclotrons from Japanese universities and have them dumped in the Pacific Ocean.[17] The scientific community was outraged, even the phlegmatic Rudolf Peierls told Chadwick he was 'profoundly shocked'. The escapade provided easy evidence to show that the military could not be relied upon to act in a restrained and logical way in this most dangerous of spheres. At first the Pentagon tried to pass off the incident as 'War Department policy',[17] but as Chadwick wrote to London in December:[37] 'There seems to be no doubt that the recommendation for destruction emanated from Groves' office'. Unwilling to blame his wartime colleague for such stupidity, Chadwick added unconvincingly: 'it is not so clear how far Groves himself was responsible'. He cautioned against 'pursuing the subject on our own account'.

With his considerable experience and feel for Anglo-American relations, Chadwick could already feel new rifts developing in the trans-Atlantic partnership. He warned Anderson that 'the cohesive forces which held men of diverse opinions together during the war are rapidly dissolving: any thought of common effort or even common purpose with us or with other peoples is becoming both weaker in strength and rarer to meet'.[38] Now they were no longer fighting a common enemy, the friendship between the two countries was bound to be less intense, and the cooler demands of self-interest supplanted the emotional rhetoric of battle. Chadwick, like many who had been in the thick of the alliance for the previous two or three years, was slow to recognize and to accept the new circumstances. He still believed Vannevar Bush when he told him that as the two countries worked on the issues of nuclear collaboration they were 'tying themselves together for one hundred years'.[38] In practice, Bush had for years 'opposed too close a tie to the United Kingdom for fear it might impair American chances of negotiating successfully with the Soviet Union on the all-important question of international control.'[17]

Vannevar Bush was now the most influential, nuclear scientist–politician in Washington and had, at short notice, devised the American strategy at the November meeting between Truman, Attlee and King.[17] He was not the only American to be concerned that an over reliance on the wartime partnerships would be detrimental to the future. There were noises of discontent from Capitol Hill where a liberal element felt that the Soviet Union should be involved directly in any proposals to be put before the United Nations. Secretary of State Byrnes in a speech the day after the Truman–Attlee–King declaration, played down the Anglo-American aspect and insisted responsiblity would rest with all governments.[17] He was now chairman of the Combined Policy Committee (CPC), the tripartite body established by the Quebec Agreement in 1943, which met on 4 December to consider the Truman–Attlee–King directive and the associated Anderson–Groves memorandum. Groves, still fighting a rearguard action against perceived threats to American security, pointed out to Byrnes that under Article 102 of the new United Nations Charter, every new treaty or international agreement between member countries should be reported to the UN Secretariat as soon as possible and published. The November agreements amounted to a closed alliance between the US, the UK and Canada which would share nuclear information and fissionable materials with each other, but not with the rest of the world. To publish such an arrangement would provoke the wrath of the Soviets, and probably outrage the French; it would also scupper any prospect of a United Nations Atomic Energy Commission.

According to Chadwick, this piece of intelligence allowed Byrnes to give full rein to his 'obstructive and evasive abilities'.[38] In fact the British Foreign Office had also identified the stumbling block presented by Article 102, and were less than straightforward in their proposed solution. At the next CPC meeting in April 1946, Lord Halifax suggested that instead of creating completely new treaties, they should rely on amendments to the original Quebec Agreement and the 1944 Declaration of Trust (which was designed to monitor and monopolize the world's supply of uranium and thorium). The Americans were able to reject Halifax's scheme with a clear conscience, after legal advice that it would not surmount the obstacle of Article 102. At the meeting, Chadwick was disappointed that Bush, despite his recent private assurances, remained silent and offered no support to the British. The British pointed out that the November agreement was now empty, and Chadwick wrote to Anderson, two days later, full of foreboding because there had been 'a feeling which goes beyond a mere reluctance to conclude an agreement or a desire to strike a hard bargain'.[38] Attlee immediately telegraphed Truman in an attempt to rescue the earlier agreement, but was bluntly rebuffed. The flow of information and technical assistance from the United States abruptly ceased, despite the pledge a few months earlier of 'full and effective co-operation in the field of atomic energy'.[38]

The same April CPC meeting also considered the problem of the allocation of raw materials, now that there was such demand for uranium. In the war, naturally all the uranium had gone to the Manhattan Project, but now the British needed to secure their own supply, since while they could overcome the lack of technical support from the US, they could not start a programme without raw material. 'We must by hook or by crook', explained Chadwick,[39] 'divert some of the Congo material to England.' At the CPC, the British made the optimistic bid that material should be shared on a 50:50 basis, which Groves retorted would mean that the Americans would have to start shutting down plants. The committee agreed that Chadwick and Groves should be left to reach a compromise. While it is understandable that the Americans would rely on the tough-minded Groves to fight their quarter, the choice of the reticent Chadwick, who had been beset with poor health for some months now, as his solitary opponent might seem strange. In the event, he proved himself again to be a highly accomplished negotiator. Chadwick proposed a scheme whereby the Americans should keep all the uranium landed in New York before the end of March 1946, but all subsequent supplies from the Belgian Congo (the main source of uranium ore at the time) should be equally shared. In addition, the US was to give Britain 50 tons of special oxide and 15 tons of refined uranium metal.[38] Groves railed against the injustices of the plan and also revived all the security arguments against basing a nuclear pile in Britain, but Chadwick's ideas were essentially ratified by the CPC. Without this arrangement, the United Kingdom, for better or worse, would not have become an atomic power at such an early date.

In Washington, the vacuum left by the demise of the Attlee–Truman–King agreement was filled by the emergence of the McMahon bill. This was introduced in the Senate just before Christmas 1945 as a replacement for the unpopular May–Johnson bill. McMahon was careful to meet most of the original objections raised by the scientists to the May–Johnson bill, and proposed minimal restrictions on the exchange of basic scientific information as well as a strict international framework to oversee nuclear research. The Atomic Energy Commission would have the ability to control its own research programmes, and there would be proper recognition of patent rights.[17] From the point of view of the Administration, the new legislation corrected the fatal flaw in the May–Johnson bill and guaranteed AEC accountability to the civilian executive rather than to the military. In January 1946, Truman's Secretary of Commerce testified before McMahon's committee in the Senate and endorsed the new bill as being an improvement on previous attempts at legislation, and also stressed that it should be consistent with evolving international policy. Despite his success with Truman's administration and the scientists, McMahon had an uphill task in convincing the members of his own committee to sign on to his bill.[40] They were alarmed by the unprecedented powers which would be granted to the AEC, and were not convinced about the liberal dissemination of data amongst scientists. Their security concerns were fed by War Department lobbying, and then in February 1946 came a complete bombshell.

Alan Nunn May, no doubt alerted by Moscow, which in turn had been warned by Donald Maclean, failed to appear at his scheduled liaison outside the British Museum. Since he was not going to lead MI5 deeper into the nests of Soviet espionage, he was soon arrested, as were 22 other suspects in Canada, identified from Gouzenko's papers. The news that atomic secrets had been betrayed to the Russians shook Washington and the American public to their roots. General Groves, the master of the Manhattan Project, was invited to Capitol Hill; to the annoyance of McMahon, he mesmerized the committee with personal opinions on the joint dangers of a full-time, civilian AEC, and the perils of excluding the Pentagon in this most vital and sensitive area of national defence. After a series of petulant questions, with which he failed to undermine the General, McMahon had completely lost control of his committee, several of whom later apologized to Groves for their Chairman's unfair attitude.[40]

A keen observer of these abrupt changes in fortune, Chadwick, wrote to Peierls a week after General Groves' virtuoso performance at the Senate hearing that: 'If Groves seizes his chances, he will be able to re-establish himself, after having almost lost the game. It is a very odd situation at the moment.'[41] Peierls was by this time back in England and, in turn, wrote to Chadwick about a new Association of Atomic Scientists, which had been recently formed, with Harrie Massey, an ex-Cavendish man, now at University College, London, as chairman. According to Peierls,[42] there would be an insistence from the outset that 'a proper division should be made between statements of scientific facts and opinions held by scientists, and that on the latter type of question the Association should not express any definite views or advocate any definite policies.' Perhaps anticipating a sceptical reaction, Peierls added: 'even if in your present position you would consider it unwise to become a member of it, we will, of course, always be grateful for any advice or opinion that you care to express.' Chadwick lost little time in offering his opinion of the organization and some unsolicited advice to its chairman. In a letter to Massey on 2 April 1946, he wrote:[43]

> such an Association may have a useful function to perform at the present time but I do not think that it is likely to have a long life, or should have, for there should be little need for it when the present excitement has died down.
>
> . . . I think I should not neglect to warn you of the dangers of extravagant statements and unbalanced views which may be put out by some members of the Association. These will only do harm to the objects you have in mind and they will bring discredit not only upon your Association but also on scientists generally. I know that it is hardly necessary to warn you about this aspect but we have had here a very striking example.*

More specifically, he continued:

> There is another and perhaps a more immediate danger which I should mention, that of taking too partial a view of the May case. I suggest that you should be very cautious indeed how you approach this matter. I am not suggesting that you should not act as an adviser on scientific matters to his counsel, if you are asked to do so, but I think it would not be wise to espouse his cause too warmly or to jump spontaneously to his defence without some knowledge of the circumstances. He is entitled to the best defence that can be provided for him, but I am afraid that some of his friends and colleagues may take up May's cause without reflection and from political prejudice. I have already heard the words 'scientific witch hunt' and similar phrases. This is pure nonsense. I am quite certain that proceedings would have been taken against May only after thorough investigation and out of a deep conviction that he had committed a serious offence.

---

* It seems likely that Chadwick was referring to Harold Urey, the Nobel Prize winning chemist. He had become a vociferous critic of Groves, and had described the May–Johnson bill as 'the first totalitarian bill ever written by Congress'. By March 1946, Chadwick thought Urey, although he continued to be very vocal, was losing ground.[41]

> I was a little uneasy to read that you had *volunteered* to act as an adviser for the defence. I may be unduly cautious but to me this is a very different matter from consenting to act as an adviser on the request of his counsel.
>
> I expect you know the Canadian affair in general and the case of May in particular, has produced a very strong effect on this side. Whatever the result of May's trial may be, there will remain a deep feeling of uneasiness. In every way this is a most unfortunate affair.

Nunn May was eventually sentenced to ten years' penal servitude, which was attacked as harsh by many scientists and by Labour Party insiders. The Association of Scientific Workers, a left wing Trade Union, put out a statement[44] protesting against the treatment of Nunn May and asserting that 'scientific workers fervently believe in the need to achieve freedom for the fullest discussion and interchange of scientific knowledge, knowing that such freedom is essential both for scientific progress and for international co-operation'. While Chadwick might have agreed with the proposition as stated, he would never recognize Nunn May's clandestine activities as meeting the usual terms of free debate, nor would he regard scientists who profited from the purloined work of another with sympathy. In his eyes, any form of dishonesty was inimical to science. Unfortunately, nuclear physics was no longer the innocent, intellectual, pursuit it had been before 1939, but this made it even more imperative, in Chadwick's opinion, for scientists not to assume an elevated role for themselves nor to imagine they were exempt from the laws of the land.

The British nuclear programme was making slow headway, regardless of the withdrawal of American support. The site for the research establishment at Harwell, south of Oxford, had been chosen and Cockcroft was formally appointed as its first director in January 1946. At about this time, it was decided that the Ministry of Supply should take responsiblity for the atomic energy programme outside Harwell, and Attlee pursuaded Lord Portal, the wartime Chief of the Air Staff, to become the Controller of Atomic Energy within the Ministry.[19] Portal, who reported to the Prime Minister rather than to the Minister of Supply, took a rather narrow view of his responsiblity and restricted himself to the production of fissionable materials. His appointment was greeted with scepticism by many scientists, who thought it would lead to an over-emphasis on military applications. Among the critics was Oliphant, who confided his disappointment to Chadwick. Chadwick himself was annoyed that neither he nor Cockcroft had been consulted prior to Portal's appointment, but did his best to be supportive, saying that he was 'very pleased indeed', and that Portal 'impressed me more than anyone I have met for a very long time'.[45] He told Oliphant he did not believe that Portal as Controller would concentrate exclusively on the military aspects, but did argue that 'these aspects are in the foreground at present and may remain so for some time and this is a fact that we cannot afford to ignore'. Oliphant,[46] in turn, was soon won over by Portal's charm and intelligence; he even agreed to serve on a committee with Lord Cherwell, who had become a black sheep with many, providing he was not the chairman or given a position of special authority.

Oliphant[47] remained pessimistic about progress at Harwell and reported to Chadwick that:

> most of the machinery accumulated is junk . . . it has been dumped in a hangar without greasing or oiling, and is rapidly rusting away. None of the building work showed any sign of completion. Over all . . . there hangs a dense fog of buff coloured forms and red tape.

Just as Oliphant had complained[48] that technical information Chadwick sent from Washington never reached those who needed to know it in England, Chadwick disclaimed any knowledge about events in England saying 'I am not kept in touch by anybody either about Harwell or the general organisation'.[49] He also had a more personal grievance about the confused administration of the British project—he had not been paid for some months. At the beginning of April he received US$1600 in arrears, which was very welcome but did not cover his expenses. He thanked Appleton who had arranged the payment, and commented:[50]

> Forgive me if I appear rather bitter and bad tempered about this matter, but it is not easy to live on the verge of bankruptcy unless you have the temperament of a Micawber.

He was going to write to the Ministry of Supply and if he did not obtain satisfaction would resign. In an almost apologetic, handwritten, postscript, he added: 'I do not mind so much for myself but it is most unfair to my family, especially my daughters to live under these conditions.' He was at a low ebb and wrote to Rotblat[51] around the same time: 'I cannot stay here much longer for I am completely worn out. I have had only one day [holiday] since I came here.'

There was to be neither escape nor resignation. New responsibilities flew to him and stuck like iron filings to a magnet. At the beginning of March he was named as the United Kingdom's scientific adviser to the UN's Atomic Energy Commission;[52] he was also to be the alternate representative on the Commission to Sir Alexander Cadogan, the permanent British UN delegate. The Commission was due to meet in June 1946, and Chadwick was eager to mend fences with the Americans as fully as possible before then. He was frustrated with Lord Portal's reluctance to visit Washington, since he wanted him to see the scale of the US plants and to meet General Groves. Urging him to come, Chadwick[53] warned:

> The opportunity may not be long available and may not recur for some time. I cannot swear that your visit would result in immediate agreement to full and effective collaboration but I am sure it would help for so much depends on personal contacts.

While he was right to be concerned about the unpropitious circumstances, Chadwick was mistaken in his belief that they could be improved by personal intervention. After the deadlock in the CPC, Attlee again approached Truman seeking to justify the British interpretation of their November declaration.[38] On 20 April 1946, a week after Chadwick's letter to Portal, Truman replied to Attlee[54] firmly squashing British hopes. The President was adamant that 'full and effective cooperation' applied only to basic scientific information and not to engineering and operational assistance. In Truman's view it would be unwise from the standpoint of both countries to enter into an arrangement to build an atomic energy plant in the UK. The Americans were inspired by 'the demands of people the world over that there should be some international control of atomic energy. Ever since, we have been working toward that goal . . .'

Although the visit took place too late to have any influence on US strategy, Lord Portal did come to Washington in May 1946. General Groves arranged for him to see all aspects of the American atomic effort—including the plutonium production facility at Hanford. This was the only plant to which Chadwick had been denied access in wartime, and he now asked Groves if he could accompany Portal. Groves replied that he could, but if he did 'Portal will not see very much'.[55] Chadwick thought that Groves overestimated his appreciation of technical matters, but it is likely that he would have gleaned more information than Portal, whose main impression of the gaseous diffusion plant at Oak Ridge was that it was so big that men had to ride round it on bicycles.[38] Portal did learn enough about the operational problems with the nuclear piles at Hanford, where the graphite moderators unexpectedly expanded under heavy neutron bombardment (the so-called Wigner Effect), that he asked Chadwick whether it would not be better to revert to a uranium-235 plant and to forget about plutonium. Chadwick's answer was unequivocal: 'The decision has been taken to make bombs and therefore we must make plutonium'.[23]

As soon as he believed that a British atomic weapons programme was inevitable, Chadwick began to promote the man he thought best qualified to be in charge—William Penney. While in London in the autumn of 1945, Chadwick made it his business to lunch at the Athenaeum with Sir John Lennard-Jones, an old friend, who was leaving his wartime work as Chief Superintendant of Armament Research at the Ministry of Supply.[56] At the lunch he persuaded Lennard-Jones that Penney was the right man to succeed him, and in turn Penney was appointed. Now that Chadwick had direct access to Portal in Washington, he strongly urged him to let Penney run the atomic bomb project in England, a suggestion which Portal accepted very willingly.[23] At this stage, the decision to proceed with a British bomb was still top secret: the Prime Minister had still made no public policy statement, and even Portal's discussions with the military Chiefs of Staff were 'off the record'.

Nor did the British government make any discernible planning for the forthcoming United Nations debate on atomic energy, other than to nominate Cadogan and Chadwick as representatives. By contrast, the Truman administration appointed a strategic policy group which became known as the Acheson–Lilienthal committee, after two of its main members.[57] The chairman, Dean Acheson, was the Under-Secretary of State, and David Lilienthal was the brilliant head of the Tennessee Valley Authority, a huge energy consortium. The most influential scientific consultant to the committee was Robert Oppenheimer, and the daring nature of the eventual report owed much to his powers of persuasion. In essence, the report, which was endorsed by Bush and Groves, divided atomic activities into 'safe' and 'dangerous'. All *dangerous* activities would be strictly controlled by a new international Authority which would seek to maintain a strategic balance between nations. The Authority would have a monopoly over raw materials and would designate where nuclear plants could be built. *Safe* activities could be pursued in individual nations, their industries and universities, under the Authority's permission. There would be an incremental transition to the new international system, with safeguards at each stage, and the United States would, at each step, give up an element of its technical and material advantage. Such a scheme may now seem hopelessly naive and idealistic, but it was passionately supported by the Acheson–Lilienthal committee. They had the foresight to recognize that even if there were international agreements prohibiting the direct manufacture of plutonium for weapons but allowing civil nuclear power programmes, unscrupulous countries would find it irresistible to divert the plutonium formed as a by-product in their power plants to weapons use. If the supranational Authority had the monopoly both of uranium mining and plant construction, any unapproved activity by an individual country would serve as a clear danger signal to the rest of the international community, which could take steps (unspecified) to protect their collective security. The report was forwarded to Secretary of State Byrnes and the President, and Acheson also presented it to McMahon's Senate panel, where it was generally well received.

On 16 March 1946, just as the Acheson–Lilienthal committee were putting the final touches to their report, President Truman appointed the US representative to the UN Atomic Energy Commission.[57] His choice, heavily influenced again by Secretary Byrnes, was Bernard Baruch, a legendary Wall Street investor then in his mid-seventies. Acheson was dismayed by the choice of Baruch, whose reputation as a wise adviser to Presidents he thought was 'without foundation in fact and entirely self-propagated'.[58] To Lilienthal the notion of Baruch as 'the vehicle of our hopes' seemed fantastic 'not only on acount of his age but also because of his unwillingness to work and his terrifying vanity'.[38] Baruch, though, was shrewd enough to recognize that the utopian Acheson–Lilienthal report would polarize public opinion, and he took care to distance himself from it immediately.[55] After two months of study, during which Baruch had only committed himself to four hours work per day, he decided to support the central concept of the Acheson–Lilienthal report that there should be a powerful new international atomic agency, that there needed to be explicit penalties specified for errant countries, and no nation could escape these by the use of the veto in the UN Security Council. This would become the basis of the American submission to the United Nations conference in June.

In contrast to the wide-ranging analysis of the Americans, which resulted in their revolutionary proposals on international control, according to Chadwick,[55] there was one meeting in the British Embassy prior to the UN conference between him, the Ambassador and Sir Alexander Cadogan. The new Ambassador was Lord Inverchapel, who, as Sir Archibald Clark-Kerr, had previously been the Ambassador in Moscow. While there, he had produced the most imaginative assessment of the atomic bomb's effect on the Russian psyche.[36] In his words, just as they had defeated the Germans after paying a terrible price, and national security seemed to be within grasp for the first time in generations, 'plumb came the atomic bomb'. When they realized that their Western comrades-in-arms were not going to share this weapon with them, and indeed were contemplating its use as an instrument against them, the lack of trust brought all their old suspicions to the surface and threatened them with national humiliation once more. Chadwick was most impressed by the new Ambassador's historical perspective.[55]

The other career diplomat present, Cadogan,[59] was the youngest son of one of England's wealthy, aristocratic families. He had entered the Foreign Office in 1908 when his first task had been to make handwritten copies of a letter, beginning 'My dear Brother and Cousin and Nephew', in which the Emperor Franz Joseph announced to King Edward VII the annexation of Bosnia and Herzegovina. He had served in the pre-First World War embassies at Constantinople and Vienna, and subsequently spent many years at the League of Nations in Geneva, where he patiently drafted plans for disarmament, which were rendered void by Hitler's rise to power. While not cynical, he was not at an impressionable age, and Chadwick was rather shocked by the shallowness of his plans. He recalled[55] that Cadogan blithely announced: 'Well, if this is a discussion about the future arrangements on nuclear bombs and such matters, it's really a matter for the physicists'. Chadwick disagreed completely and said it was entirely a political matter. He told Cadogan, 'the physicists wouldn't have any difficulty in coming to some kind of reasonable agreement amongst themselves, but it wouldn't mean anything whatever'.

After this unpromising start, the two men went to New York for the UN gathering which was held in the gymnasium of Hunter College in the Bronx. Whereas Chadwick was the sole British scientific adviser, in the American section were his old friends Tolman, Compton, Urey, Bacher and Oppenheimer. They all sat through Baruch's speech[40] which opened with the phrase, 'We are here to make a choice between the quick and the dead', and continued in dramatic fashion for well over an hour. To Chadwick, it seemed that Baruch saw himself 'in his sere years as Moses descending with a message for the salvation of humanity'.[38] Perhaps as a result of listening to Lord Inverchapel, Chadwick[55] did not think that 'the Russians would agree to anything'. After six months of debate and posturing, the Russians effectively stymied the American plan.

From the narrower perspective of Anglo-American relations, the sterile outcome of the UN debate mattered less than the passage of the final McMahon bill in August 1946. With each revision and amendment, the bill had become more restrictive.[40] For example, the section originally titled 'Dissemination of Information' had been changed to 'Control of Information'; the attempt to distinguish between 'basic scientific' and 'related technical' data had been abandoned and it would all be 'Restricted Data'. Chadwick had been the first to notice this hardening of American security, and had wanted to approach Senator McMahon privately[38]—a move that was vetoed by the Embassy for fear of creating worse political difficulties. In Chadwick's opinion,[60] the British were suffering 'disadvantages from our strict adherence to the spirit of the Quebec Agreement'. The Americans, on the other hand, 'have for some time refused to fulfil their obligations towards us. There is, in effect, no longer any real co-operation between us, except as regards the Combined Development Trust, from which the US have derived far greater benefits than have accrued to us.' While he did not think it was wise 'to apply pressure for a more active co-operation . . . to do nothing is to convince the Americans that we acquiesce in, and are reconciled to, the inferior position which they are trying to force upon us'.

The Foreign Office inclination, as expressed by Makins[60] from the British Embassy, was to hold fire until President Truman had appointed the US Atomic Energy Commission, for until then there was no unified authority with which to deal. Chadwick[60], while greatly respecting Makins' opinions, flatly disagreed with him on this occasion. In his view the AEC 'will not be composed of men who carry much weight.' That summer, Washington was rife with rumour about who would be chosen as the chairman of the new Commission, and the name of David Lilienthal gradually emerged from the pack, which included General Marshall and Joseph P. Kennedy.[61] While Chadwick[60] thought it possible that Lilienthal would accept the chairmanship, he believed the rest of the Commission were 'likely to be men of less repute', and would be surprised 'if any scientist of high standing' joined. 'I cannot believe', he continued, 'that such a Commission will have much influence on international policy,' and questioned whether they would be often consulted. When they were, he thought their main role would be to provide cover for Secretary of State Byrnes. He had become thoroughly mistrustful of Washington and could not understand 'why Makins is so confident that the U.S.–U.K. documents will be communicated to the Commission, for it is contrary to all our experience'.

It was true that few American politicians knew about the Quebec Agreement, and only Dean Acheson[62] seemed to be troubled that the US Government 'having made an agreement from which it had gained immeasurably, was not keeping its word and performing its obligations'. While Makins saw him as Britain's most likely ally,[60] Acheson could not seek a change in policy to ease his conscience, and the idea that Britain would be treated on an equal basis with other nations did not cause the political pulse in Washington to miss a beat.

References

While I have used original documents where feasible, the material involved is voluminous and often still classified. I have, therefore, had to rely heavily on the official UK history by Gowing and the official US Atomic Energy Commission history by Hewlett and Anderson. In both cases, relevant chapters are given as separate references.

1 Harris, K. (1995). Into power, 1945. In, *Attlee*, pp. 255–91. Weidenfeld & Nicolson, London.

2 Gowing, M. (1974). Statements by the prime minister and Mr Churchill on the atomic bomb, 6 August 1945. In *Independence and deterrence* , Vol. 1, pp. 14–8. Macmillan, London.

3 Oliphant, M.L. (1947). Rutherford and the modern world. *Proceedings of the Physical Society*, **59**, 144–55.

4 Chadwick, J. (7/9/45). Letter to H.M. Wilson. CHAD IV, 11/56, CAC.

5 Chadwick, J. (8/9/45). Letter to W. Akers. CHAD IV, 11/56, CAC.

6 Chadwick, J. (14/9/45). Minute to W. Akers re German scientists. CHAD IV, 11/56, CAC.

7 Notes on meeting of sub-committee September 10, 1943 (R.C.T.). RG 77, File 334, Box 60. NA2.

8 Chadwick, J. (20/2/46). Letter to J. Anderson. CHAD IV, 11/56, CAC.

9 Sime, R.L. (1996). Suppressing the past. In *Lise Meitner: a life in physics*, pp. 326–46. University of California Press.

10 Chadwick, J. (21/3/46). Letter to D. Rickett. CHAD IV, 11/56, CAC.

11 Aide-mémoire of Hyde Park agreement, (18/9/44). PREM3/139/8A, PRO.

12 Chadwick, J. (10/9/45). Letter to P. Moon. AB 1/485, PRO.

13 Chadwick, J. (14/8/45). Letter to C.F. Powell. CHAD IV, 2/10, CAC.

14 Peierls, R.E. and Moon, P. (5/10/45). Letter to J. Chadwick. RG 77, Box 85, file 201, NA2.

15 Peierls, R.E. and Moon, P. (29/10/45). Letter to J. Chadwick. RG 77, Box 85, file 201, NA2.

16 Memorandum by British scientists at Los Alamos (1945). CHAD IV, 11/56, CAC.

17 Hewlett, R.G. and Anderson, O.E. (1990). Controlling the atom: from policy to action. In *A history of the United States Atomic Energy Commission: the new world*, pp. 428–81. University of California Press.

18 Harris, K. (1995). *Attlee*, p. 401. Weidenfeld & Nicolson, London.

19 Gowing, M. (1974). Labour's machinery of government. In *Independence and deterrence,* Vol. 1, pp. 19–59. Macmillan, London.

20 *The Times*, 22/8/45.

21 Rickett, D.H.F. (24/8/45). Cable to J. Chadwick. CHAD IV, 10/42, CAC.

22 Chadwick, J. (10/9/45). ANCAM cable. CAB 134/6, PRO.

23 Gowing, M. (1974). Deterrence. In *Independence and deterrence* , Vol. 1, pp. 160–206. Macmillan, London.

24 Penney, W. (1945). Report on atomic bomb damage to Hiroshima and Nagasaki. CHAD II, 3/17A, CAC.

25 Oliphant, M.L. (1/10/45). Letter to J. Chadwick. CHAD I, 25/1, CAC.

26 Chadwick, J. (14/9/45). Letter to D. Rickett. CHAD IV, 10/42, CAC.

27 Chadwick, J. (5/10/45). Letter to D. Rickett. CHAD IV, 10/42, CAC.

28 Newton, V.J. (1991). *The Cambridge spies*. Madison Books, New York.

29 Rhodes, R. (1995). *Dark sun: the making of the hydrogen bomb*. Simon & Schuster, New York.

30 Cockcroft, J.D. (1945). Memo on Dr Nunn May. CHAD IV, 6/29, CAC.

31 Intelligence report (1945). Estimate of the knowledge of Metlab operations possessed by Evergreen scientists. RG 77, Box 60, file 334, NA2.

32 Groves, L.R. (3/2/45). Notes on meeting with J. Chadwick, 20/1/45. RG 77, Box 85, file 201, NA2.

33 Boyle, A. (1979). *The fourth man*, p. 65. The Dial Press, New York.

34 May, A.N. (29/8/44). Letter to J. Chadwick. CHAD I, 25/1, CAC.

35 West, N. (1989). *Molehunt: searching for Soviet spies in MI5*, p. 160. William Morrow, New York.

36 Gowing, M. (1974). External policy: brief hope of interdependence. In *Independence and deterrence* , Vol. 1, pp. 63–86. Macmillan, London.

37 Chadwick, J. (19/12/45). Letter to D. Rickett. CHAD IV, 11/56, CAC.

38 Gowing, M. (1974). External policy: hopes are dupes. In *Independence and deterrence* , Vol. 1, pp. 87–130. Macmillan, London.

39 Chadwick, J. (29/4/46). Letter to Lord Portal. AB 1/353, PRO.

40 Hewlett, R.G. and Anderson, O.E. (1990). The legislative battle. In *A history of the United States Atomic Energy Commission: the new world*, pp. 482–530. University of California Press.

41 Chadwick, J. (6/3/46). Letter to R.E. Peierls. CHAD I, 24/2, CAC.

42 Peierls, R.E. (12/3/46). Letter to J. Chadwick. CHAD I, 24/2, CAC.

43 Chadwick, J. (2/4/46). Letter to H. Massey. CHAD I, 24/2, CAC.

44 Association of Scientific Workers (1946). Press release re. Dr Nunn May. CHAD IV, 6/29, CAC.

45 Chadwick, J. (21/3/46). Letter to M.L. Oliphant. CHAD I, 24/2, CAC.

46 Oliphant, M.L. (23/3/46). Letter to J. Chadwick. CHAD I, 24/2, CAC.

47 Oliphant, M.L. (30/3/46). Letter to J. Chadwick. CHAD I, 24/2, CAC.

48 Oliphant, M.L. (2/1/46). Letter to J. Chadwick. CHAD I, 24/2, CAC.

49 Chadwick, J. (4/4/46). Letter to M.L. Oliphant. CHAD I, 24/2, CAC.

50 Chadwick, J. (11/4/46). Letter to E. Appleton. CHAD I, 24/2, CAC.

51 Chadwick, J. (22/3/46). Letter to J. Rotblat. CHAD I, 24/2, CAC.

52 *New York Times*, 11 March 1946.

53 Chadwick, J. (12/4/46). Letter to Lord Portal. AB 1/193, PRO.

54 Truman, H.S. (1956). *Years of trial and hope*, p. 12. Da Capo, New York.

55 Weiner, C. (1969). Sir James Chadwick, oral history. American Institute of Physics, College Park, Maryland.

56 Gowing, M. (1974). *Independence and deterrence* , Vol. 2, p. 6. Macmillan, London.

57 Hewlett, R.G. and Anderson, O.E. (1990). Last best hope. In *A history of the United States Atomic Energy Commission: the new world*, pp. 531–79. University of California Press.

58 Acheson, D. (1969). *Present at the creation*, p. 154. W.W. Norton, New York.

59 Dilks, D. (1972). *The diaries of Sir Alexander Cadogen*. G.P. Putnam's Sons, New York.

60 Chadwick, J. (1946). Relations with the U.S.A. on atomic energy. Undated report. CHAD IV, 10/42, CAC.

61 Hewlett, R.G. and Anderson, O.E. (1990). A time of transition. In *A history of the United States Atomic Energy Commission: the new world*, pp. 620–55. University of California Press.

62 Acheson, D. (1969). The Quebec Agreement. In *Present at the creation*, pp. 164–8. W.W. Norton, New York.

18 Liverpool and Europe

The sojourn in America had lasted three years when Chadwick returned to his post as Professor of Physics in Liverpool. He had been trying to extricate himself from his post in Washington and the UN Atomic Energy Commission for some months, and finally 'just packed up and left'.[1] He had worked virtually every day during those tumultuous times, and unlike a returning soldier he could not simply unload his wartime responsibilities on returning to civilian life. He and Aileen sailed from New York on the *Queen Mary* in the high summer of 1946. The luxury liner had been used as a troop ship and was still not refitted to its prewar state. Sir James and Lady Chadwick were assigned a small cabin below decks with two bunks. The slight, unassuming, middle-aged, scientist had experienced many uncomfortable journeys during his extensive travels on the North American continent, and this was certainly preferable to another crossing by RAF bomber. Aileen was outraged on his behalf that the Embassy had not made better arrangements; she stormed off to find the purser, who was informed in no uncertain terms that her husband, the Nobel Prizewinner and leading nuclear physicist, Sir James Chadwick, would not be returning to England in such cramped quarters. Chadwick smiled meekly at the steward who returned with Aileen to move them into a refurbished stateroom. On their arrival at Southampton they were met by a car and driver from the Ministry of Supply. As they approached London, the car ran out of fuel. The driver had no coupons to purchase more petrol so that the Chadwicks had to complete their journey to Euston station by bus. It was the perfect introduction to the austerity of postwar Britain.

Chadwick's friends in Liverpool were shocked to see how much his wartime efforts had taken out of him. He called on the new Vice-Chancellor, James Mountford, who noted in his diary that he had never seen a man 'so physically, mentally and spiritually tired'[2] as Chadwick was at that time. Mountford sensed that Chadwick 'had plumbed such depths of moral decision as more fortunate men are never called upon even to peer into', and as a consequence suffered 'almost insupportable agonies of responsibility arising from his scientific work'. Other Liverpool academics were more self-righteous, and one member of the Physiology Department wrote to criticize him for showing so little sense of responsibility in helping to create such an awful weapon. In his reply, Chadwick[3] did not attempt to justify his own contribution to the Manhattan Project, but restated his belief that the problems posed by atomic weapons could only be solved by international statesmen acting in concert. He pointed out that he and many of his colleagues had expended a great deal of effort in bringing these difficult issues before their political masters.

Although Chadwick was undoubtedly deeply affected by the advent of nuclear weapons and shared in the remorse which to some degree affected all those scientists who had worked on their development, he never contributed to public debate on the subject. In part, this was because he was an intensely private man who would never disclose his emotions even to close friends. There was also his official position as a government adviser, for which he took the ordinance of a civil servant and considered it improper 'to state his own views at length'.[3] While he did not hesitate to tell ministers and generals that an independent bomb was a necessary component for the British defence programme, he would have been perfectly content to be contradicted by them. Others might, and did, see his reserve as an abrogation of moral responsibility, but it was a position born of genuine humility, reinforced by innate shyness. When he was awarded an Honorary Doctor of Laws from Liverpool University in 1947, the orator, who was perhaps acquainted with Chadwick's opinion, remarked: 'He has released the genie from the bottle; whether it works in our service or for our annihilation is for us, and not for him, to decide.'[4]

Although Chadwick had not been physically present in Liverpool for three years, he had attempted to maintain an interest in university affairs. His friend Sir Arnold McNair, who was Vice-Chancellor for most of the war years, would write to him in Washington with news about the Physics Department, the Liverpool Cancer Commission and occasional tittle-tattle. In return, Chadwick would send incisive opinions and suggestions about future university policies. He wrote to McNair[5] in July 1944 following a dinner conversation with James Conant, the President of Harvard University, on the subject of university admissions. Conant had told him about Scholastic Aptitude Tests (SATs), which Chadwick thought would be a useful supplement to scholarship exams. He sent a package of materials to Liverpool, saying that he thought SATs would be especially good for assessing students from country schools, where the 'syllabus and teaching did not reach the standards of the good private schools'. Chadwick sent such material through the diplomatic bag and was annoyed that it was subject to the same censorship as Tube Alloys communications. He pointed out that he was forced to use the diplomatic channel because ordinary mail took eight weeks to cross the Atlantic, which would effectively cut him off from Liverpool. He was worried that his university correspondence would be unduly restricted, making it 'quite impossible for me to be any use to my university'.[6] McNair wrote to him,[7] at about the same time, trying to reassure him:

> You really must not worry about your duty to the University . . . I am quite certain that the prestige which will in future, when the veil of secrecy is lifted, accrue to you and your Department, will amply repay any inconvenience which the University may have sustained by reason of your absence.

It will be remembered that Chadwick in 1944 was receiving monthly reports on Tube Alloys work from several British universities, as well as attempting to quell a virtual insurrection against Feather at the Cavendish: he was also dealing with regular university paperwork sent by Stanley Dumbell, the Registrar at Liverpool. In April 1944, Chadwick criticized[8] the Faculty of Science report for being an unco-ordinated document, where each department seemed to be plotting a separate future. He was disappointed that it contained no provision for a Chair in Theoretical Physics, 'which, apart from its desirability *per se*, would form a link between Physics and Mathematics and, in time, Chemistry'. Regarding the Physics Department, he considered 'the present laboratory and facilities are unworthy of our University. They do not rank with those offered by Manchester, Birmingham, Leeds or Bristol, say, but rather with one of the Welsh colleges.' Sounding, for once, like a latter day Rutherford, Chadwick had no doubt that after the war his department could become very prominent: 'I have the men—and more will come; all they need is apparatus and free time.'

The over-confidence was short-lived, and it soon became apparent after the war that there were more posts available than men of experience to fill them.[9] It was proving difficult to bring some of the best scientists back from America, where they could command salaries at least double those in England. Even the illustrious Cavendish was having difficulty with recruitment; Sir Lawrence Bragg, Rutherford's successor, contacted Chadwick in January 1946 about the Swiss physicist, Egon Bretscher, whom he wished to secure as a lecturer, but who was proving elusive. Bretscher was still at Los Alamos, where he was working on the 'super' or hydrogen bomb, and Chadwick found him very unsettled about his future. Setting himself up as a mediator, he wrote to Bragg[10] that Bretscher could be troublesome on occasions 'owing to his hypersensitivity to imaginary slights, neglect, etc., but he is undoubtedly very likeable and very able'. He thought the root of Bretscher's uneasiness was that he had never had a position commensurate with his ability nor any settled future. 'In addition', Chadwick continued, 'he has suffered by being a foreigner; in England we are too slow in absorbing good men.' He suggested that Bragg offered him a Readership rather than a lectureship.

The following month, Chadwick heard from Bretscher,[11] who said he had received a completely unsatisfactory offer from Bragg, from which he concluded that Cambridge were not really interested in him. As if to exemplify his hypersensitivity, Bretscher expounded his jealousy of Norman Feather, who, in his opinion, had 'advanced with astonishing speed to great heights of distinction and influence though . . . he had hardly contributed anything to our Project'. He went on to accuse Feather of reaping the glory of his ideas on plutonium as a fissionable isotope. In reply, Chadwick agreed[12] that the salary of £1000 was inadequate and sympathized to some degree with Bretscher's criticisms about Feather, but pointed out to him that Rotblat had thought of the plutonium idea in June 1940, before either of them, and had never published it. Chadwick advised temperance and said he would try to repair the situation. In the meantime, Bragg wrote to Chadwick again,[13] saying that he had tried to extract the maximum from the University in terms of salary and seniority, and regretted that this had not been enough.

Bretscher's employment intentions became almost a *cause célèbre* in the small world of British nuclear physics. Oliphant heard of his discontent and wrote to Chadwick:[14] 'I know that he is an ass, and quite unreasonable, but can we afford to lose him?' The same concern was bothering Chadwick since Bretscher, along with Klaus Fuchs, was one of the few non-Americans who were working on the superbomb. Chadwick wrote to Bretscher[15] about this in April 1946, pointing out that the diplomatic relationship with the Americans was deteriorating steadily. In a handwritten postscript to the letter, in which he also held out the prospect of a job in Liverpool, Chadwick said 'Collect anything you can and let me have it at any available opportunity'. The opportunity came two months later with the departure of Fuchs from Los Alamos. Bretscher had made some calculations on deuterium–tritium reactions, which were central to the hydrogen bomb. Although he knew this would be regarded as highly classified information by the Americans, he took the view that Britain was entitled to information produced by her own scientists and unwittingly asked Fuchs to take the calculations to Chadwick in Washington. Fuchs was delighted to oblige, breaching the security of Los Alamos for the last time.[16]

The Liverpool University Senate approved Chadwick's request for a Chair in Theoretical Physics and it was eventually ratified by the University Grants Committee. In May 1946 Chadwick received a rather startling cable in Washington from Dumbell: 'WOULD YOU APPROVE SCHROEDINGER IF AVAIL FOR CHAIR THEOR PHYSICS'.[17] Erwin Schrödinger[18] shared, with Dirac, the 1933 Nobel Prize for Physics for his wave equation which described the permissible motions of electrons within atoms—a piece of work which had become an indispensible tool in quantum mechanics. Now nearing sixty, he had escaped from his native Austria in 1938 and was working at the new Institute for Advanced Studies in Dublin, founded by de Valera, the President of the Irish Free State, who also happened to be a talented mathematician. In 1944 Schrödinger wrote a small, philosophical, tract entitled 'What is Life?', which had not found favour with the Roman Catholic Church, which was now calling for his departure from Dublin. Chadwick's immediate reaction was one of delight that such a preeminent figure might be attracted to Liverpool. He then received reports collated by Rotblat,[19] which essentially pointed out that Schrödinger was a difficult man who took an idiosyncratic approach to his own work and found it uncongenial to collaborate with others—even with his own students. After reading these confidential reports, Chadwick came to the conclusion that Schrödinger was not the right man to create a school of theoretical physics, and wrote to the Vice-Chancellor,[20] with some reluctance, that it appeared Schrödinger's best work was behind him.

Schrödinger remained in Dublin and de Valera attempted to recruit another Jewish émigré, Herbert Fröhlich,[21] who had spent the war working in the physics department at Bristol. While there he had made many valuable contributions to various aspects of physics, including work with Heitler on nuclear forces. This had brought him to Chadwick's attention, and soon after the invitation came from Dublin, Chadwick offered him the Chair in Theoretical Physics at Liverpool. After establishing that the theoretical department would be an independent entity under his direction, Fröhlich accepted and occupied the chair for 25 years, making a series of brilliant revelations in solid state physics and the theory of superconductivity in particular.

Steps were also taken to create a position for Bretscher at Liverpool. When Rotblat found out about this, he was surprised that Chadwick had not informed him directly. While he thought Bretscher would be excellent for the laboratory, he had shared an office with him for a year at Los Alamos and warned Chadwick[22] that he would find it intolerable if Bretscher came in a position senior to him. Chadwick apologized both to Rotblat and Dumbell for forgetting 'the ways and methods of our University',[23] and in the meantime received a characteristically hypochondriacal missive from Bretscher,[24] in which he was unenthusiastic about the prospect of coming to Liverpool. On the general subject of retaining staff, Rotblat[25] also warned Chadwick that the new establishment at Harwell was luring men with offers of higher salaries. Word had already reached him in Washington that Harwell was attempting to poach Michael Moore, the cyclotron engineer, from Liverpool. This led to a warning from Chadwick that he would 'oppose very strongly any effort to rob universities for the benefit of Harwell'.[26] In reply he was assured that Harwell would 'never dream'[27] of making Moore any offer without his consent, to which Chadwick retorted[28] that he felt sure the rumour which had reached him 'must be a gross distortion of the facts', and he was much relieved to get that true story.

It is not surprising that Chadwick himself was approached to join in the game of academic musical chairs, which was taking place in physics departments across England. Bragg was conscious of the need to preserve the great tradition of nuclear physics at the Cavendish, even though he was steering the laboratory in new and fruitful directions. The Jacksonian Chair, occupied in Chadwick's time at the Cavendish by C.T.R. Wilson, was vacant and was now offered to Chadwick. Replying to the invitation in March 1946, Chadwick[29] admitted to Bragg:

> In spite of the strong ties I have with Liverpool, I cannot help but feel attracted towards Cambridge, for my associations with the Cavendish and with Caius are too long and too intimate to be disregarded, and form a large part of my life.

Aside from personal considerations, Chadwick was concerned to avoid the over-concentration of expertise in the Cavendish which he thought would be bad for the country, and he was not convinced in his 'innermost heart' that it would be right for him to return there. He discussed the offer with Mountford, the Vice-Chancellor at Liverpool, who was sympathetic,[30] and said the decision must depend essentially on where Chadwick felt he could best advance his subject. After several months of vacillating, he resolved to stay in Liverpool.

When he finally arrived back in the city that summer, Chadwick found a department which was still physically cramped and had undergone considerable damage and dilapidation. The 37 inch cyclotron was, in Chadwick's words, a 'war casualty'[31] having done more work than any outside Los Alamos. It was rebuilt, largely by Michael Moore, who after prolonged visits to Berkeley and Los Alamos, was now one the world's leading cyclotron engineers. Pickavance, the senior lecturer, moved to Harwell early in 1946, but other staff members remained, eager to restart fundamental nuclear research. When all the men returned from North America, there was an academic staff of 14 in the Physics Department, supported by about twice that number of technical staff.[32] Whereas only a deacade earlier, Liverpool was a research backwater, many of the staff had now worked, intensely, with the world's leading nuclear physicists; this experience had immeasurably raised their collective self-confidence and expectations. They had come to regard themselves as being at the leading edge of the subject. The teaching side still lagged behind, and this was not helped by the example of Professor Chadwick, who still hated lecturing. At the start of the year, he always insisted on being included on the lecture schedule, but would invariably find a pressing engagement at Senate House or the Royal Society which prevented him from giving his talk. On one occasion, he was asked to talk about the discovery of the neutron to the student Physical Society. On the day of the talk, he called a young lecturer to his office to say that he had to go to a meeting that evening and asked him to give the neutron story with the aid of his own meticulous handwritten notes. The lecturer had no alternative, but felt rather put upon and not a little inadequate as a substitute. When he arrived in the lecture theatre, he looked up to see the grinning face of Chadwick, who had come to hear his presentation.[33]

It had been realized since 1937 that there was an inherent limit to the energy of accelerated particles that could be produced in a conventional cyclotron.[34] As the speed of a particle begins to approach the speed of light, relativity predicts that the mass of the particle will increase. When the mass increases by more than a few percent, the time it takes for the particle to travel round the cyclotron becomes longer so that the particle gets out of phase with the electrical potential that is accelerating it and it slows. In practice, the limit for a conventional cyclotron is protons of approximately 25 MeV energy, which travel at nearly a quarter the speed of light and experience an increase in mass of about 3%. In 1945, a solution to this problem was proposed independently by an American and a Russian, which was to vary the accelerating high frequency voltage in order to compensate for the relativistic mass increase. This came to be known as the principle of phase stability and it opened a new era in nuclear physics, where particles could be accelerated to much higher energies in a machine known as a *synchrocyclotron.*[35] Chadwick, with enthusiastic support from Rotblat and Moore, decided that Liverpool should have one, and he was in an unrivalled position to ensure that it would come to pass.

At an early meeting of the Advisory Committee on Atomic Energy,[36] Patrick Blackett called for a Nuclear Physics Sub-Committee (NPC) to be set up under the chairmanship of Sir James Chadwick to oversee the relationship between Harwell and the universities, and to co-ordinate research work done in the universities. The suggestion was immediately adopted, and Chadwick found himself at the head of an august group that, besides Blackett, included Cockcroft, Sir George Thomson, Oliphant, Dee, Peierls, Feather and Sir Charles Darwin. Their brief was to 'make recommendations regarding the programme of nuclear physics to be pursued in this country as a whole'. At its meeting on 12 November 1946[37] with Chadwick in the chair, the sub-committee considered research programmes put forward by the universities. The ambitious proposal from Liverpool to build a frequency modulated cyclotron with 120 inch diameter pole pieces was not only approved, it was suggested that Chadwick should investigate the possiblity of an even larger magnet. Much of the detailed planning for the machine was delegated to Joe Rotblat, and it was he who travelled to the Department of Scientific and Industrial Research (DSIR) in London to seek funding for this grandiose project.

The suppliant in front of Rotblat[38] at the DSIR was Nevill Mott, from the University of Bristol. He emerged shaken after three-quarters of an hour with the committee and told Rotblat that he had never had such a grilling in his life. Rotblat enquired about what he had been requesting—two or three new microscopes for a few hundred pounds. Trembling, Rotblat was shown into the room where the committee had broken for tea, and he was introduced as coming from Chadwick's department in Liverpool. There followed a lengthy period of social chatting with various members—Rotblat wondering all the while when they would start to consider his application and how he would deal with hostile questions. Eventually the meeting reconvened and the chairman explained that Rotblat had submitted a request to finance a synchrocyclotron for Liverpool. It seemed to him that the application was satisfactory and unless anyone had any comments (which they did not), the DSIR was happy to approve a grant for £200 000 over five years.[32] Such was the perceived potential of nuclear physics to the British Government in the immediate post-war period. Nor was Liverpool the only university contemplating a nuclear programme; the Nuclear Physics Sub-Committee[37] had received notices of intent from Glasgow, Edinburgh, Cambridge, Bristol, Birmingham and Oxford as well. It was just that in this field dominated by Rutherford's old boys, Chadwick was *primus inter pares.*

Moore and Robtlat were a dynamic team. Months before the project had been approved by the NPC, Moore had been in correspondence with the City Electrical Engineer requesting an increase in supply to a load of 1000 kW.[32] The magnet would require 1500 tons of steel, which was in short supply. A choice had to be made for the magnet's coils between copper, which was expensive and difficult to work, and aluminium, which was subject to corrosion by water. There were major questions about the shielding necessary—how thick and what material—for radiation protection of personnel. One point was quite obvious: the synchrocyclotron would not fit into the basement of the George Holt Laboratory and a new site must be found. There was a derelict plot, partly cleared by the Luftwaffe, close to the university. For many years it had been the location of the workhouse, but in the 1930s was intended to be the site of the new Metropolitan Catholic Cathedral. Work had started on the crypt, but was interrupted when Archbishop Downey belatedly submitted Lutyen's blueprints to the Vatican for their approval. This was withheld when it was realized the new building would be larger than St. Peter's, and construction came to a standstill.[32] Walking over the triangular patch of land, Rotblat had the idea that the sunken crypt and slope of the ground would help solve the radiation protection problem. By December 1946, there was a scale map in the Physics Department, showing the enormous planned footprint of the cathedral and a small area in the southeast corner now earmarked for the nuclear physics laboratory. The land was consecrated, and the University Estates Department were instructed to enter negotiations with the Catholic Church with a view to obtaining a lease.

In January 1947, Chadwick was recalled to the UK delegation to the UN and sailed to New York on the *Queen Elizabeth.* He expected to stay for two months and hoped, amongst other things, to visit Lawrence at Berkeley to pick up what information he could about their 184 inch synchrocyclotron. He found that the political climate in the US had changed dramatically in the six months he had been away, with the Republicans having taken control of Congress for the first time since Roosevelt had won the White House in 1932. The international talks on atomic energy at the UN had reached no concrete agreement after six months. At midnight on 31 December 1946, the civilian Atomic Energy Commission appointed by Truman had come into being with sweeping powers over all domestic production of fissionable materials and nuclear research.[39] The five members of the AEC had to be confirmed by Congress, and although they were in no way partisan appointees of the President, this proved to be a lengthy and controversial process. Chadwick was anxious to establish a relationship with the AEC chairman, but as he wrote to Lawrence,[40] he felt that he could not approach Lilienthal 'before the Senate has confirmed his appointment, for it might make things even more difficult for him'. While he was in New York, Chadwick was disturbed to learn that the question of relations with France was again under active discussion, spurred in part by French patent lawyers. He wrote to the Cabinet Office in London[41] warning of many pitfalls.

> I would agree that we should recognise some obligation to the French for their action in 1940; not so much for what they actually accomplished, either in France, which was practically nothing, or in England, which was little more, but for their good intentions.
>
> . . . We should be tender with the French if they approach us but on no account should we approach them. This could only give rise to serious misunderstandings with the U.S., as past events have amply demonstrated. Apart from the general undertakings we have given the U.S., we have to remember that the Americans are suspicious, to say the least, that information given to the French quickly reaches Russia. [I must say here that Groves did not accuse Joliot himself of disclosing information, but said he had definite proof that some men in Joliot's laboratory transmitted information to Russia. Groves did not show me the evidence but I am inclined to believe him.]
>
> The U.S. view is that French co-operation is of no great value; they have so little to put in the pool. I agree with this, although I would not put it so bluntly . . . the advantage we get from co-operation with France is imponderable rather than tangible, more political than scientific. The French, on the other hand, have much to gain and little to lose.

While he would not argue against co-operation with the French under any circumstances, Chadwick wished to induce utmost caution in any dealings with them. He thought many of their patent claims were specious and should be resisted. His advice to the Foreign Office would be to be to say that Britain must await the outcome of the UN Atomic Energy Commission before entering any bilateral arrangements—exactly the argument that had been made on Capitol Hill the previous year, to the infuriation of the British.

After two months kicking his heels in the Empire State Building, Chadwick could see no end in sight to the Senate hearings and returned to England without meeting Lilienthal. Nor was he able to visit Berkeley, but he did send Lawrence some photographic plates from Liverpool to test with fast neutrons from the new 184 inch synchrocyclotron. These plates were coated with a new concentrated emulsion developed by Ilford Ltd in England, as a result of input from a small panel, chaired by Rotblat and set up under the auspices of the Ministry of Supply. With the new emulsions, it was possible to detect particles which produced very little ionization, and as Chadwick told Lawrence, 'alpha [particle] tracks are thick black lines and even in fast proton tracks the grains are almost contiguous. This makes detection of rare events much easier and enormously reduces the strain of observation.'[40] The chief exponent of photographic techniques in nuclear research was Cecil Powell in Bristol, and with the new emulsion he and his group soon detected a new, unstable, radioactive particle called the *pion*, π- or *pi-meson* in cosmic rays. The half-lives of these particles, which have a mass about 300 times that of an electron are measured in nanoseconds (billionths of a second); they are formed by the collision of protons in cosmic rays with atmospheric nuclei and very few of them are present at sea level. Powell[42] made his historic discovery by taking his photographic plates to mountain tops and sending them to the upper atmosphere in home-made hydrogen balloons. It was quickly accepted that the π-meson was the particle of intermediate mass that a Japanese physicist, named Yukawa, in 1935 had predicted should exist in association with nuclear fields. Yukawa hypothesized that the π-meson was responsible for the strong interaction between protons and neutrons, by being constantly exchanged between them. Thus Powell's 1947 discovery in a sense brought to a satisfactory conclusion research on the nature of the forces holding the nucleus together that Chadwick had begun at the Cavendish during 1921 with his assistant Bieler. Cecil Powell would be awarded the Nobel Prize in 1950.

When Chadwick sent the photographic plates to Ernest Lawrence in March 1947, he included some instructions[43] from Rotblat on the processing technique used in Liverpool. Chadwick may have also suggested that once the plates had been exposed at Berkeley, they should be returned to Liverpool for development, which was a tricky process. In the event, one of Lawrence's team, Eugene Gardner[44] performed some experiments and nothing much was found. He returned the developed plates to Liverpool, and although Chadwick thought they should contain interesting material, they had been ruined by poor processing. The following year, Gardner was joined at Berkeley by Lattes, one of Powell's team, who showed him how to develop the plates properly and in March 1948 they were the first to report the artificial creation of π-mesons using the synchrocyclotron.[45] They had bombarded carbon and other targets with 380 MeV α-particles, and in 30 seconds had produced 100 times as many π-mesons as Lattes had been able to photograph in 47 days of cosmic ray observation. The implications for particle physics research were profound—copious supplies of mesons could be produced by laboratory-based accelerators, offering a novel means for studying the forces at play in the atomic nucleus and the fundamental properties of matter. The random and rare events produced by cosmic rays did not need to be harnessed, provided that accelerators of sufficient energy were constructed. Even before the exciting report from Gardner and Lattes, Chadwick had decided the Liverpool machine needed to be bigger than first planned.[32]

The original 120 inch synchrocyclotron was intended to produce protons of about 250 MeV, which would have been above the threshold for meson production, but not comfortably so. Rotblat[32] made enquiries and ascertained that the maximum length of steel plate that could be machined in Great Britain was 33 feet, and with skilful design, this could be utilized for a magnet with pole pieces 156 inches in diameter. A synchrocyclotron of these dimensions should produce protons of 400 MeV, which would be satisfactory for fundamental particle research. On 1 December 1948, Rotblat communicated these facts to the Nuclear Physics Sub-Committee. He was now in charge of the project, since Chadwick left Liverpool in the summer of 1948 to take up the Mastership of his old Cambridge college. Rotblat explained to the committee that the capital expenditure was expected to rise from the £165 000 he had presented to the DSIR in 1946 to approximately £330 000. He was now truly begging on a large scale, to use Chadwick's aphorism, and since the DSIR were still well disposed, the project proceeded.

Viewed from the continent of Europe, nuclear physics, in which the French, Danes, Italians, Austrians and Germans had such fine traditions, was beginning to look like an Anglo-Saxon preserve. The UK government, in part due to Chadwick's warnings, had pointedly failed to entertain overtures from France and, still mindful of the Quebec Agreement, they were not about to irritate the Americans by collaborating with other European countries. After the war, Lew Kowarski, the Russian who had brought heavy water to England in 1940, returned to France from Montreal and became a director of the French atomic energy commission (CEA), as well as a member of the French delegation to the UNAEC. In late 1949, he wrote a seminal report[46] on European atomic co-operation in which he concluded that in the nuclear field 'continental Western Europe could only count on its own resources'. While he saw the British as being, to some extent, in thrall to American policies, he also recognized the real lead which they held over the rest of Europe through programmes such as Harwell and the Liverpool synchrocyclotron, which was a disincentive to their co-operation. He wanted to see a supra-national programme in Europe that could provide a major laboratory with a really big particle accelerator. He saw two main obstacles to such a project. First, the fact that 'all national resources in the atomic field are inextricably connected with defence secrets and with a mentality which is in no way open-minded to large projects and supra-national realizations'. There was, though, a political will to build Europe, as manifested in the Organization for European Economic Co-operation and the Council of Europe. Kowarski's second point or 'trap' was that individual countries, still aspiring to bilateral collaboration with the USA or UK would hold back rather than commit themselves to a nascent multinational organization that might not have the approval of the two established atomic powers. He thought the news that the Russians had exploded an atomic bomb in the summer of 1949 should make the Americans less averse to international co-operation since it must now be obvious, even to members of Congress, that there were no atomic secrets.

Six months after Kowarski's report, the concept of a European nuclear research centre received an unexpected boost at a UNESCO Conference in Florence. Isidor I. Rabi, the 1944 Nobel Prizewinner, was a member of the US delegation and put forward a resolution calling for the organization of regional research centres to aid the search for new knowledge 'in fields where the effort of any one country in the region is insufficient for the task'.[46] Acknowledging that 'the United States and perhaps the United Kingdom held almost a monopoly', Rabi called for a European centre devoted to high-energy physics, biology and computing. In December 1950, a small meeting was held in Geneva (to which no British representative was invited) which resulted in ambitious recommendations for the scientific programme and funding of a European nuclear research laboratory.

The reaction of British scientists[47] on hearing the outcome of the Geneva meeting, tended to involve the epithets 'crazy' and 'high-flown'. Ever circumspect, Chadwick wrote to Cockcroft[48] in March 1951 commenting that the opinions of British physicists 'about the European Centre were somewhat diverse but the balance is definitely against our taking any active part in it'. 'At the same time' he added, 'I do not see why we should not help when we can and especially in the preparatory phases'. By that summer, a split had become apparent amongst the European scientists over the proposed scale of the centre and in particular the size of the accelerator to be constructed. Pierre Auger, the driving force behind the 1950 Geneva meeting, clung to the original plan that the accelerator should be the biggest the world had seen, eclipsing those under construction in the US at Brookhaven and Berkeley. A more modest faction was beginning to emerge around the Dutchman, Kramers, who suggested instead that the new laboratory should be grafted onto the Bohr Institute in Copenhagen.[49] In the middle of August, Kramers travelled to Copenhagen to put the idea to Bohr. By chance, Chadwick was staying with Bohr and was asked for his opinion—which was to endorse Kramers' proposal immediately. It seemed to him to be eminently practicable to base the new centre on existing facilities, where there was already a pool of talented men, not to mention Bohr's inspiring presence. Auger was not reconciled to this proposed reduction in scale of the original proposals, and the tension between the different camps grew.

An intergovernmental conference was scheduled for December 1951 in Paris, during which it was expected a consensus would emerge. This time the British were invited, but Chadwick, the original delegate, withdrew citing other business;

Sir George Thomson went in his stead. This was a cause of regret to Kowarski, who thought Chadwick's 'experience of nuclear physics and nuclear diplomacy would be of greater use than G.P.'s robust and well meaning, common sense'.[50] In the days before the conference, Kowarski discerned three factions, all of whom favoured some form of European collaboration. First, there was the original Rabi or 'Big Brookhaven' school, which included the French, the Belgians and perhaps the Italians. Second was the 'modest Brookhaven' school who wanted a machine that would produce π-mesons as soon as possible—the Swiss, Norwegians, and Dutch (with the exception of Kramers) fell into this camp. Finally, there was the Bohr–Kramers axis to expand existing institutions, which had the support of Denmark and Sweden. Chadwick told Kowarski[51] that he thought the UK had little to gain from the joint European programme, but should support it in principle. To this end he thought there could be some affiliation with Liverpool, but there was no question of giving the synchrocyclotron to Europe. He repeated his opinion that the 'Big Brookhaven' group were 'quite unrealistic' and thought with Bohr's agreement that Copenhagen was a 'pre-eminently suitable site'. As an alternative, he thought the idea of building a new 300–400 MeV accelerator was feasible, and could provide interesting work within a reasonble time; the Europeans could then plan the next step—'a really big machine'.

At the meeting, Thomson threw his lot in with the Bohr–Kramers party and held out the carrot of access to the new accelerator in Liverpool; he also made it quite clear that the British Government would not make any further investment into nuclear research.[52] A compromise was eventually reached where the European effort would be based in Copenhagen for the foreseable future, and use would be made of the Liverpool machine, but planning would also continue for a much larger European accelerator, rather along the lines suggested by Chadwick to Kowarski. Whereas only a few months before, Chadwick expressed himself as not being 'strongly attracted to the idea of an international centre',[53] he now found himself advising Bohr[54] on how to approach the Liverpool University authorities so as to finalize agreement on using the synchrocyclotron. The notion of renewing the prewar links with Europe was growing on him, as was the prospect of exchanges with Bohr's group in particular.

In order to guide the United Kingdom's involvement, the Royal Society set up a new Advisory Committee on the Nuclear Research Centre in January 1952, with Chadwick as chairman.[47] Abandoning the previous policy of aloof detachment, the Committee agreed at its first meeting that the provision of facilities at Liverpool qualified Britain to join the governing Council of CERN* and that she should take up her place. This recommendation was, however, rejected by the Cabinet Steering Committee on International Organisations (IOC). The standing Government policy was not to join any new international organizations unless they were shown to be absolutely necessary, and there was also the suspicion that to join would inevitably entail significant extra expenditure. So at the next meeting in Geneva

---

* CERN—European Council for Nuclear Research.

in February 1952, Britain did not sign the Agreement constituting the Council of CERN.

By now the British scientists were so caught up by the momentum of the new European venture that they were not going to be prevented from active representation without a fight. Chadwick's Advisory Committee met again in May 1952, and decided to lobby members of the IOC. Chadwick wrote to his old colleague from the British Embassy in Washington, Sir Roger Makins,[55] now back at the Foreign Office.

> I understand that opinion in the F.O. is that there are already too many of these formal international organisations. I think that most, and possibly all, of the English scientists who are concerned in this matter that an agreement of the kind adopted at Geneva was too formal and too inflexible and that something looser would have satisfied all necessary purposes. But the decision has been made: we cannot change it at this juncture. The only question we have to decide now is to sign or not to sign.
>
> I believe that we have more to lose by not signing than by signing.

Chadwick stressed that the Geneva agreement was only a temporary one and did not commit its signatories beyond the next 18 months. Showing a naivety towards investment, Chadwick also sought to rely on the temporary nature of the agreement as a safeguard against much greater expenditure. 'But the chief loss', he concluded, 'would be that we should be unable to influence the course of development of the European effort and to steer it along the lines we desire.' Makins, who had also received representations from Cockcroft and Thomson, used his influence with the IOC to persuade them to reverse their original position, which they did on 11 June 1952, but two weeks later their call for British participation was rejected by the Chancellor of the Exchequer.[47] This application of cold water seems to have given Chadwick second thoughts because he wrote to Makins again,[56] on 17 July, expressing doubt about the need to sign the agreement after all.

There were now divisions appearing within Chadwick's advisory group and amongst the wider community of British physicists.[47] Cockcroft had visited Brookhaven in March and had seen the new Cosmotron, a 3000 MeV accelerator, in action. Still an electrical engineer at heart, Cockcroft had naturally been impressed and now considered the question of large machines in a new light. There was, he found, a high level of interest amongst the younger members of staff at Harwell, and co-operation with the Europeans seemed to be the most likely way to obtain such an accelerator on their side of the Atlantic. Whereas Cockcroft was starting to believe that big machines offered some economic advantage in terms of cost per MeV, Chadwick was worried that the country was close to bankruptcy and should not be drawn into such huge expenditures. When approached on the issue by old colleagues like Peierls and Pickavance, he could be as rigid in his opposition[57] as Rutherford had been towards the idea of the cyclotron, two decades earlier. But Chadwick's attitude seems to have been constantly changing. At the beginning of October 1952, he wrote to Bohr[58] describing 'the growing opinion in this country, especially among the younger men' in favour of building a big machine, which made it 'a critical point' in the development of a national policy. Chadwick indicated that he thought there would be a decision in favour of co-operation with Europe, but warned Bohr of likely dissent about the siting of a laboratory in Denmark. He had recently spoken to Lord Cherwell, an influential figure again, now that Churchill was back in power, and he had 'expressed himself strongly against location in Denmark on the grounds that Denmark was too vunerable to Russia. Arguments that the laboratory would have no military significance, etc. were of no avail; much as I regret it, his view is likely to be shared by our government.'

By the autumn of 1952, the question in the collective mind of British nuclear physicists was whether a very-high-energy accelerator should be undertaken as a domestic project or in conjunction with CERN. An augmented meeting of Chadwick's Advisory Committee took place at the Cavendish on 1 November.[57] It was unanimously agreed that British nuclear physicists should have access to a proton synchrotron and that this would be best achieved through CERN. In the opinion of the physicists, the Government should sign the agreement with CERN and contribute £250 000 per annum for eight years. Chadwick finally seemed convinced by the positive arguments of his colleagues, but he still harboured some misgivings that CERN would have a detrimental effect on research in the universities, which he believed would remain the most fertile places for innovative ideas. He wrote to the DSIR on 8 December urging that any convention signed by the United Kingdom should contain an explicit definition of the aims of the new organization. 'It should be made clear', he suggested, 'that the purpose of the Laboratory is to supplement existing Laboratories by providing facilities for work in the special field of high energy particles, facilities which no single Laboratory can afford. It is to feed other Laboratories, not to rob them or to bleed them to death.'[59]

As he explained to Bohr,[60] while there was now strong support amongst the community of British nuclear physicists and the Royal Society, the final decision was likely to depend on money, and he warned that there might still be objections from the Chancellor of the Exchequer. If the Chancellor's difficulty related to the precedent of signing an intergovernmental agreement Chadwick thought 'we should be able to overcome his objections; but if, as I fear, it is a question of money, the decision is hardly open to argument.' In fact the Chancellor, R.A. Butler, was impressed by the 'clearly expressed view of British scientists that our physicists should have access to the more powerful machines now being planned' and gave his final approval to the UK joining CERN on 29 December 1952.[57]

This convoluted story marked the end of Chadwick's second career as a nuclear diplomat. While there were times during the long, intertwined debates over big accelerators and membership of CERN when he appeared reactionary, he was in the end persuaded by the opinions of younger men and then supported them wholeheartedly. Indeed the conversion of Chadwick was probably noted in government circles, and may in some small way have weighed in the final decision to join CERN. Certainly, the promise of the Liverpool synchrocyclotron and his deep friendship with Niels Bohr helped shape the mould for postwar nuclear physics in Europe.

References

1 Chadwick, J. (27/11/46). Letter to E.O. Lawrence. Bancroft Library, University of California, Berkeley.

2 Mountford, J. (1974). Obituary of Sir James Chadwick. *University of Liverpool Recorder*, **66**, 14–5. University of Liverpool Archives.

3 Chadwick, J. (24/4/46). Letter to C. Brunton. CHAD I, 24/2, CAC.

4 Oration for LL. D. (1947). CHAD II,3/9, CAC.

5 Chadwick, J. (14/7/44). Letter to A. McNair. CHAD I, 25/1, CAC.

6 Chadwick, J. (10/7/44). Letter to D. Blok. CHAD I, 25/1, CAC.

7 McNair, A. (22/8/44). Letter to J. Chadwick. CHAD I, 25/1, CAC.

8 Chadwick, J. (10/4/44). Letter to S. Dumbell. CHAD I, 25/1, CAC.

9 Chadwick, J. (21/3/46). Letter to S. Dumbell. CHAD I, 24/2, CAC.

10 Chadwick, J. (15/1/46). Letter to L. Bragg. CHAD I, 24/2, CAC.

11 Bretscher, E. (20/2/46). Letter to J, Chadwick. CHAD I, 24/2, CAC.

12 Chadwick, J. (26/2/46). Letter to E. Bretscher. CHAD I, 24/2, CAC.

13 Bragg, L. (12/3/46). Letter to J. Chadwick. CHAD I, 24/2, CAC.

14 Oliphant, M.L. (30/3/46). Letter to J. Chadwick. CHAD I 24/2, CAC.

15 Chadwick, J. (27/4/46). Letter to E. Bretscher. CHAD I, 24/2, CAC.

16 Moss, N. (1987). *Klaus Fuchs: the man who stole the atom bomb*, p. 116. Grafton, London.

17 Dumbell, S. (13/5/46). Cable to J. Chadwick. CHAD I, 24/2, CAC.

18 Erwin Schrödinger, 1887–1961. In *Nobel lectures: physics, 1922–41*, pp. 317–19. Elsevier, Amsterdam.

19 Rotblat, J. (3/6/46). Letter to J. Chadwick. CHAD I, 24/2, CAC.

20 Chadwick, J. (10/6/46). Letter to J. Mountford. CHAD I, 24/2, CAC.

21 Mott, N. (1992). Herbert Fröhlich, 1905–91. *Biographical memoirs of fellows of the Royal Society*, 37, 147–62.

22 Rotblat, J. (30/4/46). Letter to J. Chadwick. CHAD I, 24/2, CAC.

23 Chadwick, J. (9/5/46). Letter to S. Dumbell. CHAD I, 24/2, CAC.

24 Bretscher, E. (2/5/46). Letter to J, Chadwick. CHAD I, 24/2, CAC.

25 Rotblat, J. (9/5/46). Letter to J. Chadwick. CHAD I, 24/2, CAC.

26 Chadwick, J. (26/4/46). Letter to H. Skinner. CHAD I, 24/2, CAC.

27 Skinner, H. (8/5/46). Letter to J. Chadwick. CHAD I, 24/2, CAC.

28 Chadwick, J. (14/5/46). Letter to H. Skinner. CHAD I, 24/2, CAC

29 Chadwick, J. (11/3/46). Letter to L. Bragg. CHAD I, 24/2, CAC.

30 Mountford, J. (10/5/46). Letter to J. Chadwick. CHAD I, 24/2, CAC.

31 Gowing, M. (1974). *Independence and deterrence*, Vol. 2, p. 220. Macmillan, London.

32 King, C.D. (1993). The Liverpool 156" synchrocyclotron (1945–54). Unpublished M.Sc. thesis, University of Liverpool.

33 Edwards, D. and King, C.D. (1992). Interview with Professor J. Rotblat. Physics Department, University of Liverpool.

34 Bethe, H. and Rose, M. E. (1937). The maximum energy obtainable from the cyclotron. *Physical Review*, **52**, 1254–5.

35 Mersits, U. (1987). From cosmic-ray and nuclear physics to high-energy physics. In *History of CERN* (ed. A. Hermann, J. Krige, U. Mersits and D. Pestre), pp. 3–62. North-Holland, Amsterdam.

36 Minutes of Advisory Committee on Atomic Energy (1945). CAB 134/6, PRO.

37 Minutes of the Nuclear Physics Committee (1946). CHAD I, 5/3, CAC.

38 Brown, A.P. (1994). Interview with Joseph Rotblat, London. Physics department, University of Liverpool.

39 Hewlett, R.G. and Anderson, O.E. (1990). The inheritance. In *A history of the United States Atomic Energy Commission: the new world*, pp. 1–8. University of California Press.

40 Chadwick, J. (13/2/47). Letter to E.O. Lawrence. Bancroft Library, University of California, Berkeley.

41 Chadwick, J. (12/2/47). Letter to D. Rickett. CAB 126/36, PRO.

42 Frank, F.C. and Perkins, D.H. (1971). Cecil Frank Powell, 1903–1969. *Biographical memoirs of fellows of the Royal Society*, **17**, 541–55.

43 Chadwick, J. (4/3/47). Letter to E.O. Lawrence. Bancroft Library, University of California, Berkeley.

44 Gardner, E. (2/4/47). Letter to J. Chadwick. Bancroft Library, University of California, Berkeley.

45 Gardner, E. and Lattes C.M.G. (1948). Production of mesons by the 184" Berkeley cyclotron. *Science*, **107**, 270–1.

46 Pestre, D. (1987). The first suggestions, 1949-June 1950. In *History of CERN* (ed. A. Hermann, J. Krige, U. Mersits and D. Pestre), pp. 63–96. North-Holland, Amsterdam.

47 Krige, J. (1987). Britain and the European laboratory project: 1951–mid 1952. In *History of CERN* (ed. A. Hermann, J. Krige, U. Mersits and D. Pestre), pp. 431–74. North-Holland, Amsterdam.

48 Chadwick, J. (13/3/51). Letter to J. Cockcroft. AB 6/912, PRO.

49 Pestre, D. (1987). The period of conflict, August–December 1951. In *History of CERN* (ed. A. Hermann, J. Krige, U. Mersits and D. Pestre), pp. 147–78. North-Holland, Amsterdam.

50 Kowarski, L. (15/12/51). Letter to J. Chadwick. CHAD I, 1/11, CAC.

51 Chadwick, J. (18/12/51). Letter to L. Kowarski. CHAD I, 1/11, CAC.

52 Pestre, D. (1987). The establishment of a council of representatives of European states, December 1951–February 1952. In *History of CERN* (ed. A. Hermann, J. Krige, U. Mersits and D. Pestre), pp. 147–78. North-Holland, Amsterdam.

53 Chadwick, J. (14/9/51). Letter to G.P. Thomson. CHAD I, 1/19, CAC.

54 Chadwick, J. (3/1/52). Letter to N. Bohr. CHAD I, 1/3, CAC.

55 Chadwick, J. (2/6/52). Letter to R. Makins. CHAD I, 1/6, CAC.

56 Chadwick, J. (17/7/52). Letter to R. Makins. CHAD I, 1/6, CAC.

57 Krige, J. (1987). Britain and the European laboratory project: mid-1952–December 1953.

In *History of CERN* (ed. A. Hermann, J. Krige, U. Mersits and D. Pestre), pp. 475–522. North-Holland, Amsterdam.

58 Chadwick, J. (3/10/52). Letter to N. Bohr. Niels Bohr Archive, Copenhagen.

59 Chadwick, J. (8/12/52). Letter to B. Lockspeiser. CHAD I, 1/11, CAC.

60 Chadwick, J. (12/11/52). Letter to N. Bohr. Niels Bohr Archive, Copenhagen.

19 The mastership

Well into his fifties, Chadwick was now showing some disenchantment with the field of nuclear physics. While he did not express any regret about the military applications, he found it hard to come to terms with the new approach to the subject that had been hastened by the war. Even by the early 1930s, there was a discernible difference between the benchtop experiments long favoured by Rutherford and the large-scale endeavours of Lawrence at Berkeley. In the aftermath of the Manhattan Project, 'Big Science' was going to be the order of the day, not just in the USA, but in Europe too. It held no appeal for Chadwick, whose creative ideas in the past might occasionally have been discussed with a mentor like Geiger or Rutherford, but would never have been rehearsed in a team meeting or group discussion. For him, research was essentially a private activity—a scientist 'must have quiet and solitude in which he does his thinking'.[1] While his original work was all behind him by the mid-1940s, it is difficult to imagine that a young Chadwick would have achieved so much in the new era. Writing a reference for a young Liverpool physicist who, he said, possessed considerable ability and overflowed with energy and enthusiasm, the following opinion revealed Chadwick's[2] disapproval of the modern methods:

> His development has been hampered by too much team work, and he would have benefited by working on his own.

He concluded the reference by citing 'impetuosity' as the man's worst fault—something which resulted from not thinking 'enough or deeply enough, before plunging into experiment'. Another trend he deplored was for 'half a dozen men working together and all putting their names on the publication'[1] because it made the contribution of any individual very difficult to assess. Equally, Chadwick recognized that youthful impetuosity keeps a laboratory alive. He was stimulated by 'young men coming in, often with quite hairbrained ideas, but nevertheless full of enthusiasm, full of desire to do something',[1] even if he never showed it. He also took to heart a maxim of Lord Rayleigh, the second Cavendish Professor, that by the time a scientist has reached the age of 60, his opinions on science are not worth very much, and he should make way for younger men.

Just as times had changed in nuclear physics, life in Liverpool was different too. The genteel life of the city's upper middle classes had vanished. Domestic staff were no longer plentiful or affordable, and in economic terms the city was well launched into a downward spiral—a decline made more depressing by the ravaged physical surroundings. Chadwick,[3] whose personal streak of pessimism never needed much encouragement from external forces, complained to Norman Feather:

> How difficult it is nowadays when you wish to do anything. Sometimes I wish I had stayed in the States or in Canada. The people of this country have forgotten what it is to be free agents, forgotten how to enjoy themselves, many have never known the pleasure we used to get out of simple things.

While this may be a timeless example of an old man's lament, the main fruits of Attlee's socialist government, such as nationalization and the new welfare state, brought no joy to Chadwick, and Aileen took them almost as a personal attack on her and her kind. And then came the unexpected opportunity to escape back to the tranquil and unspoilt life of Cambridge.

In 1948, Gonville and Caius College celebrated its sexcentenary.[4] The Master, John Cameron, an amiable Scotsman who had been a fellow since 1899 and Master since the death of Sir Hugh Anderson in 1928, decided it was time to retire. Since the earliest days of the College, the Master had been elected by the body of fellows. In the sixteenth century statutes, there were some uncompromising restrictions such that no man be elected Master who is 'deformed, dumb, blind, lame, maimed, mutilated, a Welshman, or suffering from any grave or contagious illness, or an invalid, that is sick in a serious measure'. In the nineteenth century, the statutes became more positive in tone, and the fellows were encouraged to elect a Master 'best qualified to preside over the College as a place of education, religion, learning and research'. In 1948, the fellows decided, after gentle persuasion from one of their senior members, Sir Arnold McNair (recently the Vice-Chancellor of Liverpool and now the British representative on the International Court of Justice at the Hague), that Sir James Chadwick, himself an Honorary Fellow of the College, would be the best choice. His selection was not a complete surprise: in June 1945, Mark Oliphant had visited Cambridge ('that hotbed of rumour') and found the place 'seething with the story that you [Chadwick] would go to Caius as Master'.[5] Now he had been chosen, it was the first time in living memory that a Master had been selected from outside the body of fellows:[6] a situation which resulted from the lack of a suitable inside candidate rather than from Chadwick's irresistable claim to the post. The President of the College, 'Chubby' Stratton, was dispatched to Liverpool to invite his old friend to return to the Mastership.

Whereas, just two years earlier, Chadwick had eventually refused the Jacksonian Chair at the Cavendish because he felt he ought to remain in Liverpool, by 1948 the future of that department as a centre for nuclear research seemed bright. The initial funding of £200 000 from the Department of Scientific and Industrial Research had been secured, the Roman Catholic Church had agreed to lease land to the university and building on the Nuclear Physics Research Laboratory started in late 1947. Rotblat, who was responsible for the day-to-day planning of the synchrocyclotron, had proved himself more than capable as an interim head of department at the end of the war, and would no doubt do so again. While Chadwick was genuinely reluctant to give up the intellectual stimulation of life as a research scientist, he would certainly not miss teaching. Sir Charles Darwin,[7] head of the National Physical Laboratory, did write to Chadwick to question whether, after his departure, Liverpool was still the most appropriate site for the new cyclotron. Chadwick assured him that it was, and told Darwin[8] that the answer was to appoint a successor who could carry on the project.* In Chadwick's eyes,[1] the balance of duty now tilted towards Cambridge, especially when Stratton told him that the fellows had been unable to agree on a new Master from their own ranks. It was the college, through Hugh Anderson, who had taken him in after the First World War, when he was almost destitute and in poor health, and now was his opportunity to repay that debt. He had little trouble in persuading Aileen to leave Liverpool; her loyalty to him was unswerving, and the prospect of a triumphal return to Cambridge as a Master's wife was deeply satisfying. After informing Mountford, the Liverpool Vice-Chancellor of his decision, Chadwick indicated to Stratton that he would accept, and on 18 May the fellows held a formal election to appoint him.

The traditions of Caius,** like those of any ancient college at Cambridge or Oxford, are rich and idiosyncratic. Its founder, Edmund Gonville, was a Norfolk cleric in the reign of King Edward III, and in its earliest times the College partly depended on tithes collected by the churches of East Anglia. These links to the Church and Norfolk persisted into the twentieth century; the medieval courts of the college and above all the chapel are constant reminders of its ecclesiastical roots. As an institution, Caius was more secular than many other Cambridge colleges,[4] but it was only in 1860, for example, that the requirement for fellows to remain celibate was ended, and after 1871 they were no longer bound to be believing members of the Church of England. It is a self-governing society, with the ultimate power residing in the whole body of fellows, but essentially controlled by the College Council, a body of 13 headed by the Master, which can trace its own origins back to Tudor times. Under Anderson's and Cameron's benign but effective administrations, the real power had been concentrated even more into the hands of an unofficial inner cabinet.[6] Chadwick may have felt that this arrangement gave Hugh Anderson the flexibility to take him on as Wollaston scholar in 1919, as a favour to his friend Rutherford (although his credentials, by that time, were outstanding).

Apart from its spiritual and political aspects, the college is primarily a place of learning with its own academic distinctions. The college was refounded and re-endowed in the 1550s by a London physician, John Caius, and 'he set a medical stamp upon his College which has never faded.'[4] In the closing years of the

---

* Chadwick was succeeded as head of department in Liverpool in 1949 by Herbert Skinner, who had been Cockcroft's deputy at Harwell. Rotblat resigned a few months later to take the Chair in Physics at St. Bartholomew's Hospital, London—a move actively opposed by Chadwick, who thought it a great loss to nuclear physics. He tried everything to dissuade Rotblat from going, including warning him that it meant he would never be elected a Fellow of the Royal Society. This prediction was invalidated in 1995 when Rotblat was elected FRS at the unprecedented age of 86.

** Gonville and Caius College is familiarly known as 'Caius' (pronounced KEYS) and for brevity I will follow this convention.

sixteenth century, William Harvey had been a scholar, and just a few years before Chadwick's arrival in 1948, the pre-eminent neurophysiologist, Sir Charles Sherrington had returned to Caius, in his mid-eighties, as a long-term guest of the Camerons in the Master's Lodge. The Camerons' generosity also serves to remind us that Caius is a social family, where it is common for generations of the same biological family to pass through or indeed to have long, successful, careers as dons. One of the aspects of the college which had not changed during Chadwick's time away was that it was still almost exclusively a male preserve—the exceptions being the College bedmakers and the Master's wife.

It was with a mixture of relief and anticipation that the Chadwicks moved into the Master's Lodge in the late summer. The undergraduates had not yet returned from the long vacation so the courts were largely deserted, although Chadwick could meet many of the fellows at lunch every day in the Combination Room. The Lodge itself is a maze of a building, beginning within the medieval fabric of Gonville Court and extending outwards in Georgian and Victorian additions to overlook the Master's private walled garden. It was not in good decorative repair, and although the Chadwicks lived in this magnificent setting rent free, he did not receive much of an allowance for entertaining, which was one important part of the job. But, as Chadwick's successor as Master, Sir Nevill Mott, has written 'the Master's duties were not entirely clear'.[9] The Master is undoubtedly the political head of the college, and as such enjoys a honeymoon period with the electorate before they start expecting to see results or at least to be given an idea of his overall policies. On a more mundane level, he is the manager of the college, ultimately responsible for personnel issues amongst the staff, as well as for the finances, which became a considerable burden for Chadwick, as we shall see. He is a figurehead for the students, a man of admirable character and perhaps great academic achievements. He owes the undergraduates a duty of pastoral care, and his level of involvement with the students will ultimately depend on his and his wife's personalities.

Even by the standards of the late 1940s, the Chadwicks were a formal couple. There was an annual intake of about 120 undergraduates, whom the Chadwicks did their best to meet individually. Every student was invited at least once to the Master's Lodge on a Saturday evening for sherry. Chadwick also instituted an annual matriculation dinner for all freshmen to welcome them to the College and to tell them about some of its traditions. He had become a devout student of Venn's biographical history of the College and would regale the students in his after-dinner speech about the virtues of the sixteenth century Master, John Caius. The majority of freshmen were educated at fee-paying public schools, with less than a third gaining admittance from state grammar schools, although this fraction showed a steady increase from the time of Chadwick's mastership.[4] Quite a number of men came up to Caius after completing their two years' national service, and were immensely grateful for the peace and civility the college offered. Unlike their counterparts a decade or so later, they were not politically active and have been characterized as generally showing 'meritocratic ambitiousness'.[4]

Chadwick had not had the same opportunity to mix with young students in Liverpool, and found himself impressed at their qualities. He remembered some of their fathers as undergraduates, in times of greater student affluence, and found their sons, if less adventurous, 'a very good lot'.[1] Sport played an even greater role in College life in the 1940s and 1950s than it does now, and the Master was expected to be on the banks of the river for the bumps races and on the touchline at rugby games—both alien sites to James and Aileen, but ones which they took to with genuine enthusiasm and enjoyment. In order to fix a student in his mind, Chadwick kept a little notebook[10] in which he would jot a few lines after talking to him. The thumbnail sketches would list the student's secondary school, with a note of any national service and a mention of their degree course. There would then be an often surprisingly detailed account of sports achievements (e.g. full-back 2nd XV and captain of rowing), activities such as music or chess, and general interests, with Chadwick seeming to make particular record of students sharing his predilection for 'country pursuits'. The whole, which never amounted to more than six lines of notes, was rounded off by a cryptic assessment which could be as simple as 'good boy' (a highly favourable rating) or more equivocal as in: 'to go into father's business—not very bright but quite nice boy'. If there were any distinctly negative impressions, they do not seem to have been committed to paper.

There seems little doubt that Chadwick's main ambition as Master was to enhance the academic quality of Gonville and Caius, making it attractive to the brightest students. Perhaps underestimating the need as head of the institution to articulate his policy, Chadwick did not make a song and dance about his intentions, but in his quiet way set about bringing them to fruition. The gradually increasing proportion of students accepted from state schools led to occasional reproachful letters[11] to Chadwick from fathers who were 'dismayed and not a little hurt' that their old college had chosen one of 'the spate of state scholars' in preference to their son. Chadwick displayed the same lack of prejudice, and overriding concern for the candidates' merits in fellowship elections. Soon after the unmasking of Klaus Fuchs as a Russian spy in February 1950, (an event, of course, which Chadwick found deeply shocking), he emphatically supported the appointment of Peter Bauer, a Hungarian by birth who had recently been naturalized and spoke English 'with a thick accent';[12] this was despite the fact that the college already had a teaching fellow in economics; it was done in such a way as to show his contempt for the wave of xenophobia sweeping the country. In a similar move, he also supported the election of a young Chinese biochemist, Ts'ao T'ien-Ch'in, at the height of the Korean War in 1951, a few days after the Gloucestershire Regiment had been defeated in battle.[6] He was especially good at spotting academic talent in his own subject, of course. In 1953, Sam Edwards,[13] the theoretical physicist and future head of the Cavendish, returned from a spell in the United States to find a letter from Chadwick waiting. In it he was asked whether he would like to be a candidate for a research fellowship at Caius. Edwards was indeed tempted, but had already been offered a job by Rudolf Peierls in Birmingham. He went to discuss his dilemma with Chadwick, who immediately relinquished any claim, and firmly advised Edwards to go to Birmingham and work with Peierls, whom he regarded as the best theoretician in the country. Having delivered this selfless piece of advice, Chadwick announced 'I have an important funeral' and disappeared, leaving Edwards to make his own way out of the labyrinthine lodge.

While contact with the undergraduates brought Chadwick nothing but pleasure, his relations with some fellows soon soured. There is no evidence that the fellows set out to undermine Chadwick as Master, but their machinations (which have become enshrined in College lore as the Peasants' Revolt) did come to cause him deep unhappiness. Indeed, it seems likely that the aims of the Master and the Peasants, as far as the desired quality of the whole fellowship, were similar, but their attempts at reform caused a deep rift between the parties. Whether Chadwick refrained from publicizing his policy because he wished to avoid a statement of the obvious, or was just following a lifelong habit of never explaining his own thoughts, the effect was the same.

As mentioned before, during Cameron's time, the inner cabal of Caius consisted of the Master, the Bursar and three tutors. This oligarchy had been a benevolent one, but it guaranteed a conservative governance of college affairs and became in itself an obstacle to innovation. Over the years, the influence of such a small number of college officers had come to be resented by some members of the Senior Combination Room, who felt their own opinions to be ignored. When they muttered about the iniquities of the Caius system, as they had no doubt done over their post-lunch coffees for many years, they now found an audience who were quite easily moved to the conclusion that something should be done. The influx of young fellows, some of whom had fought in the war, brought a fresh group, who expected to have a say in their own futures and were not prepared to let the cosy *status quo* continue unchallenged. There was also an element of academic idealism among the younger fellows, to whom it seemed that excellence in research counted for less in the upper echelons of the College than being a sound teacher or administrator. They wanted recognition for men like Joseph Needham, a polymath whose history of Chinese civilization and science many fellows regarded as one of the great sustained works of twentieth-century scholarship.

Matters came to a head at the general meeting of the Master and fellows on 6 October 1950—the first meeting after the long vacation. These meetings were always held on Friday afternoons, and on the preceding Wednesday a group of eight fellows, which included two very senior men, met in college rooms to discuss their strategy. They decided they would field two candidates in the council election and in particular, they would stand against H.E. Tunnicliffe, who had been a tutor at Caius since 1929 and had occupied a seat on the Council, unopposed, for many years. The two candidates, selected by drawing straws, were Michael Swann, a research fellow in zoology, and the young economist, Peter Bauer. Word spread around the College over the next few days and several senior fellows indicated that they would vote for the two new men. The name Peasants' Revolt was bestowed on the movement by Patrick Hadley, Professor of Music, as a pun on Bauer, the

German for peasant. News of the plan also reached the Master, and his first concern was for Tunnicliffe, an old friend. It was not in Chadwick's nature to mount a counter-campaign on his behalf, nor would it have been politically astute to do so, but he was concerned that Tunnicliffe should be forewarned. He was assured that this would be done, but perhaps predictably, Tunnicliffe arrived at the meeting as one of the few fellows who did not know what was afoot. He was defeated and both Peasants* were elected. When the result was announced, Tunnicliffe swept out of the meeting, carrying his papers, but recovered his equanimity later that term after being granted a short leave of absence by the sympathetic Master. A year later, another tutor of long standing and personal friend of Chadwick, Henry Deas, was also voted off the council, by a much larger majority than Tunnicliffe. He took the defeat much more to heart and withdrew from the social life of the College, deeply disenchanted and 'feeling that something had gone from the society he cared for so much'.[6]

The following year it was the turn of Francis Bennett, the Senior Tutor, to come under the Peasants' scrutiny. He had replaced Chubby Stratton as President at the start of Chadwick's mastership and would normally expect to hold the position, unchallenged, until he decided to relinquish it or died. He had been a well-loved fixture at Caius since the First World War and still occasionally wore a black jacket and striped trousers, the uniform of a grander age. He also wrote with a quill pen and, together with his aging clerk, was visibly losing the battle against the incoming tide of paper to his office. A young fellow drew the short straw and was sent to explain to Bennett that there were other very distinguished senior fellows such as R.A. Fisher, the statistician and geneticist, and Joseph Needham, who had never received official recognition in college life, and that he should not be affronted if one of them were elected to replace him. In the event, the fellows re-elected Bennett, but again, bad feeling was engendered.

The other key figure in the college administration, the Bursar, E.P. Weller, whose influence was also resented by the Peasants, became seriously ill, and Chadwick found himself taking on his role as well. Weller had handled the college's financial affairs for two decades and had originally been appointed because of his expertise in estate management. The major part of the College's wealth still derived from agricultural land, which had been acquired over the centuries, although there were also valuable properties in London and Hertfordshire.[4] The accounts included a number of venerable trusts, and the rules governing investment were intricate. While the college could hold gilts, it was not allowed to invest in stockmarket equities. The war had been disastrous for the British economy, and although Chadwick was not well versed in investment strategy, he was advised by college economists that gilt-edged stock was 'a very quick way of losing money'.[1] He presented this question to Sir Arnold McNair, whose judgement he had come to value highly

---

* Both later entered the House of Lords, completing the remarkable odyssey from Peasant to Peer. Lord Swann was ennobled after serving as Chairman of the BBC, and Lord Bauer in recognition of his distinguished career as an economist, specializing in underdeveloped countries.

over the years. McNair would help him out during his visits to Cambridge from the Hague, and agreed with the economists' opinion. Chadwick decided that the investment rules must be liberalized so that Caius could invest in equities. The first step was to change the college statutes, which required a two-thirds majority of fellows, and it took Chadwick a year to engineer this, even though it was uncontroversial and supported by virtually the whole council. He then had to present the case to the university, followed by the Privy Council. At times Chadwick felt out of his depth in these matters, and they certainly came as an unexpected burden. Within a few years, the change in policy brought about a marked improvement in the financial health of the College

In the back of his mind, Chadwick might have imagined he would like to renew his contact with the Cavendish on his return to Cambridge. Certainly there were some hopes amongst the nuclear physicists there that his proximity might lead to an improvement in their status, which had declined under Bragg. Denys Wilkinson opened a sweepstake in 1948 on how many days it would be before Chadwick was seen back in the laboratory. There were about 20 contestants whose guesses ranged from one day upwards, peaking at five or six days. One derided wit said a year, but even he was too eager: Chadwick's only visit was for the meeting about CERN in November 1952. Apart from the unexpected travails of his mastership, Chadwick was mindful not to encroach on another man's territory. When Bragg resigned from the Cavendish Chair in 1953, Chadwick was one of the Board of Electors who chose his successor. The chair went to Nevill Mott from Bristol, which must have brought Chadwick pleasure, not only because he and Mott had published a paper together back in 1930, but because Mott was a Caian. After his appointment, Mott[9] came on a preliminary visit, when he stayed in a bitterly cold guest room in Caius. He discussed the prospects for the Cavendish with Chadwick who, after a lengthy period of contemplation, confided to him, 'I think, perhaps, taking everything into account, we may have done the best thing.' Mott, who was used to Chadwick's deadpan manner, took this as high praise.

Apart from his continuing interest in the development of European nuclear physics, Chadwick had little contact with colleagues from his former life. There was a trickle of American visitors to Caius, and he was always pleased to renew wartime friendships. Edward Teller came in 1949 and was interrogated by Aileen through dinner about what had become of all the people from Los Alamos.[14] Hans Bethe spent time as a temporary visitor at Caius in the mid-1950s. The radiochemist, Harold C. Urey, who won the Nobel Prize for his discovery of heavy water, also paid Chadwick a visit. After the war, he became interested in the planets and the origin of the universe. He told Chadwick about this line of research with great enthusiasm; it was soon after the Russians had exploded their first atomic bomb, and Chadwick commented lugubriously that it might be more profitable to concetrate on the end of the world. When Urey later fell foul of McCarthy's Committee on un-American activities, Chadwick thought he was very badly treated.[15] He also expressed sympathy for J.R. Oppenheimer during his hearings; he repeatedly said that while Oppie may have been foolish, he was never wicked.

The Masters of the individual colleges were eligible to serve as Vice-Chancellor of the university. Chadwick was proposed in 1950 to take over the office the following year for a two-year term. He was undecided about accepting the offer because of his poor health and asked if it could be postponed.[1] The University replied that Vice-Chancellors were required to be less than 60 years of age on taking up the position, so that in Chadwick's case it was now or never. Recognizing that it would be important for Caius, Chadwick accepted, despite personal uneasiness. In addition to the strain of the mastership, Chadwick had now accepted another high-profile position of enormous responsibility. The prospect began to overwhelm him, and he came to believe that the extra work would kill him. At Aileen's insistence, they travelled back to Liverpool to consult his old physician, Lord Cohen. Cohen examined him and could find no organic cause for his lassitude. He advised Aileen that James should not take on the additional burden, and on his return to Cambridge, Chadwick asked to withdraw. He also stood down from the Council of the Senate and took to his bed for several months.

It is interesting to speculate on what impression Chadwick might have made as Vice-Chancellor. There were some like Oliphant, who thought after the war that Cambridge badly needed 'some relief from the stranglehold of second-rate men in high places, in University affairs as well as in the colleges'.[5] Chadwick had certainly been effective on the governing body at Liverpool University, and indeed had been groomed as Pro-Vice-Chancellor there. To Chadwick's mind a succesful Vice-Chancellor needed the trust and confidence of the Senate. When advising Liverpool University on a successor to McNair, he had said that it was more important to show depth of intellect than breadth—'a comparatively common quality [which] sometimes carries with it an inactivity of mind and an indecisiveness. You can't grow good crops by merely scratching the soil'.[16] His personal standing was such that he would have commanded respect from the other members of the university hierarchy, and none would doubt the depth of his intellect. His experience of the upper echelons of the Civil Service and personal contacts in the Government would have been valuable at a time when the universities were starting to receive more funding from central sources. He had not taken part in any public debates on the future of higher education, but certainly had thought about the subject at length, and had written several times to Sir Henry Tizard about the country's policy on training scientists. His views, coming from a Cavendish physicist, were somewhat heretical. He wrote in February 1951:[17]

> Our leading position in fundamental science certainly brings prestige, but, as far as I can see, little else; if we were leaders in Engineering etc., we should get perhaps less prestige . . . but real and tangible benefits. (e.g. Whenever there is money to be made out of Physics it is called, and becomes, Engineering.)

He suggested to Tizard that the British universities should try to attract foreign students for their engineering courses, on the basis that they would return home as leaders of industry and would use British goods and services. He also thought that industry must be given adequate representation on the governing boards of university departments so that they could 'take part in, and share the responsibility for, the formulation of policy etc. of these departments'.[18] He was concerned that 'the learned world tends to offer second hand scraps of information, and second-hand knowledge leads to mediocrity'. This was a particular problem 'with what so often passes for a literary education', whereas 'the peculiar merit of a scientific education is, or should be, that it bases thought upon first-hand observation and knowledge'.[19] The philosophy inculcated by Rutherford remained as vital as ever.

Even without the Vice-Chancellorship, the mastership continued to test his emotional and physical reserves. In 1951, the Transport and General Workers Union tried to recruit the college servants as members.[20] This produced some hilarity in various quarters, but was staunchly resisted by Chadwick. During the same period, one of the domestic staff had a grievance which she wanted to bring to the Master's attention. On hearing about this, Chadwick again took to his bed for a prolonged period, until the woman finally despaired and left the college's employ. The same year did see the appointment of the first female college officer as Steward, which was a successful innovation.

The provision of extra money as a result of the financial reforms, rather than easing the tensions within the college tended to exacerbate them. There was now the prospect of hiring extra research fellows, a development welcomed by Chadwick and the junior fellows alike. The research fellows existed mainly on a modest stipend from the college; the Peasants contended that these stipends should be increased and that research fellows should be enrolled in the university pensions scheme. Although he was pleased about the increase in their numbers, which he saw as an important step towards his main goal of enhancing the academic quality of the college, Chadwick was determined to husband the college's new wealth carefully and to direct it towards long-term projects. He was implacably opposed to any creeping inflation in the stipends of existing fellows, however meagre these might be. Chadwick also took the view that these were essentially temporary positions, which should not qualify for pensions rights. Such an unyielding position was never likely to prove popular, but Chadwick was not given to be conciliatory on what he regarded as important points of principle. In former days, the word of the Master was likely to be accepted by men who had grown up in Caius with a deeply engrained respect for his authority; but now the fellowship was a more diverse body, whose members were drawn from other colleges and universities, and carried no such emotional loyalty. The gulf widened as 'new men entered a society in which there was all too much temptation to align themselves either with those who in the main supported the Master or with those who regarded him as insensitive to the need for necessary change'.[6]

The fortnightly council meetings on Friday afternoons became heated affairs. While Chadwick was an accomplished committee man and a master of self-expression on paper, the instantaneous cut and thrust of verbal debate among such clever men was beyond his control. There were some famous exchanges, one of the most memorable concerning the provision of extra space and a bathroom for a fellow's set of rooms. Perhaps remembering Mark Twain's warning that soap and education are not as sudden as a massacre, but more deadly in the long run, Chadwick resisted the proposal for many hours, holding it to be unreasonable. Eventually, the President, Francis Bennett, intervened to say 'It may be unreasonable, Master, but I do not think we shall get to bed tonight until we accept it'.[4] Recognizing the weight of this argument, Chadwick finally backed down with a smile of resignation. On another occasion, Chadwick was presiding at a council election. Sir Ronald Fisher, the renowned statistician and geneticist, asked Chadwick if it were permissible to vote for oneself.[21] There was already an air of expectation, which was heightened by the lengthy pause that now ensued as the Master considered the question. He eventually responded that Fisher must follow his conscience in the matter. Fisher replied sharply that his conscience would never permit him to vote for the Bursar. It was a prime example of what has been described as Fisher's 'mastery of the elegantly barbed phrase [which] did not help dissolve feuds.'[22]

While Chadwick was wounded by the behaviour of some of the fellows, he was long suffering by nature, and never publicly complained or issued rebukes. Aileen was less forgiving and made little attempt to disguise her anger, which tended to increase the antagonism between the Master's Lodge and the Senior Combination Room. Her more combatative nature was plainly seen in spirited exchanges with at least one eccentric young don, who had his set of rooms on H staircase, overlooking the Master's garden. He had recently become obsessed with the music of Wagner, and used to play recordings of *The Ring* for hours on end. The beginning of each 78 r.p.m. record would stimulate the Chadwicks' dog to bark, at which signal the volume of the Wagner would be increased until there was a deafening cacophony. Aileen sent him a number of brusque notes, complaining about the disturbance he was creating, with his unspeakable 'music'. Unabashed, the don would take these to Professor Hadley, in mock outrage, asking him to refute Aileen's criticisms of Wagner.

By the mid-1950s many of the small band of original Peasants had moved on to other universities or had become less interested in college affairs, but the legacies of their revolt continued. In 1956, Francis Bennett's second and final term as President came to an end. Fisher was duly elected President, which served to isolate the Master's position further. Fisher was an eloquent man, whose obvious brilliance captivated many fellows in a way which drew attention to the Master's apparent aloofness and forbidding nature. Chadwick soldiered on, often feeling unwell, becoming an increasingly remote figure to the majority of fellows. There were still occasional flashes of his mordant wit. When one of the dons was commiserating with him about his evident weariness, Chadwick sighed and said pensively: 'Anyway, it will soon be coming to an end.' The fellow took this to mean Chadwick was going to step down as Master, or possibly expire, and began to tell him how sorry they would all be to lose him, only to be cut short by the quiet rejoinder: 'I was merely referring to the end of term.' Many came to realize that there was human warmth, beneath the exterior shell. Peter Gray, who had learned to appreciate Chadwick's kindly nature after one or two minor infractions as an undergraduate and research fellow,[21] returned to Cambridge one summer with his wife and two small children. Mrs Gray had been doubtful when he suggested visiting the Chadwicks, but they found themselves warmly welcomed, and Aileen assured them that Sir James would 'stand on his head to entertain young children'. Members of the college would have been equally amazed to see Chadwick escape to the cinema with his own daughter to watch a cowboy film with all the excitement of a young schoolboy, aiming his fingers at the screen and ducking imaginary gunfire.

March 1958 saw the Quatercentenary Celebration of John Caius' endowment of Gonville Hall and the refoundation of Gonville and Caius College. The celebration began with a service in the chapel, conducted by the Bishop of Norwich, and was followed by a dinner in hall, where Norwich was joined by the Bishops of Ely and Salisbury. The Vice-Chancellor of Cambridge University proposed the toast of the college to which Chadwick, as Master, replied. Other feasts followed during the early summer at which Chadwick presided with evident pleasure. The festivities provided an incongruous backdrop to what were the final unhappy months of his mastership.

The abrupt end came about as a result of an entanglement of personalities and points of principle, difficult for an outsider to comprehend so many years later. In 1958, Tunnicliffe and Deas, the two dons deposed from the Council in the original Peasants' Revolt, were due to be re-elected as tutors.[6] The position of tutor in a Cambridge college was not an academic office, but more a pastoral one to the undergraduates. Contact between a student and his tutor could be limited to one meeting at the beginning of term ('Had a good vac?'), and applying to him for an exeat at the end of term ('Have a good vac').[4] So the duties were not strenuous, and Tunnicliffe and Deas both enjoyed their roles. In keeping with the notion of increasing the rotation of college officers, the College Council suggested the compromise that both men should be reappointed for limited terms—until the academic year of their sixtieth birthday. Chadwick was against this departure from tradition in their case because he knew how hurt Deas in particular had been over his earlier ousting. In the event, both men took the council proposal as a vote of no confidence and stood down as tutors.

There was now some urgency to find replacements for the next academic year, and Chadwick was asked as Master to nominate new candidates to be confirmed by the council. Chadwick's nomination was also a departure from tradition—he suggested a university teacher who was a Caian, but not a fellow. He did discuss the proposal with McNair, Stratton and Weller, but made no attempt to canvas the wider body of fellows or even other council members.[6] While the suggestion was made in good faith as Chadwick's way to introduce new blood into college government, and was not in breach of the statutes, it came to be viewed with suspicion. The general fellowship thought it could lead to an expansion of the Master's powers and to the foisting of unknown quantities into their midst. Accordingly, when Chadwick's nominee was put before the council for ratification, he did not gain a single vote. Chadwick thereupon refused to make any further nomination on his own initiative. It is a measure of the intricacy of the College statutes that such an eventuality is anticipated, and the council has the authority to appoint whom it sees fit, even without the Master's agreement. To try to preserve amity, the council asked Chadwick to nominate two candidates of its own choosing, who were fellows of Caius. With some reservations, Chadwick did so, and one was formally appointed, while the other declined.

As he reflected on these events, Chadwick decided that they amounted to a desire on the part of the fellows 'to limit the powers and responsibilities of the Master'.[1] A month later, with no explanation or recrimination, he tendered a letter of resignation. He continued to fulfil his duties punctiliously during the Michaelmas Term; as late as November 1958, he warned Bohr[23] that he might not be able to attend his Rutherford Lecture in London because he had a College Council meeting on that afternoon. At the end of December, Chadwick moved out of the Master's Lodge, with a heavy heart. The fellowship was deeply divided, and did not elect a successor, Nevill Mott, until two months after Chadwick had departed.

It is easy to bring forth reasons for Chadwick's discomfiture as Master. Some of them are rooted in his personality—a painfully shy man, required to carry out a job which depends to some extent on the holder being able to dispense a certain geniality. He was a man of complete integrity and high principles, without an ounce of expediency. His whole career to date had been spent in the pursuit of fundamentally important matters, and he sometimes found it hard to concentrate on the banal issues which now came his way, such as selection of new curtains for the college rooms. There is no doubt that he missed the real academic world of research, which had been his *milieu* for so long and where he was an acknowledged master. It was particularly galling to him that his duties as Master clashed with council meetings of the Royal Society so that he had to resign as Vice-President in 1949, after only one year of a two-year term because he had not satisfied the attendance requirements.[1]

In Rutherford's laboratories, there was one unquestioned leader. This did not imply that there was active repression of new ideas or innovation, but it was generally accepted that there was one final arbiter. Men like Chadwick, Kapitza and Oliphant came to have influence with Rutherford, but rarely showed open dissent if their opinions were disregarded. Chadwick, in turn, always treated the newest recruits to the Cavendish with natural respect as fellow scientists, never as inferiors. There were occasional men of exceptional character, notably Blackett, who could not tolerate Rutherford's autocracy, but in general the atmosphere was a happy one. When Chadwick came to Liverpool he followed the Rutherford model, without the personal flamboyance, and his standing was so high amongst his colleagues there that his authority was accepted almost without question. During his wartime service on the Maud Committee and the Manhattan Project, his ability and judgement became so widely appreciated that again it was rare for his decisions to be challenged. None of this served to prepare him for the heat and intrigues of local politics.

As a young research fellow in the Caius of the early 1920s, Chadwick had developed an especial fondness for the Master, Sir Hugh Anderson, who in turn was far warmer towards him than his own father had been. There is no doubt that Chadwick would have liked to continue the same paternal style as Master that Anderson and then Cameron had achieved, but he was not they and the times had changed. The pre-existing oligarchy was untenable in the postwar years. Chadwick's adherence to the spirit of the college that he had known, and his devout interpretation of the nineteenth century statutes came to be encumbrances in the transition to a more open, and at least in the short term, more contentious power structure.

The turmoil occasioned in his mastership should not obscure the real successes that also accrued. His active promotion of the best available candidates from other colleges and universities bore fruit both at the post-doctoral level and also in professorial fellows. There was an unprecedented increase in the number of research fellowships, from 31 to 49. The expansion in numbers was made possible by the improved financial standing of the college, which also allowed the construction of Harvey Court on undeveloped college land to house the increasing population of students. In the early years of his mastership, a brilliant but erratic physicist from the Cavendish named Francis Crick was given a toehold in Caius while he completed a Ph.D. in biology. During the course of this, he and Watson discovered the double helix of DNA.

During his mastership, Chadwick became deeply absorbed in the history of the College, and developed an almost mystical veneration for John Caius, the co-founder and sixteenth century Master. Caius had his own quarrels with a lively group of young fellows, not in the least because of his own interpretation and revisions of the statutes. During the height of their disagreements in December 1565, the Archbishop of Canterbury recorded his impressions of the state of affairs in Gonville and Caius College. On balance he wrote: 'I do rather bear with the oversight of the Master . . . in respect of his good done and like to be done in the College by him, than with the brag of a fond sort of troublous factious bodies.' Four centuries later, Sir James Chadwick took some solace in the fact that Caius too had resigned the mastership, and this represented a historical link between them.

References

Christopher Brooke's elegant monograph (ref. 4) is the source of most of the historical aspects of the college quoted here. Some of the anecdotes are just that, and cannot be attributed.

1 Weiner, C. (1969). Sir James Chadwick, oral history. American Institute of Physics, College Park, Maryland .

2 Chadwick, J. (1949). Reference re. F.C. Flack. CHAD II, 1/4, CAC.

3 Chadwick, J. (29/8/48). Letter to N. Feather. Feather's papers, CAC.

4 Brooke, C. (1985). *A history of Gonville and Caius college.* Boydell, Woodbridge.

5 Oliphant, M.L. (20/6/45). Letter to J. Chadwick. CHAD IV, 12/66, CAC.

6 Skemp, J.B. (1978). Mastership of Sir James Chadwick. *Biographical history of Gonville and Caius College,* **7**, 485–502. Cambridge.

7 Darwin, C.G. (27/9/48). Letter to J. Chadwick. CHAD I, 30/2, CAC.

8 Chadwick, J. (11/10/48). Letter to C.G. Darwin. CHAD I, 30/2, CAC.

9 Mott, N. (1986). Master of Caius 1959–66. In *A life in science*, pp. 121–29. Taylor & Francis, London.

10 Notebook. CHAD III, 1/7, CAC.

11 Gaskell, A. (22/2/57). Letter to J. Chadwick. CHAD II, 1/5, CAC.

12 Interview with Lord Bauer (1985). *The Caian*, Gonville and Caius College.

13 Edwards, S. (1991). Sir James Chadwick, 1891–1974. *The Caian*, 59–61. Gonville and Caius College, Cambridge.

14 Teller, E. (13/11/51). Letter to L. Strauss. Lewis L. Strauss papers, Herbert Hoover Library, West Branch, Iowa.

15 Cohen, K.P., Runcorn, S.K., Suess, H.E. and Thode, H.G. (1983). Harold Clayton Urey, 1893–1981. *Biographical memoirs of fellows of the Royal Society*, **29,** 643.

16 Chadwick, J. (16/3/45). Letter to S. Dumbell. CHAD I, 25/1, CAC.

17 Chadwick, J. (6/2/51). Letter to H. Tizard. CHAD I, 3/5, CAC.

18 Chadwick, J. (26/5/48). Letter to H. Tizard. CHAD IV, 10/46, CAC.

19 Chadwick, J. Undated notes. CHAD II, 3/7, CAC.

20 Chadwick, J. Undated notes. CHAD II, 1/6, CAC.

21 Gray, P. (1988). Interview. *The Caian*, 19–48. Gonville and Caius College, Cambridge.

22 Gridgeman, N.T. (1972). R.A. Fisher. In *Dictionary of Scientific Biography*, **5**, pp. 8–9. Charles Scribner's Sons, New York.

23 Chadwick, J. (5/11/58). Letter to N. Bohr. Niels Bohr Archive, Copenhagen.

20 Final reflections

When Chadwick became Assistant Director of Research at the Cavendish in the mid-1920s, he was enrolled into a pension scheme run by the Department of Scientific and Industrial Research. At the time he had to chose between two life assurance policies, one which would mature at the age of 60, the other at 65 years. Chadwick chose the latter, and the man from the DSIR confided in the Bursar of Gonville and Caius that he thought Dr Chadwick had been rather unwise.[1] On leaving the Mastership, he was in his sixty-eighth year, and, while prey to recurring minor ailments and depression, had never spent one day in hospital in his life. He was disillusioned by the circumstances of his resignation, and took his leave of the college which meant so much to him with no warm words of farewell.

The Chadwicks had not made prior arrangements for his eventual retirement, and only when he concluded it would be wrong to continue as Master did they buy a house. They decided to move to North Wales for the beautiful countryside and to have easy access to friends and family in Liverpool. Wynne's Parc, the house they chose, was a good-sized property in about two acres of ground, on the outskirts of Denbigh. Arriving in the dead of winter, they found that many repairs and improvements needed to be made. Niels Bohr thoughtfully sent a case of Carlsberg lager from Copenhagen,[2] and Chadwick's spirits were gradually restored in the spring, when he was able to take country walks and to go fishing. He and Aileen also set about creating a magnificent garden, where Chadwick employed some tips he had learned in the 1890s from his grandfather, who had been a gardener in Bollington. They made new friends and entertained old ones from Liverpool; Chadwick enjoyed renewing closer contact with colleagues from prewar days. Norman Feather, who was now Professor of Natural Philosophy in Edinburgh, became a copious correspondent once more. He was writing several papers on the history of the Cavendish, Lord Rutherford and the discovery of the neutron. Chadwick encouraged him, but took great pains to correct what he thought were glib or misleading passages in Feather's writings. He also passed on to Feather some remarks about Rutherford from another, unidentified, writer, who was under the mistaken impression that Chadwick was preparing a new Life of the great scientist. The unnamed correspondent ventured the opinion that he fully understood the type of brain Rutherford possessed: 'He assimilated everything and it became so much part of his brain that he quite thought it was his own creation.' To this, Chadwick added the enigmatic comment: 'Well, well'.

To mark the fiftieth anniversary of Rutherford's revolutionary insight into the structure of the atom, a celebratory meeting was held in Manchester in September 1961. This gave Chadwick the opportunity to reminisce with old colleagues such as Marsden, Andrade and Darwin, who were fellow members of Rutherford's Manchester school. Chadwick was spared the ordeal of giving a lecture himself, but he did spend a considerable amount of time helping Niels Bohr with his speech, which was the centrepiece of the programme. In his letters to Bohr,[3] he was punctilious about ensuring credit in various ways for Russell, Cockcroft, Allibone and other students of Rutherford. With regard to himself, he asked Bohr to cross out his name where he referred to 'chemical investigations' by Russell and Chadwick, because while it was true he did the work, the ideas and responsiblity were all Russell's. In a similar vein, he suggested that 'Chadwick's cloud chamber pictures' should read 'Dee's c.c. pictures'. His modesty forbade him to point out that the credit Bohr[4] gave to Ellis for the 'demonstration of the continuous spectral distribution of the electrons directly emitted from the nucleus' should rightfully belong to him. After the meeting, Niels and Margrethe Bohr travelled back with the Chadwicks to stay at Wynne's Parc. The next month, James and Aileen were guests of honour at a Cavendish dinner to mark his seventieth birthday.

The following year saw the thirtieth anniversary of Chadwick's own discovery of the neutron. He was invited to a symposium at Cornell University in Ithaca, New York, for which he prepared '*Some personal notes on the search for the neutron*'. As a postscript to the story, Chadwick[5] concluded with the following words:

> It is unnecessary to record my pleasure and satisfaction that the long search for the neutron had at last been successful. The decisive clue had indeed been supplied by others. This, after all, is not unusual; advances in knowledge are generally the result of the contributions of many hands and minds. But I could not help but feel that I ought to have arrived sooner. I could offer myself many excuses—lack of facilities, lack of technical assistance, and so on. But beyond all excuses I had to admit, if only to myself, that I had failed to think deeply enough about those properties of the neutron which would furnish evidence of its existence. I consoled myself with the reflection that it is much more difficult to say the first word on any subject, however obvious it may later appear, than the last word—a commonplace reflection, and perhaps only an excuse.

Chadwick was concerned that his description of Rutherford's reaction to the Joliot-Curie paper (which triggered his own successful experiment) could be interpreted as criticism or as a 'dig'[6] at his old chief. For this reason, he softened that part of his talk and also presented others 'in a rosier light than they perhaps deserve'. As a result, he told Feather that he regarded the lecture as 'a bit of gossip rather than as a contribution to history'. Predictably, Chadwick never made the journey to Cornell to deliver his words in person. He succumbed to a bout of depression 'which had been threatening for some time', and told Feather[6] that he was 'very low'. A month or so later, he was able to write: 'I am now relieved, or been relieved, of all formal duties and commitments. And it is a relief—indescribable. Nothing now weighing on my mind.

I shall now be able to give some time to putting my books and papers in order.'

Many of these papers related to Lord Rutherford, as Chadwick had been editing a collection of scientific works for some years. This was a project which he had had in mind as early as 1940, but for which he had no time until the closing years of his mastership. The first fat volume, covering Rutherford's output in New Zealand, J.J. Thomson's Cavendish and Montreal appeared in 1962.[7] In his foreward, Chadwick credited Dr Paul Rosbaud with the conception of publishing all Rutherford's scientific papers together with his general articles and formal lectures. The scientific papers would occupy three volumes and the remaining material would constitute a fourth. Apart from serving as a memorial to Rutherford, Chadwick intended that it should inspire succeeding generations of young scientists 'to see what he did, to follow the development of his ideas, and to get at first hand some idea of the magnitude of his contribution to our knowledge of the physical world.' He was concerned that to 'a young man of the present day it all looks so easy and so obvious', when Rutherford had been a towering genius in an age of extraordinarily talented scientists. Rosbaud, a distinguished, German, science editor, had known Chadwick well for 30 years and took much of the responsibility for assembling the material. During the war, he had been an inestimable spy, code name *der Greif* (the Griffin),[8] operating at unimagineable peril in Nazi Germany; he reported to the British as early as 1942 that the German scientists had failed to grasp the principles of the atomic bomb. He died from leukaemia in January 1963, just as work on the second volume was completed. Chadwick undertook all the 'donkey work' for the third volume, which comprised Rutherford's output as Cavendish Professor. It appeared in 1965, and its production sapped Chadwick's last reserves of energy. In doleful mood, Chadwick wrote to Feather,[9] 'whisky seems to do me much more good than medicine or pills, it revives me when I am exhausted and gives immediate comfort. But I can't afford it.' He had insufficient reserves of stamina left to embark on his own memoirs. He had particularly wished to record his own account of the Cavendish under Rutherford, and 'his story of the beginning of work on the nuclear bomb, which is rather different from anything that has been published'.[1] As he began to think about the errors in the accounts of the Manhattan Project that he read, he came to realize that he too had a limited view of the whole undertaking; the concern that he might make serious omissions restrained him from ever writing his own version of history.[1]

The fourth volume of Rutherford's miscellaneous lectures and writings was never compiled, and Chadwick probably threw away much of the material. It was his scientific output which defined Rutherford's greatness, and Chadwick probably sensed that any interest in his peripheral activities would be slight and transient. One characteristic which Rutherford shared with Newton and Niels Bohr, in Chadwick's eyes, was the faculty of 'being able to concentrate his mind on a matter for long periods at a stretch without getting tired or being bored'.[1] When one recalls the tenacity with which Chadwick undertook his own experiments at the Cavendish, when he seemed oblivious to those around him, one can suppose that he also possessed this quality to some degree. While Chadwick revered Rutherford, he had 'an enormous respect and admiration' for Niels Bohr, and regarded their time together in Washington and Los Alamos as one of the highlights of his career. The regard was mutual—after Bohr's death in 1962, his son Aage wrote to Chadwick, conveying how highly his father had valued Chadwick's friendship, which was invested with 'deep affection and admiration'.[10]

As we have seen, Niels Bohr was concerned about the dangers of nuclear proliferation and issues of international control from the time he first discovered the existence of the Manhattan Project in 1943. During the war, all Chadwick's energies were employed in making sure that the project came to fruition. He remained proud of the British contribution to the Manhattan Project which, although small in relation to the totality of the American effort, was crucial in many respects. In Chadwick's opinion, the greatest influence was exerted throughout 1941 'by the profound conviction of British workers, backed up by the experimental and theoretical work that was already in an advanced stage by January of that year, that the production of an atomic bomb of tremendous explosive force was entirely feasible to any country with sufficient industrial resources.'[11] It was exactly this point regarding the potential for nuclear proliferation that he made at his Washington news conference in the week after Hiroshima and Nagasaki, but he made almost no subsequent public pronouncements on the subject. While he regarded these gravely important questions as challenges for statesmen, he seemed to accord Bohr special status and wisdom, which qualified him, among all scientists, to participate in the debate.

The third friend, who also attained greatness in Chadwick's estimation, was General Groves. Chadwick may have been unique amongst the scientists from the Manhattan Project in this assessment—most of them loathed Groves and thought him a military brute who would have fitted quite easily into the Nazi regime. While Chadwick thought he was an insensitive bully, he also realized, through his close association, that Groves' toughness and determination were essential to the success of the project. He admired Groves' incisiveness and judgement, which on occasion he recognized had saved the project months of time and millions of dollars.[1] Chadwick's strong support of Groves made a lasting impression on Edward Teller,[12] the father of the H-bomb, after he had made a mildly disparaging remark at dinner at Caius in 1949. Teller's comment was 'along the lines which, unfortunately, is not uncommon among nuclear scientists'. He was taken aback by the emotion and volubility of Chadwick's response, since before then he had found the English don did not utter 'more than five sentences in an hour'.

> He made an extremely deep impression on me. There was no question but that he was deeply concerned about the fact that General Groves had a very great merit in getting Atomic Energy development completed in good time and that he was not given credit for his achievement. Chadwick said that the scientists who were in administrative charge of the Manhattan Project would certainly not have been able to complete the work since they had neither the initiative nor the drive nor the conviction for such an accomplishment. What made Chadwick's attitude so

> very interesting to me was the fact that General Groves was always opposed to cooperation with the British and that he voiced his feelings on the subject frequently in a somewhat tactless way. Of this Chadwick was fully aware, yet he was most anxious to convince me of Groves' merits and he did it not in the manner of a man who feels that merely justice should be done but with a warmth which made it quite clear that, in some way, not only justice or wisdom but also his heart was invloved in the matter.

A few days after the conversation between Chadwick and Teller, the Russians announced the explosion of their first atomic bomb. Teller often wondered why Chadwick was so passionate in his defence of Groves, and came to believe that 'his words and his insistence were connected, not only with the past record but also with questions which involved our future safety.'

Unlike Bohr or Groves or Teller, Chadwick did not become a major voice in the international arguments about nuclear weapons. He did reflect privately on the changed nature of the world, which by his science and public service he had helped to bring about. While never recanting the use of the bombs against Japan, he came to harbour grave misgivings about the development of the hydrogen bomb. In his mind, there was a qualitative difference between the horrors wrought by the fission bombs and those posed by the next generation of hydrogen bombs. In undated notes, Chadwick[13] reflected:

> The original atom bomb is a weapon not very different from other bombs—of the same kind but more powerful. The H-bomb can hardly be classed as a weapon at all. Its effect in causing suffering is out of all proportion to its military effect. The H-bomb does not offer any improvement in the waging of war, and it brings with it a risk of making the world uninhabitable.
>
> Most people would agree that it would be morally wrong to use the H-bomb—the only possible exception could be to use it in retaliation, even then only to the extent necessary to make the enemy stop its use.

His strategic outlook was clearly coloured by the Cold War, and on balance he took some comfort from the deterrence provided by nuclear weapons.[14] He found it 'unthinkable' that the US would start a nuclear war and 'unbelievable' that the Soviet Union would start one 'deliberately'. Further, he considered it to be 'unlikely that either country would promote or encourage a major war with conventional weapons, for a major war would almost inevitably lead in the end, when defeat appeared imminent, to recourse to nuclear weapons.' It appeared to him, therefore, that 'both countries will try to avoid major conflicts with arms, will try to confine wars to local conflicts, and will try to find other ways of settling major disputes and differences.' Continuing this note of optimism, he surmised 'the development of these fantastic weapons of destruction means the end of the great wars we have known in the past. That is the brighter side of the picture.' In 1957, he received a letter from Linus Pauling,[15] the American Nobel Laureate in chemistry, who was trying to organize an international petition amongst the world's most famous scientists against the testing of nuclear weapons. He told Chadwick that he saw this as the first step toward a more general disarmament.

It seems improbable that Chadwick would have agreed with either proposition at the time, and he would certainly not have wished to lend his name to a statement that he could not wholeheartedly support. Rather than attack the idea as naive or voice his substantive reservations to Pauling, Chadwick was instead inclined to hide behind the cloak of officialdom, saying he would have to consult the UK Atomic Energy Authority, since they still retained him as a part-time consultant.

In the war, Chadwick always stressed that the military aspects of atomic energy must take precedence over any eventual civilian use, although the latter should always be borne in mind. While he must have been tempted to believe that the neutron could be harnessed to produce almost unlimited quantities of clean, cheap, energy, he steadfastly maintained a realistic appraisal of the possibilities and never succumbed to some of the wide-eyed optimism which was prevalent in the years after the war ended. In his 1946 article for the *Liverpool Daily Post*,[16] he considered popular predictions that 'our present electric power plants are now out of date; that ships will soon be running on atomic power instead of coal or oil; that in two or three years a tiny engine fed by atomic energy will propel our cars without need of replenishment for years; that, indeed, an economic and social revolution is already in sight'. He cautioned his readers that 'statements such as these are largely rash speculation and are likely to do more harm than good.' While it was already possible to release atomic energy in a controlled manner at a low temperature, to be useful for industrial purposes the heat energy must be obtained at a high temperature. He did not doubt it would be possible within a few years to obtain steam from burning atomic fuel which could then be used for heating or electricity production. But the question would then arise 'as to the economy of the atomic plant as compared with one based on coal or oil, and in this connection it should be remembered that the cost of fuel in an electric power plant is far less than the cost of distribution, so that even if the fuel were free the reduction in the cost of electricity to the user would be small.' It seemed to Chadwick 'very unlikely that any big change will be brought about in industry within the next ten years. We may see the erection of an electric power plant based on atomic energy, but it seems that this would find an important application only in places where ordinary fuel is exceedingly scarce or where transportation is very difficult.' He did predict that warships might be driven by atomic power so as to avoid the need for refuelling in wartime, but thought that this was a special case of mobile atomic power units which would generally be impractical because of the weight of radiation shielding required.

During a long tour of Canada in 1953, he gave another lecture on 'Prospects for atomic power' at the University of Toronto.[17] In his introduction, he noted that 'the consumption of power both for industrial and for domestic purposes is increasing rapidly in most countries, and the signs are that it will continue to do so for many years. The population continues to increase; the consumption per head increases as standards of living rise'. He thought the increase in demand was such that 'new large sources of energy will be needed within a few decades'. While this was a global trend, he recommended that the need for atomic power should be considered on a country by country basis. In the UK, coal reserves should last two to three centuries and considerable economies in the use of coal were certainly possible. The price of coal, however, was rising; since the UK had no oil (North Sea oil was as yet undiscovered) and little water power, the British Government was prudently setting up a large atomic power plant for experiment and investigation. By contrast, Canada, although it had abundant water power, coal and oil, faced great problems and expense from transportation and distribution. Its rich supplies of uranium made it favourable for the development of atomic energy. With Canada in mind, Chadwick then examined the two fundamental questions: (1) whether the use of atomic power would be feasible on a large scale and (2) whether it would be economic.

In his closely argued talk, Chadwick completely overlooked two of the main objections to nuclear power raised by modern critics: the proliferation of plutonium and the problem of radioactive waste disposal. Indeed, he thought the production of plutonium in a natural uranium–heavy water reactor was a positive advantage in that it provided fuel for a fast breeder reactor, which offered the prospect of large scale power supplies far into the future and slower depletion of the reserves of natural uranium. For the immediate future, he suggested Canada should embark on a programme of large natural uranium–heavy water reactors so that they could build on experience already gained at Chalk River. With expected improvements in efficiency over time, Chadwick thought it 'quite likely that atomic energy will, in one way or another, be able to compete with coal and oil'. He did not think that atomic power could meet all Canada's future energy needs.

After ten very happy years in Wynne's Parc, the Chadwicks found it too hard to go on living in Wales without help. They moved back to Cambridge to be near their daughters in 1968. Now in his late seventies, Chadwick found it 'difficult to realise that I am so old and to reconcile myself to it'.[18] As one would expect for a scientist of his distinction, he had received many academic and civil awards over the years, in addition to his Nobel Prize and knighthood. There were the Copley and Hughes Medals of the Royal Society, the Faraday Medal and the Franklin Medal. He held the highest awards that both the United States and Germany can bestow on foreign citizens—the Medal for Merit and Pour le Mérite respectively. He held honorary memberships and degrees from dozens of British and European universities and scientific associations. Many of these prizes were to mark his academic career and wartime service—there had been no public recognition of his postwar services, either to successive governments or to Liverpool* and Cambridge Universities. Then, in the New Years honours list for 1970, the Queen made him a Companion of Honour. His old friend and colleague Sir Rudolf Peierls wrote to congratulate him, 'not that your reputation needs the recognition'.[19] While his accomplishments were known to those that he cared about, some thought his contributions deserved wider appreciation. Dean Mackenzie,[20] the wartime President of the Canadian National Research Council, was one of this school:

* Liverpool University did name its new Laboratories after him in 1959 and later created a Chadwick Chair in Physics.

> ... those of us who were in a position to observe what you did in Washington in connection with those difficult negotiations in the Atomic Energy field, have always been annoyed that you did not receive the public recognition you deserved . . . my opinion grows that had you been in charge of negotiations from December 1941 on, it would have been a happier and more successful world for all of us.

Chadwick was delighted with the CH, and went to Buckingham Palace in the spring for the investiture. Prince Charles was at the time an undergraduate at Trinity College, and Chadwick talked to the Queen at some length about the history of the college. After the General Election of that summer which saw the Conservatives return to power, against some opinion poll forecasts, Chadwick wrote to Norman Feather 'not such "a damn close run thing" but what a pleasant surprise. I hope now to be able to afford to live a few more years.'[21] In truth, he never held any great political allegiance, but a Tory victory certainly pleased Aileen and would, therefore, enhance the general atmosphere in the household.

During his eightieth year, he became progressively more frail and exhausted, and experienced a significant deterioration in his eyesight. The Royal Society were planning a meeting in 1971 to mark the centenary of Rutherford's birth. Chadwick hoped not only to attend, but to deliver a short talk. The anticipation of this event, particularly the ordeal of giving a speech to a large audience, began to overwhelm him. Aileen eventually had to write to Feather[22] to explain that he would not be able to go to London, and that he was very distressed about missing the meeting. Fortunately, he did still have an opportunity to pay one last tribute to his hero, when Mark Oliphant asked him to write the Foreword to his book '*Rutherford—Recollections of the Cambridge days*'.[23] Chadwick concluded a few pages of elegant and revealing prose with a quote from Thomas Hardy: 'If I ever forget your name let me forget home and heaven . . . I never can forget 'ee; for you was a good man and did good things'. He was able to make one last visit to Liverpool on the occasion of his eightieth birthday. He toured the expanded Physics Department with great enthusiasm and declined all attempts to get him to rest. To Mike Moore, who was now Head of the Engineering Division at the Daresbury Nuclear Physics Laboratories, Chadwick 'had changed little over the years, tall, wiry and energetic; still with the same sideways glance, the same twinkle in his eyes, and the same ability to speak to the point with never a wasted word.'[24] He was introduced to Professor Charles Johnson and remarked that he was the fifth professor of physics he had met that day—in his time there had been one. When Johnson pointed out that he was the Lyon-Jones Professor, Chadwick smiled broadly. At a small dinner in the evening, where he knew he was among good friends, Chadwick spoke movingly.[24]

He enjoyed receiving letters from old friends and students from around the world. He was especially thankful that Peter and Anna Kapitza could now visit England again, and saw them several times in Cambridge. His health became so delicate that Aileen would not leave him alone in their flat. He spent many hours listening to recorded music—German *lieder* being his favourite. He was a lifelong atheist and felt no need to develop religious faith as he approached the end of his life. About three months before he died, he said to his daughter Joanna, quite calmly, 'I am not going to live very long. I am going to die in the next few months. It doesn't matter, I have nothing left to do. The only thing that matters is that Aileen is looked after.'

He died in his sleep, with Aileen by his side, on 24 July 1974.

There were lengthy obituaries in the leading newspapers of the English-speaking world. Curiously, *Nature* failed to carry one—his old colleague Lord Blackett had died on 13 July, and once again Chadwick was partially eclipsed. The most touching remembrance came from the woman who had loved him for almost 50 years and cared for him selflessly all their married life. Aileen wrote, 'James was shy, sensitive and endowed with that rare attribute, the Grace of Humility, with abounding wisdom and knowledge of many things, and the ability to understand people. He put his Duty, whether to his country, University or College before everything.'[25] No one else knew him so well.

References

1 Weiner, C. (1969). Sir James Chadwick, oral history. American Institute of Physics, College Park, Maryland .

2 Chadwick, J. (7/2/59). Letter to N. Bohr. Niels Bohr Archive, Copenhagen.

3 Chadwick, J. (1961). Letters to N. Bohr. Niels Bohr Archive, Copenhagen.

4 Bohr, N. (1962). Reminiscences of the founder of nuclear science and some of the developments based on his work. In *Rutherford at Manchester* (ed. J.B. Birks), pp. 159–61. Heywood, London.

5 Chadwick, J. (1984). Some personal notes on the discovery of the neutron. In *Cambridge physics in the thirties* (ed. J. Hendry), pp. 42–5. Adam Hilger, Bristol.

6 Chadwick, J. (1962). Letters to N. Feather. Feather's papers, CAC.

7 Chadwick, J. (ed.) (1962). *The collected papers of Lord Rutherford.* George Allen and Unwin, London.

8 Kramish, A. (1986). *The griffin.* Houghton Mifflin, Boston.

9 Chadwick, J. (11/2/66). Letter to N. Feather. Feather's papers, CAC.

10 Bohr, A. (17/12/62). Letter to J. Chadwick. CHAD II, 2/2, CAC.

11 Chadwick, J. (1946). Undated memorandum for Field Marshall Wilson. CHAD I, 24/2, CAC.

12 Teller, E. (13/11/51). Letter to L. Strauss. Lewis L. Strauss papers, Herbert Hoover Library, West Branch, Iowa.

13 Skemp, J.B. (1978). Mastership of Sir James Chadwick. *Biographical history of Gonville and Caius college*, **7**, 492. Cambridge.

14 Chadwick, J. Undated notes, probably mid-1960s. CHAD IV, 13/11, CAC.

15 Pauling, L. (6/11/57). Letter to J. Chadwick. CHAD IV, 11/56, CAC.

16 Chadwick, J. (1946). The atom bomb. *Liverpool Daily Post*, 4 March. University of Liverpool Archives.

17 Chadwick, J. (1954). *Prospects for atomic power.* Pamphlet, University of Toronto Press.

18 Chadwick, J. (1966). Letter to N. Feather. Feather's papers, CAC.

19 Peierls, R.E. (1/1/70). Letter to J. Chadwick. CHAD III, 4/4, CAC.

20 Mackenzie, C.J. (19/1/70). Letter to J. Chadwick. CHAD IV, 12/66, CAC.

21 Chadwick, J. (22/6/70). Letter to N. Feather. Feather's papers, CAC.

22 Chadwick, A. (18/10/71). Letter to N. Feather. Feather's papers, CAC.

23 Oliphant, M.E. (1972). *Rutherford: recollections of the Cambridge days*. Elsevier, Amsterdam.

24 Moore, M.J. (1975). Sir James Chadwick: an appreciation. *Quest* (House Journal of the Science Research Council), **8**, 7–8.

25 Chadwick, A. (30/8/74). Letter to N. Feather. Feather's papers, CAC.

Appendix 1

Chadwick's letter to Nature announcing his discovery of the neutron (reproduced with the permission of Nature), and his handwritten note to Niels Bohr (reproduced with the permission of the Niels Bohr Archive, Copenhagen).

Cavendish Laboratory,
Cambridge,

24 February 1932.

Dear Bohr.

I enclose the proof of a letter I have written to "Nature" and which will appear either this week or next. I thought you might like to know about it beforehand.

The suggestion is that α particles eject from beryllium (and also from boron) particles which have no nett charge, and which probably have a mass about equal to that of the proton. As you will see, I put this forward rather cautiously, but I think the evidence is really rather strong. Whatever the radiation from Be may be, it has most remarkable properties. I have made many experiments which I do not mention in the letter to "Nature" and they can all be interpreted readily on the assumption that the particles are neutrons. Feather has taken some pictures in the expansion chamber and we have already found about 20 cases of recoil atoms. About 4 of these show an abrupt bend or fork (and it is almost certain that ~~this~~ one arm of this fork represents a recoil atom and the other some other particle, probably an α particle. They are disintegrations due to the capture of the neutron by N_{14} or O_{16}. I enclose two photographs one of which shows the simple recoil atom, and the other what we suppose is a disintegration. The photographs are not very good but they were printed in a hurry.

With best regards
Yours sincerely
J. Chadwick.

Letters to the Editor

[*The Editor does not hold himself responsible for opinions expressed by his correspondents. Neither can he undertake to return, nor to correspond with the writers of, rejected manuscripts intended for this or any other part of* NATURE. *No notice is taken of anonymous communications.*]

Possible Existence of a Neutron

It has been shown by Bothe and others that beryllium when bombarded by α-particles of polonium emits a radiation of great penetrating power, which has an absorption coefficient in lead of about 0·3 (cm.)$^{-1}$. Recently Mme. Curie-Joliot and M. Joliot found, when measuring the ionisation produced by this beryllium radiation in a vessel with a thin window, that the ionisation increased when matter containing hydrogen was placed in front of the window. The effect appeared to be due to the ejection of protons with velocities up to a maximum of nearly 3×10^9 cm. per sec. They suggested that the transference of energy to the proton was by a process similar to the Compton effect, and estimated that the beryllium radiation had a quantum energy of 50×10^6 electron volts.

I have made some experiments using the valve counter to examine the properties of this radiation excited in beryllium. The valve counter consists of a small ionisation chamber connected to an amplifier, and the sudden production of ions by the entry of a particle, such as a proton or α-particle, is recorded by the deflexion of an oscillograph. These experiments have shown that the radiation ejects particles from hydrogen, helium, lithium, beryllium, carbon, air, and argon. The particles ejected from hydrogen behave, as regards range and ionising power, like protons with speeds up to about $3{\cdot}2 \times 10^9$ cm. per sec. The particles from the other elements have a large ionising power, and appear to be in each case recoil atoms of the elements.

If we ascribe the ejection of the proton to a Compton recoil from a quantum of 52×10^6 electron volts, then the nitrogen recoil atom arising by a similar process should have an energy not greater than about 400,000 volts, should produce not more than about 10,000 ions, and have a range in air at N.T.P. of about 1·3 mm. Actually, some of the recoil atoms in nitrogen produce at least 30,000 ions. In collaboration with Dr. Feather, I have observed the recoil atoms in an expansion chamber, and their range, estimated visually, was sometimes as much as 3 mm. at N.T.P.

These results, and others I have obtained in the course of the work, are very difficult to explain on the assumption that the radiation from beryllium is a quantum radiation, if energy and momentum are to be conserved in the collisions. The difficulties disappear, however, if it be assumed that the radiation consists of particles of mass 1 and charge 0, or neutrons. The capture of the α-particle by the Be^9 nucleus may be supposed to result in the formation of a C^{12} nucleus and the emission of the neutron. From the energy relations of this process the velocity of the neutron emitted in the forward direction may well be about 3×10^9 cm. per sec. The collisions of this neutron with the atoms through which it passes give rise to the recoil atoms, and the observed energies of the recoil atoms are in fair agreement with this view. Moreover, I have observed that the protons ejected from hydrogen by the radiation emitted in the opposite direction to that of the exciting α-particle appear to have a much smaller range than those ejected by the forward radiation. This again receives a simple explanation on the neutron hypothesis.

If it be supposed that the radiation consists of quanta, then the capture of the α-particle by the Be^9 nucleus will form a C^{13} nucleus. The mass defect of C^{13} is known with sufficient accuracy to show that the energy of the quantum emitted in this process cannot be greater than about 14×10^6 volts. It is difficult to make such a quantum responsible for the effects observed.

It is to be expected that many of the effects of a neutron in passing through matter should resemble those of a quantum of high energy, and it is not easy to reach the final decision between the two hypotheses. Up to the present, all the evidence is in favour of the neutron, while the quantum hypothesis can only be upheld if the conservation of energy and momentum be relinquished at some point.

J. Chadwick.

Cavendish Laboratory,
Cambridge, Feb. 17.

The Oldoway Human Skeleton

A letter appeared in Nature of Oct. 24, 1931, signed by Messrs. Leakey, Hopwood, and Reck, in which, among other conclusions, it is stated that "there is no possible doubt that the human skeleton came from Bed No. 2 and not from Bed No. 4". This must be taken to mean that the skeleton is to be considered as a natural deposit in Bed No. 2, which is overlaid by the later beds Nos. 3 and 4, and that all consideration of human interment is ruled out.

If this be true, it is a most unusual occurrence. The skeleton, which is of modern type, with filed teeth, was found completely articulated down even to the phalanges, and in a position of extraordinary contraction. Complete mammalian skeletons of any age are, as field palæontologists know, of great rarity. When they occur, their perfection can usually be explained as the result of sudden death and immediate covering by volcanic dust. Many of the more or less perfect skeletons which may be seen in museums have been rearticulated from bones found somewhat scattered as the result of death from floods, or in the neighbourhood of drying water-holes. We know of no case of a perfect articulated skeleton being found in company with such broken and scattered remains as appear to be abundant at Oldoway. Either the skeletons are all complete, as in the *Stenomylus* quarry at Sioux City, Nebraska, or are all scattered and broken in various degrees, as in ordinary bone beds. The probability, therefore, that the Oldoway skeleton represents an artificial burial is thus one that will occur to palæontologists.

The skeleton was exhumed in 1913, and published photographs show that the excavation made for its disinterment was extensive. It is, therefore, very difficult to believe that in 1931 there can be reliable evidence left at the site as to the conditions under which it was deposited. If naturally deposited in Bed No. 2, the skeleton is of the highest possible importance, because it would be of pre-Mousterian age, and would be in the company of *Pithecanthropus* and the Piltdown, Heidelberg, and Peking men, all of whose remains are fragmentary to the last degree. Of the few other human remains for which such antiquity is claimed, the Galley Hill skeleton and the Ipswich skeleton are, or apparently were, complete. The first of these was never seen *in situ* by any trained observer, and the latter has, we believe, been withdrawn by its discoverer. The other fragments, found long ago, are entirely without satisfactory evidence as to their mode of occurrence.

Appendix 2

LIST OF ABBREVIATIONS

ACAE	Advisory Committee on Atomic Energy (UK, post-war)
AEC	Atomic Energy Commission (US, post-war)
CERN	European Council for Nuclear Research
CNRS	Centre Nationale de la Recherche Scientifique (France)
CPC	Combined Policy Committee (UK, USA and Canada, post-Quebec Agreement)
DSIR	Department of Scientific and Industrial Research (UK, 1920s-1940s)
IOC	International Organisations Committee (UK, post-war)
MAP	Ministry of Aircraft Production (UK)
NDRC	National Defense Research Committee (USA, wartime)
NPC	National Physics SubCommittee (of ACAE)
OSRD	Office of Scientific Research and Development (USA, wartime)
SAC	Science Advisory Committee (UK, wartime)
SIS	Secret Intelligence Service (UK)
TA	Tube Alloys (UK, wartime)
UNAEC	United Nations Atomic Energy Commission

ABBREVIATIONS FREQUENTLY USED IN REFERENCES

AB	Atomic bomb files (PRO)
AIP	American Institute of Physics
1851 Archives	Archives of the Royal Commission for the Exhibition of 1851, London
CAB	Cabinet Office files (PRO)
CAC	Churchill Archives Centre, Churchill College, Cambridge
CHAD	Chadwick's papers at CAC
NA2	National Archives 2, College Park, Maryland
PREM	Premier's papers, PRO
PRO	Public Records Office, Kew, London
RG 77	Record Group 77 (Manhattan Project, NA2)

Index